AF553796

A TEXTBOOK OF ENVIRONMENTAL SCIENCE

(As per UGC Syllabus)

Arvind Kumar

A P H PUBLISHING CORPORATION
ANSARI ROAD, DARYA GANJ
NEW DELHI-110 002

Published by
S.B. Nangia
A.P.H. Publishing Corporation
4435-36/7, Ansari Road, Daryaganj
New Delhi 110002
Ph.: 23274050
E-mail : aphbooks@gmail.com

2026

Rs. 2995/-

Printed at
Balaji Offset
Navin Shahdara, Delhi 110032

PREFACE

Man's capability to transform his environment can bring the benefits of economic development and an opportunity to enhance the quality of life. But this same power, incorrectly applied, can cause incalculable harm to the natural environment and consequently to human life.

Man in earlier times, believed in 'nature-nurture' concept. However during the past fifty years, relationships between man and environment has changed considerably. Thus Environmental Science today has emerged as an exciting subject. Clean air, clean water and suitable climate for agriculture has made us feel concerned about vital environment issues.

The importance of preserving the earth's environment and its biological diversity was amply duly focused at the meeting of the 'Earth Summit' held in June 1992 at Rio de Jeneiro in Brazil.

Virtually all religions have much to say about relationship between mankind and the earth. In the Atharva Veda the prayers of peace emphasizes the links between mankind and all creation: "Supreme Lord, let there be peace in the sky and the atmosphere, peace in the plant world and in the forests; let the cosmic powers be peaceful, let there be undiluted and fulfilling peace everywhere".

Arvind Kumar

CONTENTS

Chapter 1

Definition, Scope and Importance of Ecology

WHAT IS ECOLOGY?

The word ecology, derived from the Greek word: oikos meaning habitation, and logos meaning discourse or study, implies a study of the habitations of organisms. Ecology was first described as a separate field knowledge in 1866 by the German zoologist, Ernst Haeckel, who invented the word oeckologie for "the relation of the animal to its organic as well as its inorganic environment, particularly its friendly or hostile relations to those animals or plants with which it comes in contact."

Ecology has been variously defined by other investigators, as "scientific natural history", "the study of biotic communities", or "the science of community population"; probably the most often given: a study of animal and plants in their relationship to each other and to their environment.

WHAT IS ENVIRONMENT SCIENCE?

Environment is the sum of substances & forces external to the organism in such a way that it affects the organism's existence. In relation to man, the environment constitutes of air, land, water, flora and fauna because these regulate the man's life.

Environment is a multi-dimensional systems of complex inter-relationships in a continuing state of change.

By environment we mean not only our immediate surrounding but also a variety of issues connected with human activity, productivity, basic living and its impact on natural resources such as land, water, atmosphere, forests, dams, habitat, health, energy resources, wild-life etc.

Like other animals man depends on environment and becomes an environmental factor with respect to other members in an ecosystem.

OBJECTIVES OF ECOLOGY

Ecology is a distinct science because it is a body of knowledge not similarly organized in any other division of biology; because it uses a special set of techniques and procedures; and because it has a unique point of view. The essence of this science is a comprehensive understanding of the import of these phenomena:

1. The local and geographic distribution and abundance of organisms (habitat, niche, community, biogeography).
2. Temporal changes in the occurrence, abundance, and activities of organisms (seasonal, annual, successional, geological).
3. The interrelations hips between organisms in populations and communities (population ecology).
4. The structural adaptations and functional adjustments of organisms to their physical environment (physiological ecology).
5. The behaviour of organisms under natural conditions (ethology).
6. The evolutionary development of all these interrelations (evolutionary ecology).
7. The biological productivity of nature and how this may best serve mankind (ecosystem ecology).
8. The development of mathematical models to relate interaction of parameters and predict effects (systems analysis).

RELATION TO OTHER SCIENCES

Ecology is one of the main divisions of biology, the other two being morphology and physiology. The emphasis in morphology is on how organisms are made, in physiology on how they function, and in ecology, on how they live. These divisions overlap broadly. To appreciate fully the structure of an organ, one needs to know how it functions, and the way it functions, is clearly related to environmental conditions. The morphologist is concerned with problems of anatomy, histology, cytology, embryology, evolution and genetics; the physiologist, with interpreting functions in terms of chemistry, physics and mathematics; and the ecologist, with distribution, behaviour, populations and communities in relation to the environment (ecosystems). The evolution of adaptation and of species is of mutual interest to the ecologist and to the geneticist; bio-meteorology is a connecting link between ecology and physiology; and system analysis inter-relates ecology and mathematics. All areas, in the final analysis, are simply different approaches to an understanding of the meaning of life.

SCOPE OF ECOLOGY

To understand the scope of ecology, we consider it in relation to other branches of biology. We can cut the biology "layer cake" into small pieces into two distinct ways as shown in Figure 1.1. We

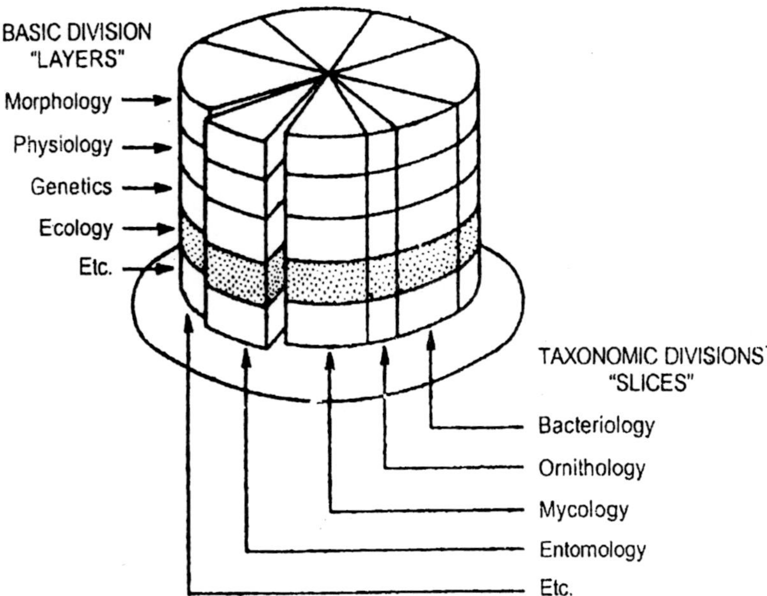

Fig. 1.1 : The biology "layer cake" illustrating "basic" (horizontal) and "taxonomic" (vertical) divisions.

may divide it horizontally into basic divisions like morphology, physiology, genetics and ecology, etc. Now we divide the cake vertically into taxonomic divisions like bacteriology, protozoology, ornithology, mycology and entomology. We find that ecology is a basic division of biology and also forms an integral part of all the taxonomic divisions.

Ecology can also be considered in terms of the concept of levels of organization. The entire biological spectrum can at the best be divided into ten levels of organization as shown in Fig. 1.2. Ecology is more or less concerned with the right hand end of this spectrum that is the levels beyond that of organism. In ecology a species or organism is defined as a physico-chemical mechanism that is self regulating and self perpetuating and is in a process of equilibrium with its environment. The environment of that organism consists in final analysis of everything in the Universe external to that particular organism. The population, is a natural assemblage of different or only one species whereas community in the ecological sense includes all the populations of an area. The community and the non-living environment function together as an ecological system or ecosystem. The portion of earth in which ecosystems operate is conveniently designated as the biosphere. An integration occurs amongst these units. Some attributes, obviously become more complex as we proceed from left to right. However, haemostatic mechanisms that is checks and balances, forces and counter-forces operate all along the line in ecosystems.

THE SUBDIVISIONS OF ECOLOGY

Ecology has been sub-divided into different fields so as to understand the subject in a more profitable way. Ecology can be commonly divided into animal ecology and plant ecology. However, two major subdivisions were preferred by ecologists. The whole subject was divided into autecology and synecology. Autecology deals with the study of individual species whereas synecology deals with the study of a group of organisms or population.

Autecology. *A* study of the individual species in relation to their environment is known as autecology. It includes the study of its geographical distribution, taxonomic position, morphological characters, reproduction, life cycle and behaviour with reference to ecological factors that might influence these activities.

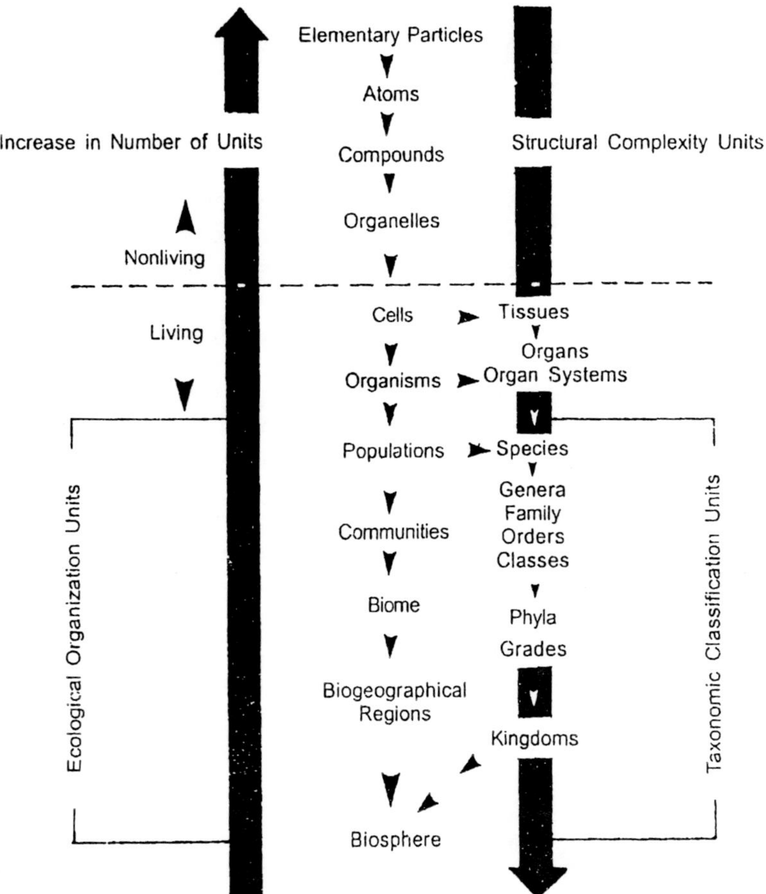

Fig. 1.2 : Levels in the organization of matter.

Synecology. A study of the groups of organisms in relation to their environment is called synecology. Here the unit of study are the groups of species. It comprises population ecology, community ecology and study of the ecosystems.

Useful subdivisions may also be made according to the habitat, taxonomic divisions and level of organization. Thus the subject can be studied through following branches of ecology:

1. ***Population ecology.*** It deals with the growth, trophic structure, metabolism and regulation of a population.

2. ***Community ecology.*** It deals with the ecology of different populations in the same habitat and same environmental conditions.
3. ***Taxonomic ecology.*** It is concerned with the ecology of different taxonomic groups, viz. microbial ecology, mammalian ecology, insect ecology and so on.
4. ***Habitat ecology.*** It includes the study of animals and plants in different habitats. According to habitat, it can be further divided into freshwater ecology, marine ecology, terrestrial ecology, forest ecology and desert ecology.
5. ***Human ecology.*** It deals with the effects of human activities on environment and vice versa.
6. ***Applied ecology.*** It deals with the application of ecological concepts to human needs including wild life management, biological control, forestry and conservation of natural resources.
7. ***Chemical ecology.*** It is concerned with the chemical affinity or preferences shown by different, organisms.
8. ***Physiological ecology (ecophysiology).*** Physiological adaptation according to ecological conditions are dealt in ecophysiology.
9. ***Palaeo-ecology.*** It deals with the environmental conditions and life of the past ages. Palaeontology and radioactive dating have aided significantly in the study of palaeo-ecology.
10. ***Evolutionary ecology.*** It deals with evolutionary problems like speciation and segregation.
11. ***Gynaecology (ecological-genetics).*** Relationship of environment with genetic variability are considered in gynaecology.
12. ***Eco-geography.*** It studies the geographical distribution of plants and animals in different environments—collectively called as biomes.
13. ***Pedology.*** It deals with the study of soil and refers to its nature like acidity, alkalinity, humus contents, mineral contents, soil types and so forth.
14. ***Ethology.*** It is the study of animal behaviour in different environments under their natural conditions.
15. ***Sociology.*** When ecology and ethology are combined it becomes sociology.
16. ***System ecology.*** When the structure and function of an ecosystem is analysed using applied mathematics, statistics or computer, it is called as system ecology.

17. ***Ecosystem ecology.*** Biological productivity of an ecosystem or nature and how it can best serve the mankind encompasses ecosystem ecology.

The essence of ecology lies in the comprehensive study of these subdivisions that form the objectives of ecology.

HISTORY OF ECOLOGY

That certain species of plants and animals ordinarily occur together and are characteristic of certain habitats has doubtedless been common knowledge since intelligent man first evolved. This knowledge was essential to him for procuring food, avoiding enemies, and finding shelter. However, it was not until the fourth century B.C. that Theophrastus, a friend and associate of Aristotle, first described interrelations between organisms and their environment. He has, therefore been called the first ecologist.

The modern concept that plants and animals occur in closely integrated communities began with the studies of August Grisebach, a German botanist, in 1938; K. Mobius, a German investigator of oyster banks, in 1877; Stephen A. Forbes, an American, who described the lake community as a microcosm in 1987; and J.E.B. Warming, a Danish botanist, who emphasized the unity of plant communities in 1895 C.C. Adams recognized and described many animal communities in his ecological surveys of northern Michigan and of Isle Royale in Lake Superior, published in 1906 and 1909. V.E. Shelford presented a classic study of animal communities in temperate America in 1913, and Charles Elton published an outstanding analysis of community dynamics in 1927. Although an appreciation that the whole community is one biotic unit, rather than one unit plants and another of animals, may be described in the writings of some early investigators, the fact has been brought to modern emphasis in the work of F.E. Clements and V.E. Shelford, especially in their Bio-ecology, published in 1939. Much interest has been stimulated in recent years by D. Ramon Margalef, Robert Mac Arthur, and others for analysing the structure of these biotic communities, particularly in respect to such phenomena as species diversity, niches, and how they come about through evolution.

Geographic ecology, in the modern sense, dates from the generalizations on the worldwide distribution of animals made by the French naturalist, Georges L.L. Buffon, and the explorations of the German naturalist Alexander von Humboldt. There was lively interest and many important contributions in this general field during the nineteenth century; notably, the life-zone concept of C. Hart Merrian needs special mention. During the present century the concept of biotic provinces is identified with L.R. Dice and the biome concept with F.E. Clements and V.E. Shelford. The broad survey of ecological animal geography made by R. Hesse in 1924 exerted considerable influence, and this treatise was later translated into English and revised by W.C. Allee and Karl P. Schmidt.

The study of population dynamics, so important in modern ecology, dates back at least to Malthus, who pointed out in 1798 the limitation to population growth exerted by available food. Darwin, in 1859, recognized the importance of competition and predation in developing his theory of evolution. Pearl, 1925, analysed mathematically the characteristics of population growth, and Lotka, 1925, and Volterra, 1931, developed theoretical mathematical equations to show the manner in which populations of different species interact. These studies led to the classic experiments of Gauge, 1935, with interacting populations of predators and prey. Nicholson's publication in 1933 stimulated much thinking concerning the factors that stabilize populations at particular levels. Andrewartha and Birch, 1954, emphasized the importance of climate and other factors on determining the size of populations.

Physiological ecology had its historical beginning in the correlation of biological phenomena with various temperature stimulated by Galileo's invention of a hermetically sealed thermometer about 1612 A.D. The French naturalist Reaumur summed the mean daily temperatures for April, May and June in 1734 and again in 1735, and correlated the earlier maturing of fruit and grain during first year with the greater accumulation of heat: A discovery of parallel significance was of oxygen in 1977 that it was an essential part of air, Claude Bernard, another French Physiologist, enunciated the principle of homeostasis in 1876. This concept originally referred to regulatory mechanisms which maintained the "internal environment" of the body constant in the face of changing external conditions. Later, the concept came to be applied also to maintenance of community interrelations.

ECOLOGICAL CONCEPTS

Ecology deals with 'the web of life' that entangles: (a) every species with the lives of others; and (b) each species with its non-living environment as a whole and each element or factor of that environment. The term organisms refers to all that is living and the term environment denotes everything external to the organisms and includes the physical environment such as climate and the biotic environment. Odum has preferred to define ecology as the 'study of the structure and function of nature'. The total physical environment of any living organism is the *biosphere,* and parts of which are of direct or immediate importance to the organism constitute the 'effective environment'. The complex constituting the organism and the physical environment has now come to be known as *ecosystem.* Ecology then, is the study of the inter-relationships within ecosystems. The environment devoted crop plants is the *agroecosystem,* where both animals and plants in the area are directly or indirectly influenced by each other. Ecosystems are not closed systems because there is always an inter-flow of matter and energy between adjacent ecosystems. The principle of the ecosystem has also brought forth very clearly that the 'whole' is more than the 'sum of its parts'. The interdependence of the organism and the environment must provide the minimum requirements and nothing that is lethal. At the same time living organisms should also be able to successfully adapt themselves to their surroundings. It must be realised that the natural pressures of the environment continue to act on the living organisms and ecological studies relate to the ways by which they–singly or collectively–respond to these pressures or changes. The inability of many animals to cope up with the surroundings, has naturally led to their extinction from the surface of the earth. It may be said that dispersal or the ability of organisms to spread from their place of origin to another area, is a basic aspect of population ecology and without proper means of dispersal, many animal populations would have been eliminated over a period of time due to modifications of the environment. In environmental studies it is usual to refer to the *micro-environment* and the *macro-environment.* Maelzer suggests that environment should be defined as the sum total of everything that directly influences the animals' chances to survive and reproduce. While the micro-environment is the intimately local and immediate surrounding of the organism, the micro-environment is the sum total of the physical and the biotic conditions existing external to the organism and its micro-environment. Both contain many factors affecting the establishment, growth and reproduction of animals and plants. Some of these factors relate to oxygen and carbon dioxide pressures, barometric pressure, wind movement, solar radiation, etc. The two divisions of the environment may differ drastically or only slightly from each other with regard to these factors; for instance, the temperature inside the burrow of a rat is different from that outside it; or within a forest the temperature beneath a fallen log or a stone would be nearly half compared with that of the air, while the temperature at their surface would be much higher. The restricted nature of such micro-habitats naturally calls for special adaptations and the term *micro-climate* is often used for the physical conditions of micro-habitats.

To the ecologist the species is his fundamental unit and the environment generally harbours more than one animal or plant species or species populations. In autecological studies the assessment of the intrinsic capacities of organisms which enable them to cope with environmental conditions is important. The term *ecological valency* has been used to refer to the adaptive range of both the individual and the species. Communities differ according to the environment and the nature of the community is an index of the environmental conditions. The place that an animal occupies in a biotic community which express its functional status, is called '*niche*'. The niche may be better defined as the place that an animal occupies in a biotic community in relation to its food and enemies. It results from the animal's peculiar structural adaptations, physiological adjustments and special behaviour patterns. Every animal exhibits a specificity with regard to its niche and no two animals can occupy the same niche permanently for a long time in the same locality. Appropriate analysis would reveal an interconnection of the different niches into a network which holds the different animals together as a community. This brings about a stratification of the community on the basis of their mode of food capture. It can easily be understood from a study of the relationship of organisms of a community to the food web in a particular environment. *Trophic level* is the term applied to such relationships on the basis of food, and through successive trophic levels the primary stock of energy as obtained by green plants is continually circulated. The chain or sequence of events in which the herbivores consume plants, carnivores feed on the herbivores (and saprophytes feed on all organisms) constitute the *food chain.*

All biotic communities are in a state of constant change resulting from the actions and the co-actions of the organisms or of the external environment. Changes occur in a more or less orderly way so that in course of time the community is replaced by another community. In general, it may be said that communities after they come into being, grow, mature, become senescent and as they die their area become inhabited by other communities. This progressive sequence of replacement of communities in a given area continues until a more or less stable community is established. This has been termed *ecological succession.*

Modern ecology relies on a variety of disciplines like physics, chemistry, mathematics, meteorology, climatology, geology, geography, etc. An essential aspect of these studies is concerned with energy transformations occurring within ecosystems. The quantity of incident energy from the sun per unit area of the ecosystem and the efficiency of conversion of this energy by other organisms is a basic principle. For the effective functioning of ecological systems energy transferring mechanisms from organism to organism as observed in food chains, are of prime importance. *Ecological energetics* deals with this aspect of ecological studies. The autotrophs (plants) and the heterotrophs (animals) utilise this energy. Such inter-conversions of energy closely follow the laws of thermodynamics and it may be said that the solar energy entering the ecosystem is equivalent to the heat energy leaving the system.

Studies such as ecological energetics provide avenues for an inter-disciplinary approach to ecological studies. Significant contributions of other disciplines towards a better understanding of ecological principles have given rise to *ecological genetics, paleoecology, ecophysiology, agroecology, chemical ecology, ecogeography,* and so on. An ecologist recognises a kind of genetic plasticity in the case of every animal. In any environment, only those animals that are favoured by the environment survive. While it is difficult to draw a line between ecology and physiology, the factors of the environment have a direct bearing on the functional aspects of organisms. All the same it may be said that ecology is concerned with supraindividual units– communities and populations–while a physiological approach generally deals with the individual animal or plant.

The *ecogeographic* approach emphasises the role of the environment in animal distribution. It is well known that *zoogeography* relates to the structural and functional relations of animals in space, which form the immediate environment of the individuals as well as the populations. The importance of faunal analysis or faunal complexes is recognised by some zoogeographers, and Uvarov considered *ecofaunas* as the lowest units of which a geographic fauna is made up of. Another aspect of ecogeography is seen in the dispersal or spacing of fauna as a result of multiple interactions between the individual and the environment. While the environment offers considerable resistance, the individuals tend to adapt themselves, exemplifying the role of selection in the evolutionary process. Dispersal within the area of distribution pertains to the concept of population ecology, while dispersal outside limits of normal range of the species is essentially a zoogeographic concept.

With the fast rate of multiplication of human population, problems of food and space will naturally arise. Man has been known to control his environment successfully to meet his needs, but indiscriminate control of insects and other animal pests through use of insecticides, the release of massive quantities of radioactive debris, discarded chemicals and industrial waste into rivers result in atmospheric and aquatic pollution, which may have short and long term ecological effects.

From very early times man has advanced his own ends by the modification of natural communities and ecosystems, sometimes in minor ways to enhance the yield of a desired commodity, but with the rise of agriculture, civilization and modern technology in ever more drastic ways. Such modifications are most likely to be productive and stable when environmental limitations of the natural ecosystem are well understood. In the following section, therefore, the structure and functioning of ecosystems are examined, including their relation to the biosphere. All life depends on how successfully man can learn to harmonize development with these environmental systems.

ECOSYSTEMS AND THEIR FUNCTIONING

Any assemblage of plants and animals able to exist within an area will in time form a biotic community. In this community the different species tend to interact with one another and to modify the conditions of life within which each exists. They therefore develop the inter-relationships and inter-dependencies which constitutes an ecosystem.

In a forest grove or woodland, for example, the larger trees–the dominants–influence the environment in which all other species live. Plants which cannot grow in their shade or tolerate the competition for soil, water and minerals provided by their root systems, or the various chemical substances that they may release into the soil, are unable to find a place within the community dominated by these large trees, or will be confined to places where such trees cannot maintain themselves. Thus, the trees exert an obvious influence on other plants. However, smaller plants may modify the environment in various ways to enhance or detract from its capacity to support the larger trees. The inability of various pine trees to thrive in the absence of the mycorrhizal fungus that attaches itself to their roots has long been known. Similarly, legumes' roots provides a home for bacteria which in turn manufacture the nitrates on which legumes depend. The animal components of such a biotic community also modify the conditions of life for the plant. The interrelationship (symbiosis) between the Yucca plant and the yucca moth is a well studied example; the plant supplies food for the moth, but is dependent on the moth for fertilization and perpetuation. Similarly, large grazing animals affect the populations of the plants on which they feed, and are dependent upon the quality and quantity of plant growth for survival. Carnivores may also limit or otherwise modify the populations and the behaviour of herbivores, and thus have a secondary impact on the kind and intensity of grazing. It is convenient to draw a distinction here between 'natural' and 'artificial' communities. The former consist of wild, naturally occurring species of plants and animals able to maintain themselves in the absence of man. Artificial communities are characterized by species introduced by man or favoured by human modification of the environment, and are unable to exist without continued human assistance or interference.

Neither the word 'natural' nor 'artificial' is entirely satisfactory, since man is part of nature, and all communities, whether strongly influenced by man or not, are also a part of nature. Furthermore, it is difficult to find any area of the earth in which some human modification of the environment has not occurred, if no more than a minor changes caused by an increased frequency of certain radio isotopes, or chemical fall-out from air pollution.

Both categories, though useful to distinguish for some purposes, always exist as part of an *ecosystem,* which is essentially a biotic community in interaction with its physical environment.

The physical environment or the abiotic component is composed of sunlight, atmosphere, water and soil or rock. All of these physical elements affect the living organisms, and all are in turn affected by them. Plants, for example, obtain their energy from sunlight, their basic food from air and water, supplemented by chemicals from the soil or rocks. Plants, however, differentially reflect, refract or absorb various wavelengths of sunlight, and thus modify it; by using carbon dioxide and by giving off oxygen and water vapour they modify the atmosphere; by removing various chemicals and by adding others they modify the soil. All plants and animals always exist as part of an ecosystem. Man, like other animals, is dependent upon the ecosystems in which he exists despite his high mobility which enables him to move from one to another, and despite the technology which permits him to create major modifications in any ecosystem.

All ecosystems must have certain component parts or functions. Thus, they must have a source of energy (usually sunlight) Biotic Component have the following organisms:

i) *Producers:–* They convert the sun's energy into chemical or food energy (usually green plants).
ii) *Consumers:–* They depend upon the producers for food. They may be herbivores, carnivores etc.
iii) *Decomposers:–* They take the chemicals built up by the green plants and break them into simpler forms in which they can

be re-used. These include the so-called reducers organisms, such as bacteria and fungi, that act in the breakdown and decay of dead plants and animals.

From the functional point of view ecosystems has two components:

(a) ***Autotrophs*** (Self nourishing)

The organisms can fix light energy and can use simple inorganic substances and build complex organic substances.

(b) ***Heterotrophs*** (Other-nourishing)

In these organisms there is utilization, rearrangement and decomposition of complex substances into simple ones.

The substances of the ecosystem depends upon the diversity of species in the ecosystem and keeping in balance the minerals between the biotic and abiotic components of the ecosystem. An ecosystem is self maintaining system except that it depends on the sun as external source of energy.

The living components of ecosystems are the plant and animal species (including micro-organisms) that make up the biotic community. Each species is adapted to a particular role in the ecosystem known as its *ecological niche.* Each depends for its existence upon the presence of a suitable habitat, comprising other species as well as necessary components of the physical environment. The functioning of the ecosystem is dependent upon the presence of a suitable combination of species each of which performs a specialized task within the total system.

In a natural ecosystem sun energy will be captured by trees, grasses, or other herbs, and converted into proteins, carbohydrates and other food components. These will be consumed by plant-eating mammals, birds, insects and a great variety of other animals, which, in turn, will provide food for meat-eating animals. All ultimately provide food for the organisms of decay which will break down plant and animal materials and restore their chemical components to the soil. Energy will pass through an ecosystem in a one-way path. Of the solar energy reaching the earth, only a portion will be stored by plants; often less than 1 per cent of the total sunlight energy falling on a vegetated area is retained as chemical energy by the green plants. Of the total energy thus available in plant tissues for consumption by animals, usually less than 20 per cent will be stored as chemical energy in an animal's body tissues.

Eventually, after supporting life for a time, most of the energy is lost to the ecosystem. Some however, is stored in log-lived organic materials such as tree trunks, or in dead organic materials preserved from decomposition–those that eventually form organic deposits of one kind or another, including peat and, over longer periods of time, oil, coal and natural gas.

By contrast, chemical nutrients flow through and ecosystem in circular pathways, from soil to plants and animals and back to soil, being reused or recycled again and again.

THE BIOSPHERE AND ITS FUNCTIONING

All of the ecosystems of the earth, together, form the biosphere. The biosphere is that portion of the earth within which life exists. It includes all oceans and freshwater, the lower layers of the atmosphere and the outer skin of the earth's crush–the rocks and soil of the earth's surface.

Within the biosphere all of the functions and structure described for anyone ecosystem exist. Green plants capture sun energy and combine it with chemical raw materials from soil, water and air. The food they produce supports all animal life, including the decay organisms which return it to the soil for plant use once more.

Man is a part of the biosphere and depends on its continued functioning for his own existence. He can modify or even destroy any one ecosystem. He cannot, however, risk major modifications of the biosphere except at the risk of his own extermination.

Thus, the continued production of plant materials, whether wild or cultivated, is the basis for the nutritional support of man as well as all other animals. The continued functioning of green plants is the source of atmospheric oxygen on which man and other animals depend. The continued functioning of the reducer organisms is the means by which the chemicals in human wastes or in the bodies of plants and animals are made available for further use by living things. A breakdown in any of these biospheric systems would imperil human survival.

Man has been able to take major risks in modifying local ecosystems. He cannot afford major risks in dealing with the functioning of the biosphere.

FACTORS INFLUENCING POPULATION GROWTH

The growth of any species population, whether it be of trees, wild animals or men, depends upon the excess of births (natality) over deaths (mortality).

For any local area or ecosystem this may be influenced by immigration, movement into, or immigration, movement out of, the area or system concerned. For the biosphere as a whole, however, there is no immigration or immigration.

Populations, unless otherwise checked, tend to grow rapidly when there is an abundance of space and of the materials that the species requires for its subsistence. They grow much more slowly when space or essential materials are in short supply. They level off or decline when space or materials become limiting.

A population growing at the maximum rate for which that species is capable is said to be growing at its biotic potential rate. Such a rate of growth is only exhibited when conditions are most favourable for the species–when natality is at a maximum and mortality at a minimum operating to prevent a species from maintaining a biotic potential rate of increase is the environmental resistance. This is the sum total of all of those factors that cause mortality or decrease natality. When biotic potential equals environmental resistance, the species population will not grow.

Most animal populations have long been established in the area which they occupy. They seldom show either a continued growth or decline unless the conditions of their environment change drastically. Most show a fluctuation around some mean level which is determined by the ability of their environment to support their members.

Population of animals (or of men) introduced into an area previously uninhabited by the species concerned but otherwise suitable for it, tend to show an increase in a manner which resembles that of an S-shaped or logistic curve. This is at first marked by a rapid (exponential) rate of increase, and then by a period of levelling off towards some relative stability. Previously established populations, when freed from some factor of environmental resistance, may show a similar growth pattern. There are many modifications of the shape of the curve, depending on the species and the conditions of the environment, but the eventual period of levelling off, through increase environmental resistance, is characteristic of all growth curves.

LIMITS TO POPULATION GROWTH

All populations must sooner or later level off their curve of growth. This levelling-off can result from the behaviour of an animal species. Some species are intolerant of crowding and produce relatively fewer young as their habitat becomes more fully stocked. Other species, however, are more directly limited by the action of factors outside of the population which cause increased mortality. Thus population growth must inevitably be limited by the environment. No environment is without limits and the biosphere is limited both in extent and resources.

If there were no other restrictive factors, ultimate limits to the increase of any species would be imposed by the amount of sunlight energy reaching the earth. In fact, however, other factors are always limiting. For plants, limits to growth would be set primarily be the extent of their ability to convert sunlight into food energy, or by the availability of carbon dioxide in the atmosphere.

In the operation of natural ecosystems, however, either water, or chemical nutrients present in soil or water, run out before the limits of carbon dioxide or photosynthetic efficiency are reached. Animal populations must ultimately be limited by the availability of the food that eat, which depends on the ability of plants to continue supporting herbivorous animals. In practice, most animal populations tend to be limited by other factors before the absolute limits of food supply are reached. These other factors are most notably behavioural, those responses of animals to the presence of other members of their species that prevent an excessive accumulation of individuals in any one area. Inter-specific predation can also be a limiting factor of some importance.

Human populations have attained comparative freedom from the limitations imposed by any one ecosystem–the exceptions being the remaining primitive groups of people. However, human populations cannot escape the limitations imposed by the resources of the biosphere. Of particular relevance is man's efficiency in converting those resources for use and in recycling waste products or by-products for ultimate reuse.

CARRYING CAPACITY

The environmental limits to the growth of any one species population determines the carrying capacity for that species. Carrying capacity is a measure of the number of individuals of any species that a particular environment can support.

Carrying capacity can be considered as having several levels:

1. an absolute or maximum carrying capacity which is the maximum number of individuals that can be supported by the resources of the environment at a subsistence level (this level can be termed the *subsistence density* for that species);
2. the level at which a species population is normally held by the influence of other species living in the same environment—those that hunt or prey upon the species, and those that cause disease or parasitic infestation (this level has been termed a *security density,* or threshold of security, because populations below this threshold are relatively secure from predation or disease; this is necessarily less than the subsistence density);
3. a level, which is generally considered more desirable by those concerned with the health or productivity of the species involved, termed an *optimum density.* At this level individuals in the population will have available an adequate supply of all essentials for existence and, in consequence, will show abundant individual growth and health not limited by shortages of any essential requirements. Such an optimum density can only be maintained by strong limitations on growth imposed by the behaviour of the species concerned (self-limitation) or by removal of individuals in excess of this density by the action of other species through predation. In the latter case optimum density and security density will be the same. These are not necessarily fixed levels, but will fluctuate upwards or downwards.

Intelligent management of domestic animals for maximum production of meat, milk, etc., involves holding their populations at or near an *optimum density.* At this level, their productivity will be at a maximum and the highest yield can be obtained. Under primitive conditions in Africa and Asia, for example, domestic herds were often held at or near a *security density* by the action of predators (human or wild animal), or by diseases and parasites. Under modern conditions, with predation and disease minimized, and where they are not properly managed, domestic herds often increase to a *subsistence density.* Under these conditions malnutrition is common, disease is more prevalent and productivity is low.

Obviously, man has similar choices with his own populations. In primitive times the operation of predation, disease, etc., often held local populations at or near a security density.

With the removal of predators, and the control of disease, populations can either increase to a subsistence density, a level where human suffering and general misery will be maximum, or they can be levelled off by self-imposed restrictions on growth, at maximum individual opportunity for health, happiness, and choice of high quality life styles and environments. These choices would normally be available to all human populations at some period in their growth. Now for the first time, however, the whole humans species faces possible population increase to levels in excess of the capacity of the biosphere to maintain them, the biosphere itself being strained by rising demands for material resources and products.

Carrying capacity is not fixed. It fluctuates naturally with weather and climate, and the operation of other natural factors such as fire, floods, earthquakes and vulcanism. In any one area it can be increased to some maximum level which is determined by the rate at which that environmental requirement which is in shortest supply can be provided. Such a requirement is known as a limiting factor. A limiting factor is that substance or quality in the environment, the supply of which is least abundant in relation to the needs of the animal or plant concerned.

LIMITING FACTORS AND THEIR OPERATION

Limiting factors are often classified in two categories: density dependent and density independent. The effect of the former on a population increases as the population increases in density. It is essentially a stabilizing influence on growth and causes a population to level off at carrying capacity. A density-independent factor, conversely, affects many or a few individuals without reference to the population level. Food supply is generally density dependent—the more there are to eat it, the less there is for each individual, and the greater the effect of food scarcity. By contrast, a flood is density independent since it may wipe out an entire population of a species, whether there are few or many. The more important limiting factors as follows :

(a) Climatic and atmospheric factors

These operate in a variety of ways to affect species populations. To begin with, any species has limits of tolerance and some optimum range of tolerance for such factors as sunlight, temperature, humidity, rainfall or wind.

If a local climate normally exceeds the limits of tolerance for a species, then the species will not occur in that area; if it exceeds the optimum limits, the species will not thrive there. If the local climate only rarely exceeds the limits of tolerance, then a species may temporarily occupy that area, but will be eliminated in those years when the climate becomes extreme.

Assuming that the climate is favourable to the establishment of a species, then a number of climatic factors may influence population growth. Changes in temperature—years that are warmer or colder than normal—may permit a species to thrive and increase, or they may cause a decrease and permit its survival only in the most favourable sites. Changes in rainfall and humidity, dry years or cycles and wet years or cycles, have major effects. Fluctuations in temperature and rainfall tend to be most severe in areas where temperature or moisture is already near the limits of tolerance for a species.

Thus, in the humid tropical lowlands where rainfall and temperature are high and near the optimum for the growth of the greatest number of plant species, there is relatively little annual fluctuation in either factor. In the dry tropics not only is moisture naturally limiting for plants, but fluctuations in rainfall from year to year may be extreme. In northern temperate regions where winter cold and the duration of low temperatures restrict the growing season for plants, one finds the greatest variation between temperatures in one year and another.

Climatic and atmospheric factors tend to have their most severe effects upon those introduced populations of plants and animals which man attempts to grow or produce in areas that have climates more extreme than the optimal climate for the species concerned.

Native species are adapted to the range of climate in the area they occupy, and although they may be affected by climatic extremes, these effects are usually much less limiting than for a domesticated species introduced outside its normal climatic range. In the African savannas, a drought will be more likely to wipe out a high percentage of the cattle than of the native antelopes.

Climatic and atmospheric factors also have more effect of species that are at, or approaching, the level of subsistence density. Well-situated animals or plants, with adequate nutrients, water and other essentials, are less likely to be affected by drought than populations which have increased and expanded into marginal areas and are already in poor condition because of nutritional limitations. Certain factors of weather and climate may have catastrophic effects on all populations in an area–hurricanes, tornadoes, unusually severe floods or droughts, but usually these are less significant than the more normally occurring changes, except in the case of these species which are highly limited in distribution. With most species, normal fluctuations in climate have their greatest effect when animal population densities are excessive, or when plants are established in sub-optimum environments.

(b) Soils

Here the limiting factors operate directly on plants and, through plants, upon animal populations (other than those of the soil fauna itself, which is often directly limited by soil conditions). The requirements of plants and animals, however, are not identical.

An abundant growth of vegetation does not necessarily mean there will be an abundant food supply for animals, nor that the food supply will have high nutritional value. Certain plants may produce abundant vegetative growth and high amounts of carbohydrates under conditions where certain soil chemicals are in short supply; however, they may produce relatively small amounts of the proteins and vitamins needed to support an abundance of animal life.

Thus, a shortage of a trace element, such as cobalt, in the soil may have little effect on plant growth. However, cattle feeding on these plants may do poorly or not survive. Addition of a cobalt top-dressing to the soil which is then taken up by the plants can open up a range-land to livestock use. For another example, a dense forest may produce an abundance of vegetation, but will support very few ground-dwelling grazing or browsing animals.

There is no lack of food for them, but plants growing on a forest floor under dense tree cover are commonly lacking in protein. Animals grazing of browsing them can receive adequate carbohydrates but not enough protein to thrive or reproduce. Thus, in many areas of dense forest, grazing and browsing animals will only be common around natural or manmade clearings.

The abundance of natural vegetation in an area often has little relation to the inherent fertility of a soil. Thus, the ferasols and acrisols of the humid tropics are notoriously infertile, yet they support the most luxuriant vegetation to be found on earth.

Examination of this situation reveals, however, that virtually all the nutritional elements required for the support of life are tied up in the vegetation and animal life of the area, and re-circulate quickly from the dead plant or animal back into a new, living individual. Below the organic surface layer the soil contains few of the elements needed for plant or animal nutrition, and if the vegetation is scraped away with a bulldozer the soil that remains is often highly infertile. By contrast, some desert soils that support little plant life, because of lack of water, may have a rich supply of the soluble mineral elements needed for the support of life. When water can be made available, they will support high levels of productivity, although there may be new ill effects such as salinization, water-logging and water-borne diseases.

(c) Water

Water is an obvious limiting factor for plants and animals. Plants are classified in relation to their tolerance for dry, medium or wet conditions as *xerophytes, mesophytes* and *hydrophytes.* Most land plants are in the intermediate, mesophytic range; desert plants are usually, but not always, xerophytes; hydrophytes grow in water or require an abundance of water in the soil. All plants require water to support their active growth and metabolism. Some plants cannot tolerate even brief periods of moisture deficiency and will wilt and die when the soil dries out. Xerophytes have special adaptations that permit them to survive under prolonged drought, but must have some moisture in order to grow and reproduce.

Where present in abundance, water limits the occurrence of terrestrial organisms and provides a habitat for aquatic forms of life. The latter will relate in their diversity and abundance to the quality of the water, its ability to supply oxygen, its salinity, temperature, velocity and other factors. All of these are capable of being affected, usually adversely, by pollution or siltation.

In relation to terrestrial vegetation, an abundance of water or in the soil limits plant growth primarily by restricting the availability of oxygen. Those plants grow in wet soils or flooded areas have the ability to derive oxygen through special structure, such as devices that permit the aeration of their roots; the pneumatophores of mangroves, or the so-called 'knees' of balt cypresses are good examples. Continuously moist or flooded soils inhibit the bacterial or fungal action that would otherwise bring about decay of organic debris and litter. Organic remains will usually accumulate on such soils in the form of organic mucks or peat.

Animals, like plants, require moisture for their continued existence, but vary greatly in their ability to endure drought. Apart from purely aquatic animals, amphibious animals such as the crocodile and hippopotamus cannot long survive away from water. Most land animals must drink water regularly in order to continue to live. However, some desert species can exist indefinitely in the absence of free water exist indefinitely in the absence of free water as long as the food which is eaten contains sufficient moisture and they can find enough shelter to hold the loss of water from their body surfaces to a minimum. A severe drought, bringing an absolute shortage of water to an extensive area, can be fatal to all animals that require water for drinking. Nevertheless, much of the mortality attributed to drought results not from a shortage of water, but from a shortage of food within reach of water.

(d) Biotic factors

The greatest number of limiting factors influencing plant or animal growth, abundance and distribution are biotic in nature. Food supply for animals is one of these, and is the most common factor limiting the growth of animal populations, either directly, through being short of requirements, or indirectly, through behavioural responses to food shortage.

The number of plant-eating animals in any area is ultimately limited by the abundance of the plants of which the animals feed. Furthermore, since the plants must survive and reproduce, there must always be a greater abundance of plants than what is actually needed to provide food to the animals. Most plants produce a surplus of vegetative growth and of seed, part of which can be used by animals. However, each plant, in order to survive, must maintain a metabolic reserve–a minimum amount of leafage to permit it to store food for its own survival, or set seed for its own reproduction.

Plant growth may be limited by competition from other plants of the same or different species each drawing on the same reserves of soil and water, or shading one another from essential sunlight. Some plants secret substances which inhibit the growth or establishment of other plants. A variety of organisms may prey upon plants, from the seed stage through the life cycle to the mature plant.

It is obvious that the number of carnivores (or meat-eating animals in any area is limited by the availability of the prey upon which they feed. Also, there must be enough prey animals to enable the prey species to survive and reproduce; the predator cannot eat all of the prey, or both would become extinct. Most prey animals have developed behaviour patterns in relation to their habitat conditions that permit some of them to escape predators.

Predators seek the most vulnerable and available prey, and usually do not hunt down and capture the more elusive individuals. Often the individuals captures are either they very young and very old, or the most diseased or genetically least vigorous stock; therefore, natural predation tends to have overall beneficial effects on the population of a prey species.

The relationship of the abundance of predators and prey is sometimes illustrated in the form of a *biotic pyramid.* Green plants occupy the base of the pyramid and must always have a greater total volume; *biomass* than the plant-eaters that feed upon them. The latter, the herbivores, form the second step in the pyramid, which must in term be wider and represent a greater biomass than the third step, representing the carnivores that eat other carnivores they must be fewer in number and contain less biomass than their prey. Each succeeding level in the biotic pyramid also represents a consistent loss of energy from green plants to herbivores to predators, since at each transfer of energy from one level to another some energy is necessarily lost. Far more calories are available at the lowest level of the pyramid than at any higher level.

The sequence of plants and animals feeding upon one another is called *food chain.* Food chains are interconnected, since more than one herbivore may feed on a plant, and more than one kind of carnivore may eat a herbivore. These interconnected chains form a *food web.*

As elements pass from soil (or atmosphere) to plants, and are then eaten by herbivores and in turn by carnivores, they are often concentrated in varying degrees by a process known as *biological magnification.*

Thus, the element iodine may be present in the soil in small quantities and taken up by plants as it goes into solution and reaches plant roots. It has little direct function in plants, but is essential to plant-eating mammals. These must obtain enough iodine from the plants they eat and concentrate it in their thyroid glands for these glands to function properly. The level of iodine in an animal's thyroid is far greater, therefore, than in either the plants or the soil.

Although the number of herbivores is limited by the supply of plants, and the number of carnivores is determined by the numbers of their prey, the reverse is also true in varying degrees. The abundance and distribution of a plant can be limited by the presence of herbivores. If a plant is introduced into an area where there are no species that feed upon it, can spread rapidly and become transformed into a weed or pest.

Similarly, the abundance and distribution of plant-eating insects and other animals is often controlled by a wide variety of diseases, insects or other predators that feed upon them. If the predators are destroyed, their plant-eating prey can increase suddenly and explosively, since its reproductive rate in nature is adjusted to the need for survival under constant predation. Sudden increases in numbers of rodents or other plant-eating mammals have often been related to extermination of the predators that formerly held them in check.

Populations of many animal species are limited by the effects of interactions among individuals with in a species. The species concerned are described as *territorial* and comprise a proportion of individuals which persistently or seasonally occupy a particular area to the exclusion of other individuals of the same species.

Commonly, territorial animals will tolerate or encourage the presence of certain other members of their species--a mate or mates, their juvenile offspring, and sometimes the members of a larger social group, a herd, troop, flock or pack.

However, they will drive off or otherwise exclude animals that do not belong to the tolerated category. Territorialism tends to limit the number of animals of any one species that occupy an area and to reserve for the territorial individual or group an adequate habitat and food supply; however, there are many kinds and degrees of territorial behaviour, and not all of them serve directly to limit population increase. Plants might also be considered to compete for 'territories' the well-established individual holds its ground against competitors. Their territorial defence mechanisms include the shading out of competitors and the secretion of chemicals which prevent rivals from germinating or growing.

Numerous studies have now been made of the effect of overcrowding on a variety of animal species under conditions where food supply is abundant, and all other essentials (except space) are present in adequate supply. All of these studies show adverse effects upon individuals in a population where numbers exceed a certain level. These adverse effects are caused by social interactions, and vary from mild behavioural disturbances to serious forms of behavioural pathology, leading to increased mortality and a decline in birth rate.

(e) interaction of Factors

Virtually all limiting factors interact and either reinforce or diminish their mutual effects. Thus the effects of animal overcrowding may be seen in the limitations imposed by diminished food supply, increased predation, greater mortality from disease and various behavioural disturbances. Similarly among plants, factors of soil, water, air, nutrients, the presence or absence of intense sunlight, all interact to influence the germination or growth of a plant.

Simple solutions to biological problems are therefore seldom likely to produce permanent, positive results. Modifying the impact of one limiting factor may simply increase the operation of others. For example, in an area of scarce water supply it may seem sensible to make available additional water in order to permit animal populations to increase.

Yet doing so has, in many observed instances, quickly led to major die-offs from malnutrition resulting from an insufficient food-supply for the increased population. Similarly, destruction of predators may cause protected herbivores to die in an epidemic and the elimination of a disease may cause a crash die-off from starvation. It is therefore always necessary in management planning to look at the operation of the total ecosystem and consider how all the factors interact.

Failure to recognize these principles has caused many economic development activities to fail and much money to be wasted.

GAIA HYPOTHESIS

Gaia is the term used for Greek goddess–Mother Earth. This hypothesis suggests that life manipulates the environment for the betterment of life. The planet Earth works as a large organism capable of self-maintenance. The Gaia hypothesis is actually a series of several hypotheses. The first of them suggests that life greatly affects planetary environment. The second idea asserts that life affects the environment for the good of life. A third hypothesis of Gaia is that life deliberately and consciously controls the global environment. It appears that the systems of positive and negative feedback that operate in the atmosphere and oceans are sufficient to explain the mechanisms by which life affects environment.

The hypothesis, as a whole, has some merit as it relates to the present and future. The future status of life may depend on actions we are taking now and those we will possibly take in the coming years. People living on earth today are making conscious decisions concerning the future of the planet.

REVIEW QUESTIONS

1. Define ecology and discuss the scope and objective of ecology.
2. Enumerate the sub-divisions of ecology.
3. Write short-notes on the sub-divisions of ecology.
4. Write short-notes on
   (a) Autotrophs
   (b) Heterotrophs.
5. Define carrying capacity. Enumerate and explain the factors that affect the carrying capacity.

Chapter 2

NATURAL RESOURCES AND THEIR UTILIZATION

WHAT ARE NATURAL RESOURCES ?

The word ***resource*** means a source of supply or support that is usually held in reserve. It includes wealth, supplies of goods, raw materials, etc., which can be used by organisms. The natural resources are the components of the atmosphere, hydrosphere and lithosphere essential for life. These include energy, air, water, soil, minerals, plants and animals. The nature of resources varies from society to society. For example, the Onge tribal society of the Andaman group of islands do not use uranium, gold and silver, hence, these are not resources for Onge society. However, uranium is an important resource for obtaining nuclear energy. Gold and silver are also important resources for those who use them.

Types of Natural resources. The natural resources are classified in different ways;

A. Depending upon their chemical composition (nature), the natural resources are of three types

(i) *Inorganic Resources.* They include air, water, ores.

(ii) *Organic Resources.* They include plants, animals, microorganisms, fossil fuels.

(iii) *Mixtures of Inorganic and Organic Resources.* They include soil.

B. Depending upon the abundance and availability of the natural resources they are classified into two categories– inexhaustible and exhaustible.

(i) *Inexhaustible Resources.* They are not likely to be exhausted by man's use, e.g., air, clay and sand.

(ii) *Exhaustible Resources*. They are exhausted by man's use. These resources are classified as renewable and non-renewable.

(a) *Renewable Resources*. They are naturally replenished after being used by man, e.g., water, soil and living organisms are the main renewable resources.

(b) *Non-renewable Resources*. They cannot be replaced after being used. These include fossil fuels, (coal, petroleum and natural gas) and minerals. It is important to note that underground water, forests and wild life are renewable resources but can become non-renewable if they are not used properly.

C. Natural resources are also classified on the basis of their presence in different countries.

(i) *National Resources*. They are confined to national boundaries e.g., minerals, lands.

(ii) *Multinational Resources*. They are shared by more than one country e.g., some rivers, lakes, migratory animals.

(iii) *International Resources*. They are shared by all inhabitants of the earth e.g., sunlight, air.

ENERGY RESOURCES

The major energy sources are fossil fuels such as coal, petroleum and natural gases. Other energy resources of energy are solar energy, hydroelectric power, nuclear power, wind power, geothermal energy, energy from garbage, etc.

Fossil fuels. As stated earlier they include coal, petroleum and natural gases. They are important source of energy. Oil sources of the world are limited and restricted to just certain areas. Moreover, their demand is increasing day by day because of urbanization and industrialization. The coal reserves of the world are higher than that of petroleum, The leading coal producing countries of the world are China, USSR, USA, U.K., India, Japan, Poland, Czechoslovakia, Kuwait, Saudi Arabia, Iran, Iraq, Nigeria, Libya, Arab Republic and Indonesia. In India the coal reserves are found in Bihar, Bengal, Orissa, M.P., Maharashtra, J & K, A.P., and Meghalaya. New reserves of oils have been discovered in sea-beds and off-shores of Bombay High. The per capita energy consumption of some countries is being given below.

Table 2.1 : Energy Consumption of Some Countries

Countries	*Per capita daily consumption (× 1000 kcal)*	*Proportion of world's energy consumption(%)*	*Proportion of world's population(%)*
USA	230	35	6
Canada	165	3	0.6
UK	145	6	1.5
Japan	40	3	3
India	6	2	15

Source: Data from Scientific American Sept. 1971 and from U.N. Demographic year book 1970.

The table shows that Americans represent about 6% of the world's population but they consume the greatest amount of energy in the world. The average American consumes 250,000 kilo calories of energy per day. That amount is double the amount used by individuals in other modern industrial nations.

Coal production is also necessary because coal is also used to produce energy. It can also be converted to liquid fuel (oil) and methane gas. Gas produces much less pollution than other fossil fuels do. Many of the coal deposits are of higher sulphur coal. When burned this coal releases large amounts of sulphur compounds into the air. These compounds are among the most important air pollutants, causing serious health and environmental problems.

Uses of Fossil Fuels. They are an important source of energy for the modern technology. Fossil fuels are used in industry, thermal plants, agricultural operations, automobiles, planes, ships, rail engines, heating, etc. They also yield certain useful materials like petroleum products.

Alternative sources of Energy. Some alternatives to the great dependence on fossil fuels are solar energy, hydroelectric power, wind power, geothermal energy, nuclear power and energy from garbage.

(i) *Solar Energy.* Sun is the unexhaustive source of energy without any pollution effect on the atmosphere. Scientists have devised means to utilize solar energy from the sunrays to heat water, **cook meals and run certain machines.**

(ii) *Hydroelectric Power.* It is generated from the kinetic energy of water falling from great heights. A number of power stations are established in India on a number of rivers and canals which generate electricity from water. Sea-tides are also used as a source of hydro-power (tidal energy).

(iii) *Geothermal Energy.* In some places, the heated water comes to the earth's surface as hot springs. Wherever available it can be used to heat buildings and produce electrical power. It is called geothermal energy or power.

(iv) *Wind Power.* Wind mills have been used for centuries as a source of power to grind grain and pump water. But, the amount of wind and its duration varies widely from place to place and from season to season. Therefore, wind mills are better suited for some areas than for others.

(v) *Nuclear Energy.* It is obtained from fusion or fission of atoms of certain elements. Nuclear fusion involves the fusion of the nuclei of two atoms instead of the splitting of atoms that occurs in nuclear Fission. The result is the release of enormous quantities of energy. Splitting of the 1 amu of uranium-235 can generate energy equivalent to that obtainable from burning of 15 metric tons of coal or about 14 barrels of crude oil. In India, Atomic power stations are located in Tarapur (Bombay), near Kota in Rajasthan, Kalpakkam in Tamil Nadu and Narora in Uttar Pradesh.

(vi) *Energy From Garbage.* It is reported that Americans produce more garbage per person than the people of any other country in the world. Half of this refuse is paper. The remainder is plastic, glass, metal, potato-peelings, bones, etc. Thus, waste can be used to produce electrical energy.

(vii) *Other sources of Energy.* Fire wood and the cattle dung are widely used for cooking and heating in rural areas of India. Indian scientists have developed a new technique, the *gobar gas plant*, or *biogas*. In this plant fresh cow dung is added to produce odourless, low pressure gas. This gas can be used for cooking and heating. The residue is used as manure.

Conservation of Energy. The energy problems facing the world are serious and may become critical. Therefore, we must realize the limitations of energy resources. Each individual or family should attempt to implement an energy savings programme.

We should change our life style. Excessive use of energy can be avoided, for example, turning off unneeded light which creates a significant saving. Over-consumption of oil should be checked. All oil installations should be well protected from Fire. Alternative sources of energy such as solar energy, hydroelectric power, wind power, nuclear power, energy from garbage should be established.

ATMOSPHERE (AIR)

Definition. Atmosphere is defined as transparent gaseous envelope surrounding the earth.

Layers. Atmosphere consists of four layers.

(i) *Troposphere.* It lies from the surface of the earth to an altitude of 10 kilometres.

(ii) *Stratosphere.* It extends between altitudes of 10 and 60 kilometres. Temperature increases with height in this layer.

(iii) *Mesosphere.* It extends between altitudes of 60 and 100 kilometres. Temperature decreases with height in this layer.

(iv) *Thermosphere.* It lies from an altitude of 100 kilometres upward. In this layer the temperature increases with height.

The outer extremely rarefied fringe of the atmosphere is called *exosphere* which is gradually merged with the outer space.

Composition. The atmosphere (air) is a mixture of several gases. Near the earth's surface it consists of 78% nitrogen, 21% oxygen and the remaining 1% includes 0.93% argon, 0.03% carbon dioxide and small quantities of hydrogen, helium, neon, krypton and traces of many other gases. Besides gases, it also contains water vapour, dust particles, smoke and salts. A layer of ozone called ***ozonosphere*** or ***ozone layer*** is present in the stratosphere.

Role of Atmosphere (i) The ozone layer protects the living organisms from the harmful radiations from the sun. (ii) Air acts as a medium for locomotion of flying animals such as insects, birds and bats. (iii) Atmosphere supplies oxygen, carbon dioxide and nitrogen to those living organisms which utilize them. (iv) Air helps in the dispersal of pollens, spores, seeds and fruits, (v) Air is used as medium for the speedy aviation transport of man and material, (vi) Many man made satellites used for communication, weather monitoring and research move around or are stationed in the

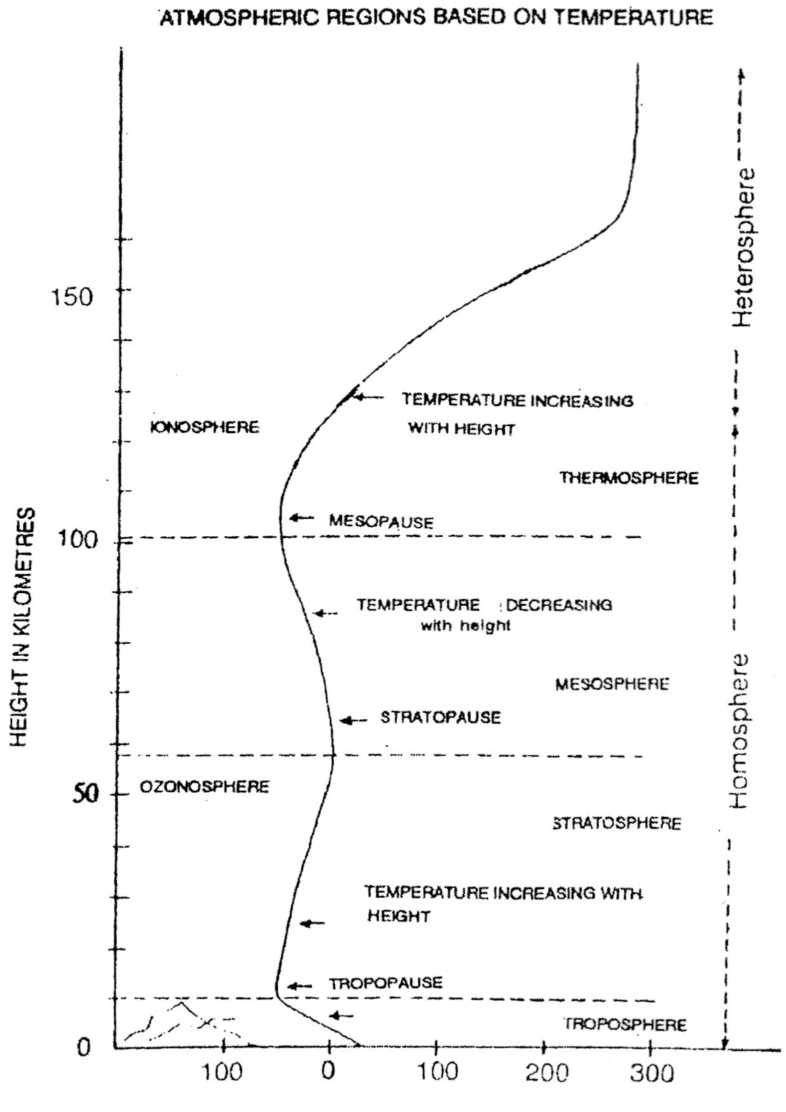

Fig. 2.1 : Different layers of atmosphere.

thermosphere at particular altitudes, (vii) The ionosphere has a large number of ions and free electrons. It reflects the radiowaves back to earth, thus, helping in long distance communication. (viii) Air transmits sound for communication among animals, (ix) Circulation of air is responsible for specific climatic conditions and water cycle.

Industries, automobiles, aeroplanes and aviation technology are polluting the air. They change the natural composition of the atmosphere. This is causing a threat to the environment, climate and finally the living organisms.

WATER RESOURCES

Four-fifth of the earth's surface is covered by water and, therefore, earth is called the water planet. Water resources consist of fresh water resources and ocean water resources.

1. Freshwater Resources. Fresh water environments range from large lakes to small temporary ponds and from large rivers to tiny streams. Fresh water is an indispensable resource which is essential for life on land as well as survival of human race. It comes to land in the form of rainfall. In India the average rainfall is about 110 cm except for semi-arid and desert areas of Rajasthan. It produces 2.75 million cubic km of water per year. Out of it only 0.6 million Km^3 seeps into ground while the rest flows into rivers. The ground water reserves of India is about 27 million Km^3 or ten times the annual rainfall. Surplus river water as well as ground water can be profitably used in irrigation. At present about 50 million hectares of land are under irrigation. This area is likely to increase with time.

Human Uses. Fresh water for human use is obtained from rivers, ponds and ground. The water is used for (i) Drinking, cooking, bathing and washing, (ii) Disposal of sewage and other organic wastes, (iii) Generation of electricity, (iv) Industrial plants, (v) Construction of buildings. (vi) Irrigation. In India, only 50 million hectares or less than 20% of land, are under irrigation. The remaining land has also to be provided with irrigation, (vii) Rearing of Fish and other aquatic organisms.

India has vast fresh water stretches in the form of rivers (27360 km), canals and irrigational channels (112650 km), reservoirs, lakes, tanks, pools, ponds, etc. Despite that, fresh water fish production is very low in India. It was only 218000 tonnes in 1951, 660000 tonnes in 1972, 748000 tonnes in 1974 and over I million tonnes in 1980.

Water Problems, (i) Deforestation, especially in the hilly areas has reduced water absorption, storage in catchment areas, greater incidence of soil erosion and floods, during rainy season. At other times the water supply is very little, (ii) Supply of fresh water to urban and industrial areas has always been a problem because of the huge amount involved. The average consumption of fresh water per person in modern society is 350 – 700 litres per day. Further, several industries consume large quantities of fresh water. The disposal of used water is still another problem. As a result there is a great amount of misuse and abuse of fresh water. (iii) Water is seldom available in the pure form because it is a general solvent of atmospheric gases and several solid substances which, therefore, get dissolved into it. Purest natural water is rain water but due to consumption of fossil fuel and other industrial activities, it is often found to contain acids formed by the dissolution of oxides of nitrogen and sulphur. The *acid rain* is detrimental to plants and has destroyed forests in many parts of the world, (iv) While seeping through the soil, water dissolves away a number of salts like chlorides, bicarbonates, sulphates of sodium, calcium and magnesium. They become part of ground water. Rapid evaporation of such water, as in semi-arid areas, brings salts to the surface and, therefore, gives rise to soil salinity. Hardness of water is due to the presence of salts. Salts present in water shorten the life of cooking utensils, water heaters, boilers, streams turbines, etc. A salt content of upto 3.5 gm/litre of water is safe for irrigation. Beyond this concentration the water becomes harmful to crops.

Conservation of water (i) Treatment of used water before passing it into irrigational channels or rivers, (ii) Prevention of wastage of water through economical use in industry and homes, (iii) Prevention of wastage of water in irrigation through brick-lining of channels, sub-surface and sprinkler techniques, (iv) Prevention of water pollution by not allowing raw sewage and industrial effluents to pass into water bodies, (v) Building of dams up-stream to store flood water for use during dry periods, (vi) Afforestation and reforestation of hills, catchment areas and slopes for increasing availability of water throughout the year. (vii) Building of tanks, ponds, etc. where other perennial sources of water are absent.

*2. **Ocean Resources (Marine Resources).*** About 70% of the earth's crust is covered by oceans. As the land based resources are limited while the oceans are still largely untapped, attention of mankind is shifting towards their exploitation.

(1) Oceans are a source of Fish and other edible animals. Over 70 million tonnes of marine fish is caught annually in the world, out of which only 1.2 million tones (1972 basis) is procured in India despite the fact that India has a long coastline. Recently mechanised boats and trawlers have been introduced for deep sea Fishing.

(2) A number of kelps (brown algae) and red sea weeds are edible. They are in use as food for man and animals to a small extent in Japan, China, Philippines, some European and American countries. The algae are rich in minerals and vitamins. Their potential remains largely untapped.

(3) Marine plants are a source of ***algin*** (from brown algae like *Laminaria, Macrocytis, Nereocytis* for stabilising emulsions in ice cream, rubber latex, shaving creams, tooth pastes, cosmetic creams, flame proof plastics, security glass, surgical threads), ***agar*** (from red algae like *Gelidium, Gracilaria* for culture medium in biology, dentistry, stabilising agent in confectionary and cosmetics, sizing of textiles), ***carrageenan*** (carrageenin, from red alga *Chondrus,* as emulsifier, stiffening and clearing agent), ***funoran*** (Funori, a glue, from red alga *Gloiopeltis),* ***diatomite*** (siliceous deposit of diatoms for filtration, as clearing agent, for sound proofing, manufacture of sodium silicate, shining paints and insulation) and some medicines.

(4) It is believed that petroleum and natural gas are oceanic in origin. Therefore, oceans must be a large reservoir of fossil fuel. U.K. has already become an exporter of petroleum by extracting the same from its offshore fields. India has also a large reserve of petroleum, and natural gas at several places along the coastline. Already good quantity of petroleum and natural gas is being extracted from offshore wells in Bombay high.

(5) Oceans possess a number of nearly pure mineral concentrates in the form of *nodules*. The mineral nodules are a rich source of manganese (hence called *manganese nodules*), copper, nickel and cobalt. They occur in pacific ocean between Hawaii and Mexico as well as along our coast. India has already chalked out a plan to extract the nodules.
(6) Ocean tides can be used in the generation of electricity.
(7) In certain parts fresh water for coastal areas is obtained through desalination of sea water. Icebergs are the other source of fresh water.
(8) Oceans are a mean of transportation of man and materials between different parts of the world
(9) Common salt is extracted from sea water through concentration in salt pans situated near the shore.

LAND RESOURCES

Area. Land forms about one fifth of the earth's surface, covering about 13,393 million hectares. About 36.6% of the land area is occupied by houses, factories, roads, railways, deserts, mountains, rocks, glaciers and polar ice marshes. About 30% of land is covered by forests. About 22% of land is occupied by meadows and pastures. Only 11% of land is suitable for ploughing. The surface layer of land is called soil. About four-fifth of the land consists of soil.

Definition of Soil. According to Buckman and Brady "Soil is a dynamic natural body, on the surface of the earth in which plants grow, composed of mineral and organic materials and living forms".

Pedology. Pedology is the science which deals with the laws of origin, formation and geographic distribution of the soil as a body in nature. Pedologist is the person who studies pedology.

Components of Soil. The soil consists of four major components:

1. Mineral matter 45%
2. Organic matter 5%
3. Soil water 25%
4. Soil air 25%

There are fine spaces between soil particles which are called *interstices*. The latter are filled with air arid water containing dissolved substances. If the volume of water decreases due to evaporation, the volume of air in the soil increases. If the volume of water increases after irrigation or rain, the volume of air decreases.

Soil also contains flora (plants) and fauna (animals).

Microflora (Bacteria)

1. Heterotrophic bacteria. A. Nitrogen-fixing bacteria (i) **Symbiotic,** *e.g., Rhizobium* (ii) **Non-symbiotic** *viz.,* (a) **Aerobic** *e.g., Azobacter,* (b) **Anaerobic** *e.g., clostridium.* **B. Non-nitrogen-fixing bacteria,** (i) **Aerboic** *e.g., ammonifers* (ii) **Anaerobic** *e.g., denitrifiers.*

2. **Autotrophic Bacteria,** (i) Nitrifying bacteria (ii) Iron bacteria (iii) Hydrogen bacteria (iv) Sulphur bacteria (v) Manganese bacteria (vi) Methane bacteria (vii) Carbon monoxide bacteria.

Fungi, (i) Yeasts (ii) Moulds (iii) Mushrooms.

Algae, (i) Green algae (ii) Blue green algae (iii) Diatoms.

Microfauna. ***A. Protozoa,*** (i) *Paramecium* (ii) *Euglena* (iii) *Amoeba* ***B. Nematodes.***

Macrofauna, (i) Earthworms (ii) Mites (iii) Centipedes (iv) Millipedes (v) Termites (vii) Snails (viii) Mice.

Types of Soil Particles. The type of soil is determined by the mineral composition of the parent rock and the size of particles present in the soil. Four types of soil particles are found (i) *Gravel Particles.* They consist of small stones with coarse sand particles as used for roads and paths, (ii) *Sand Particles.* They are porous and, therefore, well aerated. However, they have little water holding capacity. Sand particles consist largely of quartz (SiO_2). (iii) *Silt Particles.* They are carried by moving water and left at the mouth of a river. They also consist largely of quartz and are chemically inert, (iv) *Clay Particles.* They are rich in nutritive salts and can hold or retain water very firmly. Thus clay particles are chemically more active. However pure clay particles are not suitable for plant growth because they form a solid impenetrable mass. When clay soil is mixed with other constituents of the soil, the granular soil is formed, which becomes ideal soil for cultivation, because it combines the properties of porosity and water holding. It is also rich in nutritive salts.

Loamy Soil contains about 1 part clay, 2 parts silt and 2 parts sand. It is ideally suited for plant growth because it possesses good aeration, sufficient nutritive salts and good water retaining capacity.

Factors that Control Nature of Soil. The physical nature of the soil depends on its texture (physical structure), porosity and water holding capacity. The chemical properties of the soil depend on its salt content, pH and organic and inorganic nutrients such as nitrogen, phosphorus and potassium. The conditions of the soil is controlled by topography, climate and organisms inhabiting it.

Humus. It is made up of partially decayed and partially synthesized organic materials. Humus is rich in nutrients and thus promotes plant growth. Humus makes the soil granular thereby increasing its air and water holding capacity. Being black it absorbs heat and warms up the soil.

Formation of Soil. It includes two processes.

1. ***Weathering.*** It is destructive process of transformation of solid rocks into soil. Weathering is of two types.
   (i) *Physical Weathering.* It is brought about by temperature, water, wind and biological agencies.
   (ii) *Chemical Weathering.* It includes solution, hydration, hydrolysis, oxidation, reduction and carbonation.
2. ***Soil Development.*** It is constructive process that results in a soil profile. Soil development includes humification (formation of humus layer), calcification (accumulation of calcium carbonate), laterisation (silica is removed while iron and alumina remain).

Factors Reducing Soil Fertility. (i) Overcropping which withdraws minerals from the top layers of the soil, (ii) Non-rotation of crops causes deficiency of minerals at a particular level. It also increases incidence of soil borne diseases, (iii) Soil erosion due to wind or water, (iv) Deposition of silt due to floods, (v) Overgrazing, (vi) Leaching, (vii) Salination of top soil due to improper irrigation, (viii) Precipitation of minerals due to change in soil pH, (ix) Non availability of rain or irrigation, (x) Salinity of underground water.

Conservation of Soil. It requires preservation of its top layer which is most suitable for plant growth. This is achieved in the following ways.

1. ***Restoration of Fertility.***
   (i) Recovery of minerals from decomposition of fallen leaves, twigs, dead roots, dead animals and animal excreta.
   (ii) Biological nitrogen fixation.

(iii) Passing down of nitrogen, sulphur and other salts from atmosphere to the soil through rain, etc.
(iv) Addition of farmyard manure.
(v) Green manuring.
(vi) Addition of Fertilizers.

2.. ***Prevention of Soil Erosion.*** Soil erosion is caused by water and wind. Following methods are for the prevention of both water and wind erosion.

(i) *Erosion Checking Crops.* Crops such as grasses, berseem, groundnut, pulses offer excellent soil protection against erosion by water.
(ii) *Crop Rotation.* It checks soil erosion.
(iii) *Mulching.* Mulches of different kinds minimise evaporation and increases absorption of moisture.
(iv) *Contour Bunding.* Each bund holds the rain water within each compartment.
(v) *Terracing.* Terracing of slopping lands reduces the speed of water and prevent soil erosion.
(vi) *Outlet Channels.* Suitable outlet channels can carry flood water.
(vii) *Afforestation* and *reforestation* also check soil erosion.
(viii) *Regulating Grazing.* Probably the most important cause of wind erosion is overgrazing. There must be grazing regulations to restrict the use of grazing lands.

MINERAL RESOURCES

Minerals are non-renewable natural resources. Minerals occur in living organisms as components of organic and inorganic molecules and ions. *Major minerals* of the animal body are calcium, phosphorus, sodium, chlorine, magnesium and sulphur. *Minor minerals* of the animal body are iron, copper, cobalt, manganese, molybdenum, zinc, fluorine, iodine and selenium. In plants, minerals are divided into (i) *macro-nutrients* which include phosphorus, potassium, calcium, magnesium, sulphur and iron and (ii) *micro-nutrients* which include manganese, cobalt, zinc, boron, copper, molybdenum and chlorine. We have discussed the roles of these minerals in the living beings (animals and plants). Some important minerals and their other uses are given in the following table.

Some Important Minerals and some of their uses

Minerals	*Uses*
1. Metallic	
Uranium	Nuclear bombs, electricity, tinting glass
Thorium	Nuclear bombs, electricity, gas mantles
Iron	Steel
Mangarfese	Alloy steels, disinfectant
Cobalt	Alloy, catalysts, radiography, therapeutics
Columbium	Stainless steel, nuclear reactors
Chromium	Metallurgy, refractory, chemicals
Molybdenum	Alloy steels
Nickel	Over 3,000 alloys
Tungsten	Alloy and chemicals
Vanadium	Alloys
Copper	Electrical products, alloys
Lead	Batteries, gasoline, paints, alloys
Tin	Tin plate, solder, chemicals
Zinc	Galvanising, solder, die-casting, chemicals
Aluminium	Aircraft rockets, building materials, electrical wiring, utensils
Magnesium	Structural refractories
Titanium	Pigments, aircraft, alloys
Zirconium	Refractories, ceramics, metals, chemicals
Beryllium	Copper alloys, refractories, atom energy field
Gold	Monetary purposes, jewellery, dentistry
Radium	Medical and industrial uses, radiography
2. Non-metallic Minerals	
Asbestos	Insulation; textiles, roofings, glass, ceramics, gasoline, solid propellants
Corundum	Abrasives
Feldspar	Ceramic flux, artificial teeth
Fluorspar	Flux, acid, refrigerants, propellants
Phosphates	Fertilizers, chemicals
Salt	Chemicals, glass metallurgy
Sulphur	Fertilizers, acid, iron and steel industries

MINERAL WEALTH OF INDIA

India is fairly rich in mineral resource. It possesses huge reserves of iron, extensive deposits of coal and mineral oil, rich deposits of bauxite and has a virtual monopoly of mica. However, our country is deficient in some minerals like petroleum, tin, lead, zinc, and nickel, but the continued exploration of India's underground mineral wealth is yielding promising results, thus adding to the known and potential deposits of various minerals. The mineral resources of India are, however, very unevenly distributed. The Great Plains of Northern India are almost entirely devoid of any known deposits of economic minerals. On the other hand, south Bihar and Orissa situated on the north-eastern parts of peninsular India hold large concentrations of mineral deposits, accounting for nearly three fourths of the country's coal deposits as well as rich deposits of iron ore, manganese, mica, copper and bauxite, which are also found over the rest of the peninsular India and in parts of Assam and Rajasthan.

Name of some important minerals and the states where they are largely found are:

Metallic minerals	*States*
Bauxite	Bihar and Madhya Pradesh
Chromite	Orissa and Karnataka
Coal	Bihar and West Bengal
Copper	Bihar and Rajasthan
Diaspore	Uttar Pradesh and Madhya Pradesh
Gold	Karnataka
Iron	Goa and Madhya Pradesh
Lead	Rajasthan and Andhra Pradesh
Lignite	Tamil Nadu and Gujarat
Manganese	Orissa and Karnataka
Natural gas	Assam and Gujarat
Petroleum	Assam and Gujarat
Silver	Rajasthan and Bihar
Tungsten	Rajasthan and West Bengal
Zinc	Rajasthan

Metallic minerals	*States*
Non-metallic minerals	
Asbestos	Rajasthan and Bihar
Ball clay	Andhra Pradesh and Rajasthan
Barytes	Andhra Pradesh and Maharashtra
Calcite	Rajasthan and Gujarat
China clay	Rajasthan and West Bengal
Corundum	Karnataka and Maharashtra
Diamond	Madhya Pradesh
Dolomite	Madhya Pradesh and Orissa
Felspar	Rajasthan and Tamil Nadu
Fire clay	Bihar and Gujarat
Fluorite	Gujarat and Rajasthan
Graphite	Orissa and Rajasthan
Gypsum	Rajasthan and Tamil Nadu
Kyanite	Bihar and Maharashtra
Lime stone	Madhya Pradesh and Tamil Nadu
Magnetite	Tamil Nadu and Uttar Pradesh
Mica	Bihar and Andhra Pradesh
Ochre	Rajasthan and Madhya Pradesh
Pyrites	Bihar
Sulphur	Tamil Nadu
Quartz	Andhra Pradesh and Karnataka
Quartzite	Orissa and Bihar
Silica sand	Uttar Pradesh and Gujarat
Sillimanite	Maharashtra and Meghalaya

Mineral Production

Bombay High ranks first in the value of mineral production accounting to 33 per cent of the total value, followed by Bihar (15 per cent), Madhya Pradesh (11%), Assam (8%), Gujarat (8%), West Bengal (6%), and Andhra Pradesh (4%). Table 2.2 shows production of some important minerals in India.

Table 2.2 : Production of some Minerals in India (1989)

Mineral	*Quantity*	*Production*
Bauxite	Million tons	4.3
Chromite	Thousand tons	948
Copper ore	Millions tons	5.1
Gypsum	Million tons	1.4
Lead Concentrate	Thousand tons	4.5
Manganese ore	Thousand tons	1,320
Mica	Thousand tons	4.1
Zinc Concentrate	Thousand tons	123
Coal	Million tons	203
Iron Ore	Million tons	49
Lime stone	Million tons	64
Petroleum Crude	Million tons	33
Gold	Kilograms	2,000

Distribution. Every country is dependent upon other countries for the requirement of many minerals due to their unequal distribution on the earth. For example, India has abundant resources of iron, manganese, limestone, chromite, dolomite and mica but is quite deficient in lead, copper, gold, silver, nickel, potassium and phosphorus. Similarly, North America has abundant deposits of molybdenum but contains little of tin, tungsten and manganese. Parts of Asia (Indonesia, Malaysia) have the reverse of this i.e., they are rich in tin, tungsten and manganese but deficient in molybdenum. South Africa has rich deposits of gold, platinum and uranium but quite little of silver and iron.

Even for our basic need of NPK fertilizers, India is dependent upon foreign countries because of multiple reasons like deficiency of petroleum energy (for requirement of nitrogen fertilizers using Haber-Boach process), electric energy and raw material *(e.g.,* Potassium reserve). Only recently phosphate rocks were discovered in the area of Jawar Kota to Rajasthan.

It is estimated that at present rates of consumption copper, lead, zinc, tin, gold, silver and platinum will last until near the end of twentieth century Aluminium, cobalt, manganese and molybdenum will last upto about 2100-2200 A.D. and iron and chromium will last upto about 2500- 2800 A.D.

Conservation. Following steps can conserve the mineral resources.

1. The minerals should be recycled as far as possible.
2. Use of minerals as raw material should be restricted to industries producing essential materials only.
3. New technologies should be adopted in the industries using minerals so that the wastage can be checked.
4. The untapped mineral deposits of the minerals in sea beds should be exploited.
5. Proper substitutes should be found; for minerals *e.g.,* plastic wares in place of metallic containers, copper-zinc, nickel alloy in place of silver in coins; glass, plastics, ceramics and synthetic fibres in place of exhaustible minerals.

BIOTIC (LIVING) RESOURCES

They include plants, animals and microorganisms. The living resources are renewable.

(i) Plant Resources (Role of plants)

1. Many plants are cultivated for food (grains, pulses, vegetables, fruits), sugar, oil, fibres (cotton, jute), drinks (tea, coffee), medicines, decoration and fuel.
2. Timber is obtained from the trees which is used for construction work, furniture, rail sleepers and compartments, ships, boats, decks, poles, radio and television cabinets.
3. Different kinds of paner are produced from the wood pulp.
4. Certain trees provide gums, drugs, resins, tannins wood alcohol acetic acid, oxalic acid, charcoal and dry fruits.
5. Trees provide shelter and food to many birds and mammals.
6. Plants and trees provide oxygen for man and other animals for breathing.
7. Vegetation affects the rainfall by retaining moisture.
8. Vegetation benefits the soil. It prevents soil erosion. Dead fallen leaves decay and form humus that enriches the soil. Plants also check the spread of deserts.
9. The entire plant community on land can be broadly classified as *forest, grassland, shrubs* and *crop lands*.

The dry biomass produced annually on earth is 380 thousand million tonnes, in which about 325 thousand million tonnes are contributed by marine plants. However man mostly depends on the land vegetation to meat his requirements. Over 4,00,000 species of plants are found on the earth. Only a few hundred species are used for food and other needs.

Conservation. Various human activities such as environmental pollution, conversion of climax communities to crop-land, deforestation and overgrazing by cattle are causing damage to vegetation and depletion of plant life. Thus several species of plants are endangered. It is, therefore, very essential that these human activities should be checked to prevent the damage to the plant life. In addition, marine plants must be used for food and other materials. Man must explore more plant species which can be utilized as food and other human needs.

(ii) Animal Resources (Role of Animals)

1. Animals provide *food* to the man. Prawn, Fish, fowls, goat, sheep, and pigs are reared for meat, fowls and ducks for eggs, cows, buffaloes, goats, reindeer for milk and honey bees for honey.
2. Horses, mules, asses, cattle, elephant and reindeer are used for *transport.*
3. Cattle and camels are employed in *agricultural operations.*
4. Cattle, sheep, rabbits and silk worms give leather, wool, fur and silk respectively.
5. Honey bees give *Wax.* Cantharidin is obtained from blister beetles. Lac is secreted by lac insect.
6. Many animals particularly insects, help in pollination and dispersal of fruits and seeds.
7. Guinea pigs, rats, rabbits, dogs and monkeys are used as laboratory animals in research.
8. Musk is obtained from deer. Ivory is obtained from elephants.
9. Some hormones secreted by adrenal, thyroid, pituitary of some mammals are used to cure the disease of human beings.
10. Certain animals such as hyaenas, jackals, pigs, etc., are scavengers
11. Faecal matter of birds and mammals is good manure for plant growth.
12. Pearls are formed by pearl oysters (molluscs).
13. Animals help in maintaining the ecosystems.

Conservation. Destruction of habitat by deforestation, indiscriminate killing of animals for sport and poaching of animals for their products are depleting animals life. These factors have caused a great damage to several species which are threatened with extinction. About 25 percent of all the animal species are likely to disappeared in the next 25 years if immediate measures are not taken to protect their habitat and stop their indiscriminate killing. People must be educated.regarding the conservation of animal life.

(iii) Micro-organisms

They are small organisms that cannot be seen without the aid of a microscope. Micro-organisms exist almost every where on earth. They were first seen by Anton van Leeuwenhoek (1632–1723). The four major groups of micro-organisms are the bacteria, protozoa, fungi (yeasts and moulds) and viruses. Bacteria can be found as much as 5 meters deep in the soil, in fresh and salt water, and even in the ice of glaciers. They are abundant in air, in liquids such as milk and in and on the bodies of animals and plants. Most bacteria do more good than harm. Bacteria are the most active micro-organisms in breaking down the complex molecules of dead plants and animals into simpler chemicals that can be re-used by living organisms. *Escherichia coli* is harmless bacteria of the intestines of man and other mammals where it synthesizes certain vitamins. Some bacteria fix atmospheric nitrogen and provide it to plants. Some bacteria are also harmful and cause many diseases. Many fungi are useful to humans. Certain fungi are used in the preparation of bread and wine.

Penicillin, a medicine used to treat bacterial infections, is the product of a fungus. Together with bacteria, the fungi break down the biological molecules of dead organisms and therefore, play a critical role in recycling materials. Fungi can also be a nuisance and harmful.

Protozoa are harmful as well as beneficial. Harmful protozoa pollute water, reduce soil fertility and cause diseases. Beneficial protozoa provide food to many aquatic animals. The skeletal deposits of the *foraminifera* and *radiolaria* are often found in association with oil deposits. Thus, they help in the location of this important fuel. All the viruses are harmful and cause different types of diseases.

Alternations in the Environment

Man made changes in the environment have three main dimensions; changing the physical conditions, changing the biotic/ biological community and depletion and degradation of natural resources.

1. Physical conditions. Man has done much to alter the face of earth. He has disturbed the physical conditions of the environment. Man has cutdown forests for wood, farm land and for his habitations. He has utilized grass lands for the formation of villages, towns and cities. In certain places, big factories have been established on the fertile land. He has constructed broad highways and buildings. Man has also laid thousands of kilometres of railways. He has dug many canals, lakes and tanks for irrigation and transport and drains for sewage. These are great achievements but these advances have been brought at a price. The ecological balance in nature has been disturbed.

2. Biotic/Biological Communities. A biotic community is an assemblage of plants, animals and micro-organisms which live together, share same habitat and influence each other's life directly or indirectly. Man has altered biotic communities for his benefits. Originally man was a prey of carnivorous animals such as lions, tigers, leopards and crocodiles, etc., a *predator* of some herbivorous animals like goat, sheep, deer, birds and fish, etc. and a *host* for many parasitic worms and micro-organisms. When man built his houses, some animals like cockroach, lizards and mice started coming to his houses. These animals are called *inquilines.* Domesticated animals are completely dependent upon man for food, protection and reproduction. Many cultivated plants like cabbage, tomato, maize, wheat also require human help for their cultivation. These are called *cultigens.* Man has also succeeded in keeping some wild animals as pets (dogs, cats, macaws-large long tailed parrots of tropical America and parakeets – small long tailed parrots). Some rare and unusual wide plants have been collected and kept them in houses.

Man has transplanted many organisms to new habitat or a country.

Pathenium hysterophorus (the carrot grass) growing widely in several parts of India arrived as late as 1956. Man has introduced many plant and animal species to new places to meet his needs (clover and bumble bee to New Zealand, Merino sheep to Australia), as agents of biological control (cochineal insect to Australia to control *Opuntia)* or for fancy (water hyacinth) in India. Man has also changed biotic communities by destroying their habitat. Deforestation destroys the habitat of many organisms. The industrial wastes, chemical effluents, gases, etc. pollute water and air killing many fresh water and marine organisms. Insecticides are used to kill the insects but these insecticides also kill those birds that we do not wish to kill.

3. ***Depletion and Degradation of Natural Resources.*** Today the world's energy resources have reached a critical stage because we totally depend on the fossil fuels such as coal, petroleum and natural gases. It has caused such depletion that these fuels may last only another few centuries. Modern industrial and aviation technology are polluting the atmosphere (air), causing a threat to the environment and climate. Water is being polluted by human activities in various ways killing numerous fresh water and marine organisms valuable to man. Soil erosion, salinisation, water logging and acidification and alkalination of soil wrought by human actions are the great dangers to land resources. Construction of dams, roads and railways, urban encroachment and industrialisation and mining activities are also causing depletion of productive land. Depletion of almost all known mineral deposits of the world are anticipated within a few decades. There may be shortage of some elements like mercury, tin, copper, silver, gold, platinum, etc. Phosphorous is an essential component of fertilizers, its shortage may retard agricultural growth. Various human activities such as deforestation, and severe environmental pollution are causing serious damage to plant life and depletion of vegetation. Habitat destruction, indiscriminate killing of animals for sport, environmental pollution are the biggest dangers to animal life. Thus, some of the natural resources are in their serious depletion and degradation.

Consequences of Alternations

By changing natural resources man has combated with nature. But all this has caused *mental imbalance.* Therefore, the ecologists have started predicting that if man continues to disrupt his environment at the same speed, it will be a threat for his survival as a species. Man depends on natural resources directly or indirectly, therefore, his existence is not possible without them. It is essential that man must realize the limitations of his resources and environment. The present way of living imposes threat not only for the scarcity of resources but also for the ecological imbalance. Man is an integral part of the environment. He cannot completely isolate himself. It must, therefore, be realized that man should try to conserve his environment in which he is residing and should use his environment in a proper way so that the human race is not exposed to environmental hazards.

REVIEW QUESTIONS

1. What are the main factors that have made man the dominant organism of the biosphere ? Or "Man is the dominant species of the biosphere." Justify this statement.
2. How are the world's natural resources classified ?
3. What are the main sources of energy for man ? What are the problems of energy resources and their utilization ?
4. What are alternative sources of energy ?
5. Name the various layers of the atmosphere and give the range of each.
6. What is the role of atmosphere ?
7. Describe the important ocean water resources.
8. What is soil ? Explain its composition and formation.
9. (i) What is the importance of humus in the soil?
   (ii) Explain the role of soil in the life of organisms.
   (iii) Discuss nitrogen fixation.
   (iv) Name the factors that reduce the soil fertility.
   (v) Describe the measures for soil conservation.
10. Discuss the value of minerals as natural resources.
11. Explain the role of plant or animal resources.
12. Describe micro-organisms as resources of great value.
13. What steps would you propose for sustained development without depletion of available resources and ecodegradation ?
14. Explain the following terms: Exosphere, ozonosphere, acid rain, pedogenesis, food pyramid, Inquilines, Cultigens, Environmental imbalance.

Chapter 3

THE ECOSYSTEM

Ecosystem is a structural and functional unit of biosphere consisting of community of living beings and the physical environment, both interacting and exchanging materials between them. The term was introduced Tansley in 1935. An ecosystem can be temporary *(e.g.,* rain water pond) or permanent, natural or man-made *(e.g.,* aquarium), small *(e.g.,* edge of a pond, a drop of pond water) or large *(e.g.,* ocean). A very small ecosystem may be called micro-ecosystem. Ecosystems are so varied in form and stature that whatever has a distinct community of its own, would be an ecosystem, *e.g.,* rotting log, pond, lake, river, edge of a pond, village garden, orchard, park, crop field, small tract of forest, a forest, a grass land, manned spaceship, etc.

Homeostasis. An ecosystem maintains a functional balance or relatively stable state of equilibrium amongst its various components. The phenomenon is called *homeostasis* or *balance of nature*. It is not static but fluctuates within certain limits. Homeostasis or balance of nature maintained through a number of controls (= limitations = cybernetics. The important ones are: (i) *Carrying capacity* of the environment, (ii) Capacity of the ecosystem for recycling of wastes, (iii) *Self regulation* through the effect of density on reproductive potential e.g., flour beetle, tree toad. (iv) *Feed back system*. In feed back system, one living component of the ecosystem controls the population of another living component. Feedback system is of two types, *positive* and *negative*. For example, increase in plant matter would increase the population of herbivorous insects. The latter results in the increase in population of frog predator insects and insectivorous birds. This increase in population different level organisms with increase in population at the lower level food chain is called *positive feed back*. The increased population insectivorous animals would, however, soon exert a *negative feed back* on the population of herbivorous insects through excessive predation (Fig. 3.1). Thus availability of food is an important factor in maintaining population densities and hence balance of nature.

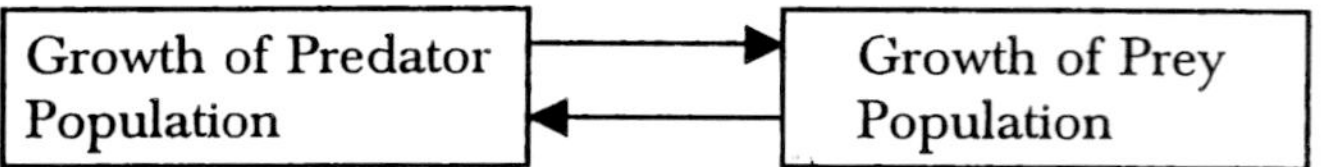

Fig. 3.1 : Feed back system for controlling population of prey and predator

COMPONENTS OF ECOSYSTEM

An ecosystem is made up of two types of components, biotic and abiotic.

Biotic Components

They constitute all the living members of an ecosystem. The different biotic components are connected through food and a number of other relations. Food *is a group of materials that contain energy and all the essential ingredients required for body building, growth and body functions.* Both matter and energy are transferred in the living world through food. Food is manufactured from inorganic raw materials by autotrophs only. Autotrophs are, therefore, also called *producers.* Other organisms which cannot manufacture their own organic food are called heterotrophs. Heterotrophs are of two main types, *consumers* (herbivores and carnivores) and *decomposers.*

1. ***Producers (Autotrophs).*** They are photosynthetic or autotrophic plants which are able to synthesise organic food from inorganic raw materials with the help of solar radiations. The process is called photosynthesis. Energy contained in solar radiations is first changed into chemical form by chlorophyll of photosynthetic plants. This chemical energy is then used in bond building of organic compounds. The energy can be released when the bonds of the organic compounds are broken down as during respiration. Producers are also called *transducers* because they are able to change radiant or light energy into chemical form. Organic compounds synthesised by producers take part in building their bodies as well as release of energy for various body activities.

All other organisms depend upon the producers for their supply of organic compounds or food. Producers also maintain 002/02 balance of nature. They pick up carbon dioxide from the atmosphere and release oxygen during the process of photosynthesis.

*2. **Consumers (Heterotrophs).*** They are animals which feed on other organisms or their parts. Consumers ingest their food. They are, therefore, also called *phagotrophs.* Consumers are differentiated into two broad categories, *herbivores* and *carnivores.* Herbivores obtain their food (and energy) directly from plants. They are called *first order consumers.* Some common herbivores of terrestrial ecosystems are deer, rabbit, mouse, squirrel, grass-hopper, some beetles, goat, cattle, etc. Similar herbivores of the aquatic ecosystem are mostly crustaceans, molluscs and protozoans.

Carnivores ingest or prey upon other animals. The carnivores which feed on herbivores are named as *primary carnivores* or *second order consumers, e.g.,* frog, centipede, birds, fishes, wild cat, jackal, fox, snakes, etc. The animals which feed on primary carnivores are called *secondary carnivores or third order consumers, e.g.,* owl, peacock, several fishes, tiger, lion, etc. Secondary carnivores may become food of tertiary carnivores and so on. The carnivores, which cannot be preyed upon further, lie at the top of food chain and are termed as top carnivores, *e.g.*. Lion.

*3. **Decomposers.*** They are saprotrophic micro-organisms which feed on dead bodies of organisms and organic wastes of living organisms. The decomposer organisms secrete digestive enzymes to digest the organic matter externally. The digested form of organic matter is partly absorbed by micro-organisms for their own assimilation. The remaining adds raw materials and minerals back into the substratum. The phenomenon is called *mineralisation.*

Decomposers are also called *reducers* because they are able to remove or degrade the dead bodies of organisms. Because of their small size they are known as *micro-consumers.*

Some workers differentiate two more categories of living beings amongst the biotic components of an ecosystem. They are *detrivores* and *parasites*. Parasites belong to diverse groups, *e.g.,* bacteria, fungi, protozoans, worms, etc. Every type of living being can be attacked by parasites. Detrivores or *scavengers* are animals which feed on dead bodies of other organisms, *e.g.*, termites, carrion butles. They are helpful in quick disposal of the dead bodies.

Differences between Autotrophs and Heterotrophs

Autotrophs	*Heterotrophs*
1. They manufacture their own organic not manufacture or nutrients from inorganic raw materials.	1. Heterotrophs do nit organic materials.
2. Autotrophs obtain only in organic materials from outside.	2. They obtain organic nutrients from outside
3. Autotrophic organisms live on inorganic medium,	3. Heterotrophic organisms live in company of or occasionally come in contact with the source of organic matter.
4. They require an external source of energy for synthesis of organic food.	4. External source of energy, other than organic matter, is not required.
5. Ecologically they are producers. They include most plants, some protists and some monerans.	5. Ecologically they are consumers. Heterotrophs many protists, include animals, fungi, several monerans and a few plants.

Abiotic Components

Abiotic component of an ecosystem consists of non-living substances and factors. They are broadly divided into *climatic* and *edaphic* factors. Climate is a complex formed by interaction of solar radiations with various constituents of lithosphere, hydrosphere, latitude and altitude, rocks, etc. to produce variables like temperature, humidity, precipitation, air currents, light, etc. The edaphic factors include those connected with soil or substratum like topography, background, mineral elements, pH, etc. Abiotic factors limit the distribution, behaviour and relationships with other organisms.

1. Temperature. Organisms generally live with a narrow range of temperature (5°-35°C) with the exception of spores, seeds, some prokaryotes and other lowly organised individuals. The latter can be found in hot springs (60°-90°C) or permafrost (-30° to -50°C). Temperature range varies in different parts of the earth. It has created different life zones—tropical, subtropical, temperate, arctic or alpine. High or low temperature causes inactivity and death of organisms. It is immediate in case of *poikilothermal* (= ectothermal = cold blooded) animals and delayed in case of *homoiothermal* (= endothermal = warm blooded) animals. Therefore, organisms show adaptations to avoid extremes of temperature.

Animal Adaptations to Low Temperature, (i) Thick coat with scales, feathers and hair, (ii) Abundant subcutaneous fat, *e.g.*,. Polar Bear, Whale, (iii) Winter eggs in case of short lived animals, (iv) Hibernation or winter sleep in warmer caves, crevices and underneath soil, *e.g.,* Frog, Polar Bear, (v) Migration to warmer areas, *e.g.,* Reindeer, Caribou, several birds and some fishes.

Animal Adaptations to High Temperature. In plains of tropics, the day temperature often crosses 45°C while in hot deserts it may touch 55°C. (i) During hotter parts of the day the animals may rest under shade in a moist place. The process is called *aestivation.* It is adopted by several animals like frog, insects, spiders, reptiles and some mammals, (ii) In deserts the animals often live in burrows during the hotter parts of the day. They become active only when it is pleasant outside—night (nocturnal), early morning (auroral) or evening (vesperal).

Plant Adaptations to Extreme Temperatures. Plants, being stationary, have to face the extremes of temperature. They develop cold hardiness or heat hardiness. For this the plants have (i) Thick cuticle, (ii) Thick corky bark, (iii) Hair, (iv) Thick leaves, (v) High solute content, (vi) Mucilage, (vii) Tannins.

*2. **Light.*** Light intensity, quality and duration control the structure, growth and activities of organisms. For this the organisms possess special *photoreceptors.* The different effects of light are as follows:

(i) *Photosynthesis.* Light is essential for photosynthesis performed by plants. Photosynthesis provides food and energy not only to plants but all other organisms as well.
(ii) *Productivity.* As compared to equator, light intensity is 50% less at 50°N. As a result, tropical forests have higher productivity than temperate forests. There is little vegetation in the tundra region. Flora influences the density of fauna.

(iii) *Growth.* In plants, light is essential for differentiation of tissues. It also influences growth. Intense light produces shorter internodes, thicker but smaller leaves, more mechanical tissues, smaller cells, more flowers, etc. Reduction in light intensity has the opposite effect like increased internode length or larger and thinner leaves.

(iv) *Pigmentation.* In strong light yellow and red tints appear on the leaves. Moderate light forms bright green leaves. Intensity of light changes skin colour in human beings as well as in several animals. Wall lizard and frog are light coloured when kept in bright light. They become dark coloured in dim light. Some mammals and birds have seasonal Colour changes. Snow Hare develops white for during autumn.

(v) *Animal Activity.* Depending upon the particular period of the day when the animals are active, the latter are of four categories: (a) *Diurnal.* Active during the day, *e.g.,* butterflies, sparrow, crow, pigeon, cattle, (b) *Nocturnal.* Active at night, *e.g.,* moths, owl, bat cockroach, (c) *Auroral.* Active at dawn only, and (d) *Vesperal.* Active at dusk only, *e.g.*. Rabbit.

(vi) *Plant Movements.* Plants show bending in response to unidirectional light. Motile algae are observed to live on the surface of water during dim light and deep in water during strong light. Many plants show folding of leaves during night (nyctinasty, *e.g., Oxalis)* or sometimes during midday (e.g., some grasses, some legumes).

(vii) *Day Length.* It controls bird migration, leaf fall, bud dormancy and appearance of new leaves. For example, leaf fall, bud dormancy and southward migration of many birds are caused by shortening day length.

(viii) *Flowering.* Photoperiodicity or periodicity of day length and darkness controls flowering in many angiosperms. Depending upon the day length required for flowering, plants are of three types—*long day (e.g.,* Radish, Spinach), *short day* (e.g., Xanthium, Dahlia) and *day neutral (e.g.,* Tomato).

(ix) *Animal Breeding.* Seasonal animal breeding is influenced by photoperiodicity – *long day animals (e.g.,* ferrets, turkeys, starlings), *short day animals (*e.g., sheep, goat, deer) or *day neutral animals (*e.g., guinea pigs, ground squirrels).

3. ***Wind (Air Currents).*** Wind affects ecosystem in a number of ways:

(i) *Pollination.* Most gymnosperms and a number of angiosperms are pollinated by wind (anemophily).

(ii) *Dispersal.* Wind is an agent of dispersal of many fruits, seeds, cysts, spores, eggs and even small animals.

(iii) *Flight Animals.* They are rare in areas frequented by high speed winds. Instead of winged insects, wingless insects are found in these areas. Some birds are, however, able to live in windy areas by building well protected nests and moving out only during calm periods. Nests are built behind sheltered rocks or stone wind screens are constructed to protect the nests.

(iv) *Wind Training.* Areas frequented by unidirectional winds develop flag trees which have branches on one side only. Trees do not grow at high altitude due to mechanical and physiological effects of wind. The plants of these areas usually possess strong spreading roots and strong but flexible shoots.

(v) *Transpiration.* Wind increases transpiration.

(vi) *Weather Conditions.* Air currents determine weather conditions including rainfall.

(vii) *Soil Erosion.* In dry areas wind causes soil erosion. Spread of deserts is caused by shifting of sand particles to fertile areas.

4. ***Humidity.*** It is the amount of water vapours present in the atmosphere. Humidity controls formation of clouds, dew, fog, etc. Epiphytes grow only in humid areas. Evaporation of water from the body of land organisms in transpiration and perspiration is regulated by humidity. Both plants and animals develop modifications for reducing water loss from their body in arid areas.

5. ***Precipitation.*** It may occur as rainfall, snow, dew, hail, etc. Periodicity and amount of rainfall determines type of forest in an area–evergreen, deciduous, chapparal, grassland, savannah, desert, etc. Animals are also adopted accordingly.

6. ***Water.*** Land plants meet their water requirements from soil. Land animals obtain the same from pools, lakes, rivers, springs, etc. Plants and animals show modifications according to availability of water in the area and requirement of conserving the obtained water. Plants of dry areas are called *xerophytes.* They develop modifications to increase water absorption, reduce transpiration and at times store absorbed water. Certain animals of the dry areas do not drink water at all, *e.g.,* Kangaroo Rat. They use water from food and its metabolism to run their body machinery. Animals of dry areas often reduce water loss by producing solid faeces and excreting solid urine.

Water is abundant in aquatic habitats. Plants of aquatic habitats are called *hydrophytes.* Hydrophytes possess *aerenchyma* or air storing parenchyma to support themselves in water. Clarity of water, salt content, depth and water waves or speed determine the growth and distribution of plants and animals. In rivers and streams, animals obtain most of their food from organic materials coming from outside the water. In ponds and lakes producers grow in sufficient strength. Organisms found in fresh water have a problem of excess internal water because of endosmosis. Organisms found in ocean or saltish water have a problem of low internal water content due to exosmosis. Some have problem of excreting excess salts obtained from outside. In oceans at a depth of more than 200 m, producers do not occur. Only consumers are found there. Deep sea animals do not possess air sacs. Many of them are luminescent.

Water currents restrict distribution of organisms in streams and intertidal areas of oceans. In streams only attached plants grow. They have dissected or ribbon shaped leaves. Animals found here are either strong swimmers, have attaching organs or live under stones, in burrows, crevices etc. Similarly in intertidal area of ocean attached plants *(Fucus, Laminaria),* sessile animals (Sea anemone and limpets), burrowing animals (e.g., *Nereis,* tube worms) or very strong swimmers are met with.

7. ***Background.*** Most animals have an adaptation to have colour, pattern and general texture similar to that of the background in which they operate. This allows them to, camouflage themselves so that they do not become conspicuous to their preys or predators. For example, elephants and rhinos have colour similar to that of tree trunks and mud. Lions and camels are sand coloured. Praying mantis, bee frog *Hyla* and grasshopper are green in colour. Jelly-fishes and sea-cucumbers

are glassy. Chameleon is able to change its colour according to the background.

8. ***Topography***. It is the surface behaviour of the earth like altitude, slope, exposure, mountain chains, valleys, plains, etc. Topography influences other environmental factors, atmospheric pressure, winds, rainfall, light intensity and light duration, temperature, water currents or wave action. It is also a causal agent for isolation, formation of new species and geographical distribution of organisms. North and south faces of hill possess different types of flora and fauna because they differ in their humidity, rainfall, light intensity, light duration and temperature regimes. It is because the two faces of the hill receive different amounts of solar radiations and wind action. Similarly, the centre and edge of a pond possess different depths of water and different wave action. Top side of a rock is exposed to wave action and light while the underside of the rock has little wave action and light. Therefore, different parts of the same area may possess different species of organisms.

9. ***Gases***. Nitrogen is present in abundance (4/5th of atmosphere) but is itself chemically inert. It forms useful salts through electro-chemical, photochemical and biological fixation. Carbon dioxide concentration of the atmosphere is always a limiting factor for photosynthesis. However, excess of carbon dioxide concentration is harmful to animals as well as climate. Its concentration increases during night but decreases during day. In water it occurs as bicarbonate and carbonate ions. Oxygen concentration is supraoptimal for C_3 plants, optimal for C_4 plants and animals except at high altitudes. In water oxygen concentration determines distribution of organisms. In the middle or intermediate stratum photosynthesis increases oxygen concentration during day but it becomes little during night depending upon population, pollution and decomposition. In deep waters, animals are faced with very low oxygen concentration.

10. ***Soil***. It determines vegetation growth and pattern, underground flora and fauna through its constitution, origin, temperature range, water retentivity, aeration, minerals, etc. Soil present on the slopes as well as the one which is uncovered are liable to be eroded by water and air respectively.

11. ***pH (Hydrogen ion Concentration).*** There is very little change in pH in oceans. Terrestrial animals are also not much influenced by pH of the substratum. However, distribution of land plants and soil organisms is determined by pH of soil. A similar control on distribution is found in fresh water habitats. Snails and earthworms do not occur in acidic soils. At this pH, *Euglena* and other flagellates are quite abundant. Animals having calcareous shells live in media having neutral or alkaline pH..
12. ***Mineral Elements.*** A large number of minerals, also called biogenic or biogenetic nutrients, are required by organism— proper growth. Deficiency or absence of any one results in abnormal growth which may lead to death. Excess, minerals are equally harmful. Abundance of some minerals favour the growth of some tolerant species. Snails occur in soils rich in calcium content. Soils deficient in nitrogen salts often possess nitrogen fixing bacteria and cyanobacteria. Plants having symbiotic relationship with these bacteria also abound in the soils. Carnivorous plants meet their requirement of nitrogen by catching small insects, worms, etc. Salinity of ocean is overcome by many animals through salt secreting glands. Similar glands occur in halophytes or plants growing in saline soils and marshes. Special adaptations are found in animals inhabiting estuaries where there are wide fluctuations in salt content, Areas having very high salt content are usually devoid of much vegetation, *e.g.*. Dead Sea, Great Salt Lake.

Trophic Levels

Trophic level is a step or division of food chain which is characterised by the method of obtaining its food. The number of trophic levels is equal to the number of steps in the food chain. The two fundamental trophic levels are *producers* and *consumers*. A fundamental similarity of all trophic levels is that it uses a part of food in body building while a major part of it is consumed in liberation of energy during respiration.

Producers belong to first trophic level or T_1. *They are autotrophic or photosynthetic organisms found in an ecosystem which synthesise organic nutrients from inorganic raw materials with the help of solar radiations not only for themselves but also for heterotrophic organisms or consumers. Autotrophs or producers possess chlorophyll for this purpose. Chlorophyll converts solar or light energy into chemical form. The chemical energy is stored in the bonds of organic materials.*

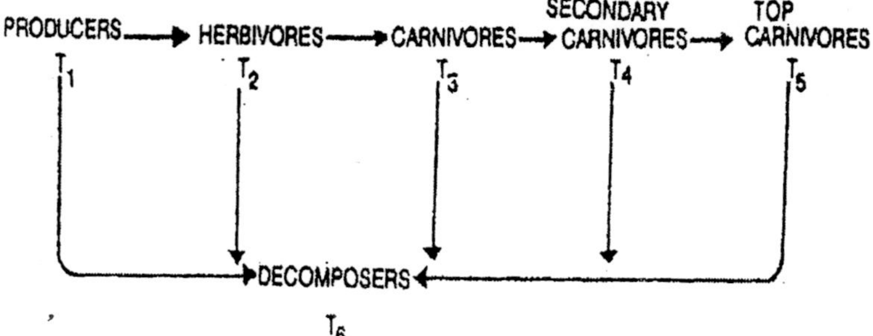

Fig. 3.2 : Trophic levels in an ecosystem.

Consumers are heterotrophic organisms which cannot manufacture their own food. They obtain ready-made organic food from outside sources. Depending upon the mode of obtaining nourishment, heterotrophic organisms are of three main types - herbivores, carnivores and decomposers. *Herbivores* or consumers of first order constitute the second trophic level or T_2. Consumers of the second order or primary carnivores form third trophic level or T_3. There may be 2-3 levels of carnivores. The ultimate or top carnivores belong to T_4 or T_5 trophic level. Decomposers form the last or detritus trophic level *(e.g.,* T_6). Parasites do not have any fixed trophic level since they feed on producers, herbivores as well as carnivores or various levels, *e.g.,* aphids, ticks, mites, leeches, mosquitoes.

Organisms of one trophic level have the same food habit, *e.g.,* herbivory. However, each trophic level may have several resources like leaves, seeds, fleshy, fruits, nectar, grasses, etc. A group of species belonging to a trophic level which exploits a common resource base in a similar fashion is known as *guild*, e.g., nectar feeding birds, grazing animals, browsing animals.

Food Chain

A food chain is a sequence of populations or organisms of an ecosystem through which the food and its contained energy passes with each member becoming the food of a later member of the sequence. A common food chain, called *predator food chain*, consists of producers, consumers and decomposers. The latter are often omitted since they operate at all the levels of the food chain. Consumers are often of 3-5 types—first order (primary), second order (secondary), third order (tertiary), fourth order (quarternary) consumers.

1. ***Producers***. They constitute the first trophic level or base of a food chain. Producers are autotrophic organisms which alone are able to manufacture organic food from inorganic raw materials in the process of photosynthesis. The energy for this process is obtained from solar radiations or sunlight. The latter is changed into chemical energy by chlorophyll. The chemical energy is used in combining raw materials into organic food. Oxygen is evolved in the process, light, chlorophyll. The chemical energy is used in combining raw materials into organic food. Oxygen is evolved in the process.

$$6CO_2 + 6H_2O \xrightarrow[\text{enzymes}]{\text{light, chlorophlly}} C6H_{12}O_6 + 6O_2$$

Organic food synthesised by producers consists of all essential ingredients like carbohydrates, fats, proteins, vitamins, etc. It contains energy. Part of the food manufactured by the producers is used up by themselves in providing energy for various body activities and in overcoming entropy. The rest is employed in their body building. Part or whole of the latter enters the food chain as food for consumers.

Producers are also known as *transducers* because they are able to change radiant or light energy into chemical form.

2. ***Primary Consumers or Herbivores***. They are animals which feed on plants or plant products, *e.g.*, grasshoppers and several other insects, rabbit, hare, field mouse, deer, antelope, elephant, zooplankton *(Paramecium, Daphnia,* some larvae), tadpoles, some fishes. A part of plant food eaten by herbivores become constituent of their body while a major part is consumed by them in production of energy for various body activities. Herbivores are also called *key industry animals* (Eiton, 1927) because they convert plant matter into animal matter.

3. ***Secondary Consumers or Primary Carnivores***. They ingest or prey upon herbivorous animals, *e.g.*, frog, predator, insects, several birds, fishes, wild cat, fox, jackal (also scavenger), etc. A part of flesh or food obtained from herbivore is used in body building by primary carnivore while the rest in consumed in providing energy for various life processes.

4. ***Tertiary Consumers or Secondary Carnivores.*** They are larger carnivores which prey upon primary carnivores, *e.g.*, wolf (feeds on fox), snake (prey upon frog).
5. ***Top Carnivores.*** They are the last order consumers or carnivores which are not preyed upon by other animals because of their size, agility and ferociousness, e.g., shark, crocodile, eagle, peacock, tiger, lion.

A food chain may terminate at the level of herbivore *(e.g.,* vegetation elephant), primary carnivore *(e.g.,* vegetation → squirrel → bear), secondary carnivore (vegetation → squirrel → wild cat → tiger), tertiary carnivore, etc. *Examples* of some common food chains are given below:

Land Food Chains

1. Grass → Grass Hopper → Frog → Snake → Peacock/Falcon.
2. Vegetation → Rabbit → Fox → Wolf → Tiger.
3. Vegetation → insect → Predator insect → insectivorous Bird → Hawk.

Aquatic Food Chains

1. Phytoplankton → Zooplanklon → Small Crustaceans → Predator Insects → Small Fish → Larger Fish → Crocodile.
2. Phytoplankton → Zooplankton → Small Crustaceans → Predator Insects → Small Fish → King Fisher/Stork.

Food Web

It is a network of food chains which become interconnected at various trophic levels so as to from a number of feeding connections amongst the different organisms of a biotic community. Food web is meant for increasing the stability of an ecosystem by providing alternate source of food and allowing the endangered population to grow in size. Normally a food web operates according to taste and food preferences of the organisms at each trophic level. However, availability of food source and other compulsions are equally important. In Sunderbans, the tigers eat fish and crab in the absence of their natural preys. Some organisms normally operate at more than one trophic level. Thus human beings are not only herbivores but also carnivores of various levels. Jackals are both

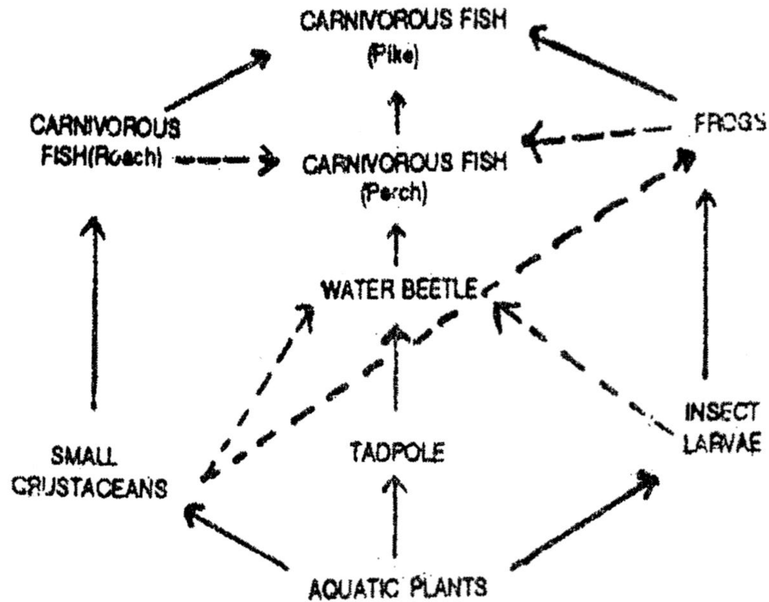

Fig. 3.3 : An aquatic food web with the interconnected food chains.

carnivores and scavengers. Snakes feed on mice (herbivores) as well as frogs (carnivores). Wild cats prey upon mice as well as birds and squirrels. A wolf eats not only fox but also rabbit and deer. Therefore, the concept of food web appears more real ecologically than the concept of a simple food chain. The mechanism of operation of food web in maintaining stability of ecosystem is given below.

A herbivore like rabbit does not get starved if its preferred plant species is reduced in quantity due to some calamity. It begins feeding on alternate plant species. The preferred one gets chance to recover from the loss. Similarly, rabbits are preyed upon by foxes, wild dogs, wild cats, jackals, etc. In case the population of rabbit decreases, the predators begin to eat mice, shrews, squirrels, etc. Meanwhile rabbits increase their population and restore the balance.

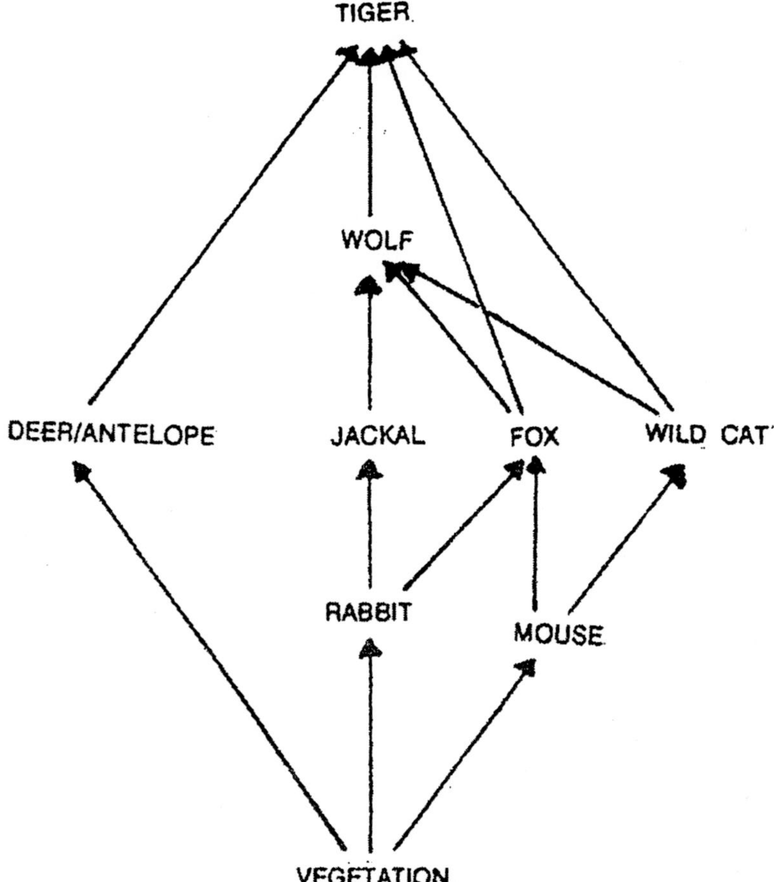

Fig. 3.4 : A terristrial food web

ENERGY FLOW THROUGH THE ECOSYSTEM

Energy flows in the ecosystem in a unidirectional manner. It comes from the sun. Different ecosystems receive different amounts of energy depending upon the latitude, slope, cloud, formation, and atmospheric pollutants. Thus Britain receives 25000 cal/cm^2/yr of solar radiations while the amount at Varanasi in India is 75000 cal/cm^2. Only a part of this energy is trapped by producers while the rest is dissipated. The energy conserving efficiency is 1.15% for grassland, 0.9% for savannah, 0.81% for mixed forest, 5% for modern crops and 10-12% for sugarcane field. Producers trap solar energy with the help of chlorophyll. The latter changes solar energy into chemical energy. The same is bound up in organic food. The green plants or producers break down a part of organic food in respiration-to obtain chemical energy for various body activities, and overcoming entropy. Dissipation of energy occurs as heat. The rest of the organic food is used in body building.

Herbivores feed on plants. The land herbivores seldom eat whole plants. There is a lot of wastage of food energy which passes into detritus (decomposer) chain. In aquatic ecosystem a major part of phytoplankton is devoured by herbivores. Herbivores digest the ingested food. The digested food is then assimilated. Major part of assimilated food is broken down to release energy in the process of respiration. A part of energy is lost as heat. The remaining is used in overcoming entropy and performing various life activities. Only a fraction of assimilated food is employed in body building.

Herbivores are eaten by primary carnivores, the latter by secondary carnivores and so on. At each step of the food chain a major part of food is broken down to release energy. Only a small part enters the process of body building. There is, thus, a decline in the amount of energy passing from one trophic level to the next. It is ultimately dissipated (Fig. 3.5).

ECOLOGICAL PYRAMIDS

An ecological pyramid is a graphic representation of an ecological parameter, like number of individuals present in various trophic levels of a food chain with producers forming the base and top carnivores the tip.

Ecological pyramids were developed by Charles Eiton (1927) and are, therefore, also called *Eitonian pyramids*. In a pyramid the various steps of a food chain are represented sequence-wise with producers at the base, herbivores above them, followed by primary carnivores and so on with top carnivores constituting the top of the pyramid. An ecological pyramid can be *upright* (with larger base and gradually tapering towards the tip), inverted (narrow base, gradually becoming broader towards the tip) or spindle shaped (narrow both at base and tip with a broader part in the middle).

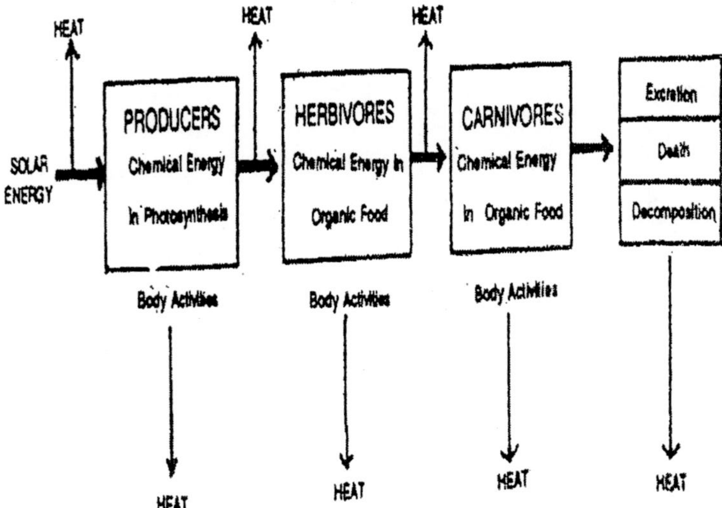

Fig. 3.5 : Energy flow in an ecosystem

Pyramids are usually prepared for three aspects of a food chain or ecosystem–number of individuals, amount of biomass and amount of energy. Accordingly there are three types of ecological pyramids.

1. ***Pyramid of Numbers** (Fig 3.6). It is a graphic representation of the number of individuals per unit area of various trophic levels stepwise with producers being kept at the base and top carnivores kept at the tip.* In most cases the pyramid of number is upright with members of successive higher trophic level being fewer than the previous one. The maximum number of individuals occur at the producer level. The producers support comparatively fewer number of herbivores, the latter fewer number of primary carnivores and so on. Top carnivores are very few in number. In a grassland, a larger number of grass plants or herbs support a fewer number of grasshoppers that support a still smaller number of frogs, the latter still smaller number of snakes and the snakes very few peacocks or falcons. A similar case is found in a pond ecosystem where a large number phytoplankton support comparatively smaller number of zooplankton (like copepods), the latter fewer number of small-sized fishes, the small sized fishes becoming food of still fewer large-sized fishes or water birds.

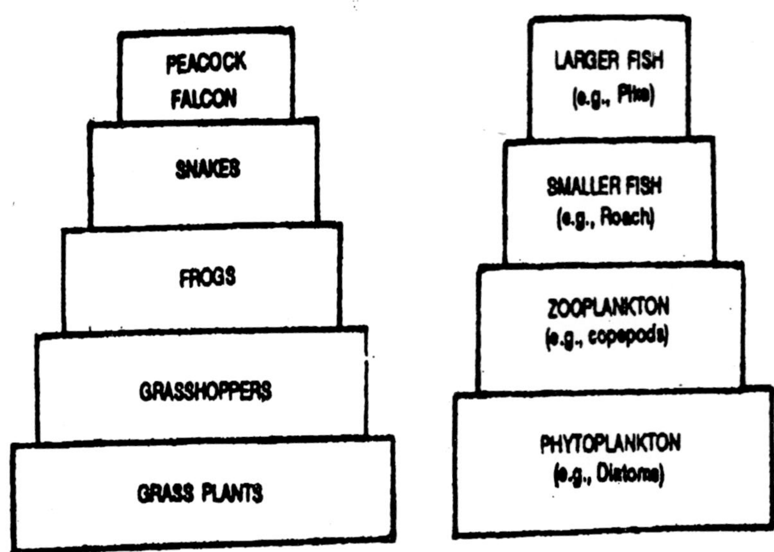

Fig. 3.6 : Pyramids of number in terrestrial and pond ecosystems

The number of pyramids in a higher trophic level is generally smaller than that of the lower trophic level because the organisms of the higher trophic level are dependent for their food and energy on the organisms of the lower trophic level. During transfer of food, about 90% of it is wasted or consumed up in respiration and only 10% becomes part of the higher trophic level. Therefore, in a food chain the members of successively higher trophic level are fewer in number. This is, however, not applicable in all the cases. A single large sized producer like tree can, however, provide nourishment to several herbivores (*e.g.*, birds). The birds may support a still larger population of ectoparasites. Such a pyramid shall be inverted. Small sized herbivorous birds are usually eaten by falcon or eagle. The number of eagles is, however, very small. This type of pyramid of numbers shall be spindle-like.

2. ***Pyramid of Biomass.*** The amount of living or organic matter present in an organism is called biomass. It is measured both as fresh and dry weight. *Pyramid of biomass is a graphic representation of biomass present sequence-wise per unit area of different trophic levels with producers at the base and top carnivores*

kept at the tip. Pyramid of biomass is more real than the pyramid of numbers because the latter does not take into consideration the size of the individual. For example, mouse, shrew, squirrel, rabbit, deer, antelope, bison and elephant are all herbivores. The same amount of vegetation will support a large number of mice, fewer rabbits, still smaller number of deer and very few elephants. However, their total biomass shall be equal and will be related to the biomass of the vegetation as well.

Maximum biomass occurs in producers. There is a progressive reduction of biomass found in herbivores, primary carnivores, secondary carnivores, etc. fig 3.7. It is found that about 10-20% of the biomass is transferred from lower trophic level of energy for giving heat, overcoming entropy and performing various body activities. Thus 1000 kg of vegetation supports a biomass of only 100 kg of herbivores which in its turn shall form only 10 kg of biomass in the primary carnivore and only 1 kg in secondary carnivore. Therefore, the cost of production of food (meat) from animals is much higher than the one obtained directly from plants. A total dependence on animal diet will be disastrous. For example, the agriculture produce of Russia is at least twice that of India while its population is less than half of India. Even then Russia imports a large amount of grains while India has become a surplus country. It is because majority of Indians are vegetarians and they obtain their food mostly from first trophic level while Russians are mostly non-vegetarians and depend upon second trophic level (herbivores) for their subsistence.

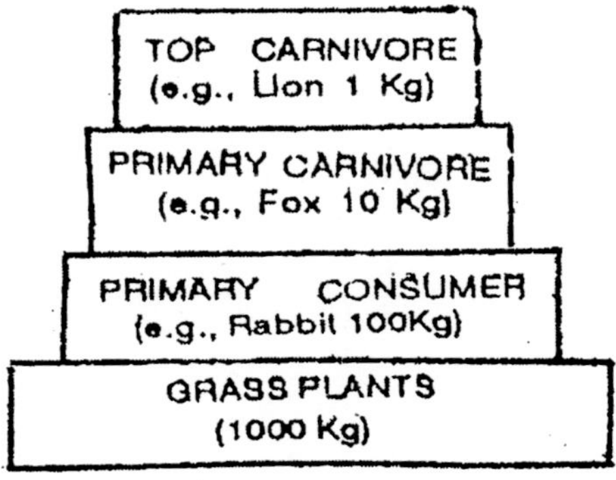

Fig. 3.7 : Pyramid of biomass

Pyramid of biomass is upright for terrestrial habitats. Inverted or spindle-shaped pyramids are obtained in aquatic habitats where the biomass of a trophic level depends upon reproductive potential and longevity of its members. Thus the biomass of phytoplankton may be smaller than that of zooplankton and that of the latter less than of primary carnivores. However, if total biomass produced per unit time is calculated, the pyramid of biomass shall always be upright.

3. ***Pyramid of Energy.*** *It is a graphic representation of amount of energy trapped per unit time and area in different trophic levels of a food chain with producers forming the base and top carnivores the tip.* The energy content is expressed as kcal/m^2/yr or kJ/m^2/yr. Maximum energy content is present in producers. They obtain the energy from solar radiations. The energy is converted in chemical form and stored inside organic matter manufactured by the producers. As the energy passes into higher trophic levels along with food, its amount decreases because of its dissipation as heat and use in overcoming entropy as well as for performing various body activities.

The data given by Odum (1971) shows that in a fish pond phytoplankton trap 31080 kJ/m^2/yr (7400 kcal/m^2/yr) of energy. Zooplankton and other herbivores dependent upon phytoplankton have an energy content of 7980 kJ/m2/yr (1900 kcal/m^2/yr). The primary carnivores (insect larvae and small fish) possess an energy content of 2100 kJ/m^2/yr (500 kcal/m^2/yr). They support secondary carnivores with an energy content of 126 kJ/m^2/yr 30 (kcal/m^2/yr). If the latter is eaten by human beings only 16.8 kJ/m^2/yr (4 kcal/m^2/yr) of energy content is stored. Thus pyramid of energy is always upright. It is more accurate than the pyramid of biomass or the pyramid of numbers.

IMPORTANCE OF ECOSYSTEM STUDY

1. Ecosystem study indicates the available solar energy and the efficiency of an ecosystem to trap the same.
2. It gives information about the available essential minerals and their recycling periods.
3. Gross and net productivity of an ecosystem are known.

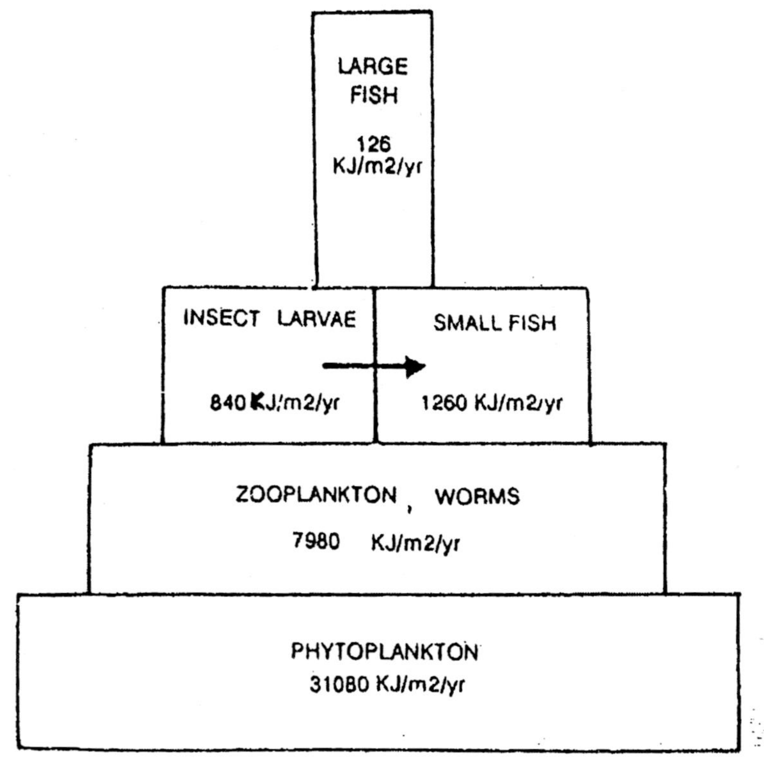

Fig. 3.8 : Pyramid of energy in a fish pond.

4. It provides knowledge about the web of interactions and inter-relations amongst the various populations as well as between populations and the abiotic environment.
5. It helps human beings to know about conservation of resources, protection from pollution and inputs required for maximising productivity.

HABITAT, MICROHABITAT AND ECOLOGICAL NICHE

Habitat. It is a specific place or locality occupied by an organism, population or community which has a particular combination of abiotic or environmental factors. Habitat is also defined as the sum total of environmental factors or conditions which determine the existence of an individual, population or community in a particular locality. The term habitat is more specific than environment because it corresponds to a particular combination of its factors. Habitat commonly refers to a large area, *e.g.*, pond, lake river, estuary, ocean, forest, desert, grassland, woodland, etc. However, it may be as small as a burrow, bark of tree or intestine of an ant.

Organisms usually live in specific habitats with specific combination of environmental factors. For example, annelid *Tubifex* lives in running and well aerated fresh water having abundant organic matter. Larvae of insect *Chironomous* occur in water devoid of oxygen. Malarial parasite lives a part of its life in red blood corpuscles of man and another part inside stomach of female *Anopheles.* Other organisms have wider choices because they tolerate changes in environmental factors. Thus fish *Hilsa* is found in both fresh and marine waters.

Microhabitat. It is part of a habitat which has a specific property of its own due to change in one or a few environmental factors, *e.g.*, forest floor, tree canopy, muddy bottom, surface of a pond, edge or centre of a pond.

Ecological Niche. Literally niche means a specific place. But in ecology it is used for the role of species/population that plays in its ecosystem. It includes several factors like (a) Type of food and its availability, (b) Type of predators, (c) Range of movement, (d) Range of tolerance, (e) Shelter, if any, (f) Microclimate, (g) Time of the day and season of the year when the organism is most active, (h) Nest building, (i) Effect on other components of the community.

In fact, ecological niche means the total interaction of a species with environment.

Examples of Ecological Niche

(i) Owls and cats feed on shrews and mice. Thus owls and cats occupy the same ecological niche in different habitats. -

(ii) Water bugs *Belostoma* (Giant water bug) and *Corixa* (water boatman) live in the shallow waters at the edge of a pond. Both have the same habitat but giant water bug is predator and water boatman is a scavenger.

(iii) *Two* birds *Ploceus melonocephalus* and *P. cotlaris* live in the same nest but have different ecological niches because one feeds on insects while the other eats seeds.

(iv) Both the arctic fox and the African hyaena are scavengers. Thus both occupy the same niche.

(v) (a) The Tadpole is herbivorous however the fox is carnivorous,

(b) Larva of rat flea is scavenger taking decaying organic matter in the soil but the adult flea is a parasite, sucking blood of its host. Thus some animals occupy different niches in succession.

Difference between Habitat and Ecological Niche. E.P. Odum, a famous ecologist, has differentiated habitat and ecological niche by saying that the habitat is organism's 'address' and ecological niche is its "profession".

Ecological Equivalents. Organisms occupying similar ecological niches but living in different regions are called ecological equivalents. Ecological equivalents possess similar types of adaptations but belong to different taxonomic groups. Succulents of american deserts are cacti while those of african deserts are euphorbias. Owls and cats both feed on mice and shrew. Owls are found in deserted or wooded area while cats occur in or around human habitations.

Biomes (Major Ecosystems)

Definition of Biome. A biomes is defined as a large natural ecosystem which is distinct in its climatic conditions and has its specific type of plant and animal life. Thus biomes are the major ecosystems of the world which are the largest ecological units.

Characteristics of a biome are determined by a number of factors. The important ones are: (i) Temperature range (ii) Latitude and altitude (iii) Intensity and duration of summer and winter (iv) Amount and periodicity of rainfall (v) Soil characteristics (vi) Geographical barriers like mountain or sea (vii) Topography (viii) Water mass.

BIOMES OF THE WORLD

Broadly speaking, biomes are of two types :

Terrestrial *Biomes terrestrial and aquatic.*

They are large terrestrial communities which are influenced by latitude, amount and periodicity of rainfall. The major terrestrial biomes are (i) tundra, (ii) taiga, (iii) deciduous forest, (iv) tropical rain forest, (v) chapparal, (vi) tropical savannah, (vii) grassland and (viii) desert.

1. **Tundra** (Rus. *tundra* – arctic hill or north of timberline). *Location.* It lies north of timberline or 60°N latitude below the polar ice. Tundra occupies some 8 million km^2 area of land mass extending across N. America, Europe and Asia. It occurs only in the arctic region and is, therefore, also called *arctic tundra.* Tundra is absent in the southern hemisphere because corresponding region is covered by antarctic region.

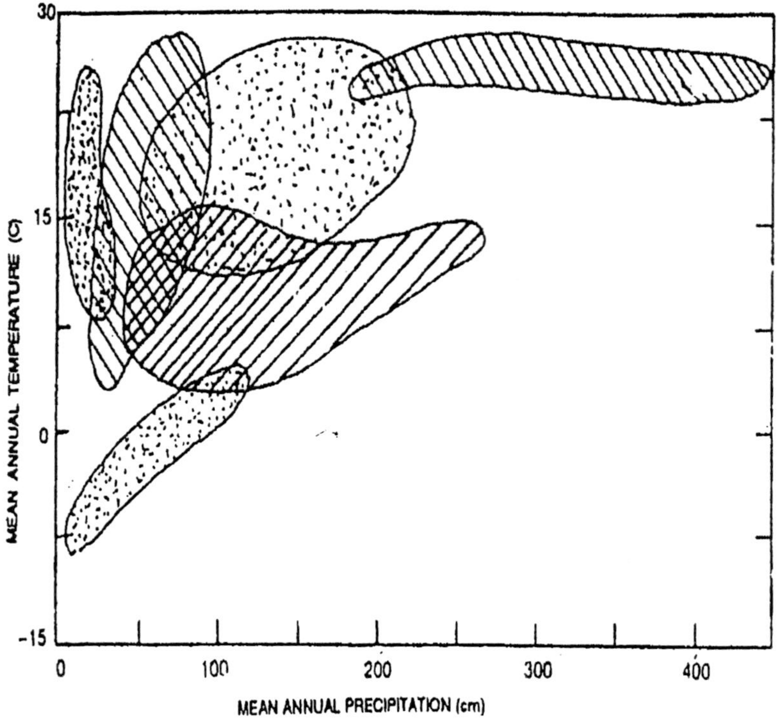

Fig. 3.9 : Terrestrial biomass governed by temperature and precipitation.

Physical Characteristics. The area receives very little precipitation, around 25 cm per year, mostly in the form of snow. The area is covered by snow for most part of the year. The climate is, therefore, extremely cold with a winter temperature – 30°C to –40°C. Strong winds and snow storms are frequent. Summer is for short duration of 45-75 days. The highest summer temperature is 10°C. It is unable to melt snow except for the upper 10-20 cm. The remaining part of the soil is in permanently frozen *(permafrost)* condition. Alternate thawing and freezing of the upper layers produces cracks. The terrain is plain or with gentle slope but has numerous depressions or cracks where ponds, pools, marshes and bogs are formed during summer.

Flora. Vegetation is thin. North tundra is often called arctic desert because it contains very sparse low growing vegetation devoid of any tree. Only those plants grow in tundra which either complete their life cycle in brief summer or can remain alive even when covered by snow for 8-10 months. They are shallow rooted as the subsoil is permanently frozen. Mosses and lichens show best development in the area. The common moss is *Sphagnum* (Bog Moss) and the most common lichen is *Cladonia* (Reindeer Moss). Other plants growing in tundra are grasses, sedges, heaths, a few shrubs, dwarf willows *(Salix* species) and dwarf birches *(Betula* species) The plants possess xerophytic characters. Leaves are often small and hairy. Their margins show folding.

Fauna. Amphibians and reptiles are absent from tundra. Arthropods and worms appear in the area during summer. Common animals of tundra are warm blooded and have protective covering like feathers (birds) and hair skins (mammals). Mammals have a thick layer of insulating fat below their skin. Important animals of the area are lemmings, polar bear, arctic hare, arctic fox, musk ox, caribou, reindeer, snow owl and snow grouse. Shore and water birds visit the area during summer. Some animals live in burrows or caves during winter and hibernate, e.g., polar bear. Others migrate in winter to less cold temperate areas, e.g., reindeer, caribou.

Tundra is a highly delicate and fragile biome. It takes a long period to recover from even a minor disturbance.

2. **Taiga** (North Coniferous Forest). *Location.* The biome occurs just south of tundra across north America, Europe and Asia. It is also found in the southern hemisphere *(e.g.,* parts of New Zealand). Roughly 10% of the land mass is under taiga.

Physical Characteristics. Precipitation is highly variable, 10-35 cm in drier parts and over 100 cm in wetter parts. It occurs both as rain as well as snow. Lakes and marshes are quite common in the wetter parts. The marshes bogs have cotton grass and *Sphagnum.* Winters are quite chilly with long dark nights. The average winter temperature does not exceed 6°C. Summers are pleasant with long hours of day light and an average temperature of less than 20°C. The growing season is for about 150 days.

Flora. Dominant vegetation consists of evergreen conifers which are able to tolerate wide fluctuations of temperature, light and soil. They are Pine, Fir, Hemlock, Spruce, Juniper, Yew, Larch, Deodar (= Cedar). Where conditions are more favourable, dense coniferous forest is present with little light reaching the ground. The ground flora consists of herbs, ferns, mosses and lichens. The latter three also occur on trees and shrubs under humid conditions. Birch and Maple are found at several places. Vines and wild flowers are common.

Fauna. Animal community of the biome is represented by mouse, wolves, otters, beavers, elk, deer, raven, rabbit, hare, squirrels, pumas, lynx, grouse, jay, many species of insects etc. During winter many animals hibernate or migrate to warmer places. The area also receives reindeer and caribou from tundra during winter.

3. **Temperate Dedduous Forest.** Location. It is found in both the northern hemisphere (Canada, eastern U.S.A., north central Europe, eastern Asia) and southern hemisphere (New Zealand, eastern Australia). The forest occurs in some 18 million km^2 of temperate areas.

Physical Characteristics. .The areas have warm summer and moderately cold winter. Annual precipitation lies between 75-150 cm.

Flora. The dominant climax vegetation consists of broad-leaved hardwood (dicot) trees like Oak, Elm, Maple, Birch, Beech,

Hickory, Magnolias, Poplars, etc. Shrubs are also abundant. The trees and shrubs usually shed their leaves with the onset of autumn (hence also called fall). New leaves are produced in early spring. A few soft-wood trees (conifers) may occur at places interspersed with hardwood trees.

Where conditions are favourable, four-storeyed forest is formed. The top stratum is occupied by trees reaching a height of 30-40 m. There is an understorey of small trees, an intermediate stratum of shrubs and a ground stratum made of herbs, grasses, ferns, mosses and lichens. Vines are found here and there.

Fauna. The animal population includes frogs, salamander, turtles, snakes, lizards, rabbits, hares, squirrels, opossums, foxes, racoons, deer, bear, thrushes, owls, sparrows and several song birds. In winter, some animals undergo hibernation or migrate to warmer areas.

4. **Tropical Rain Forest.** ***Location.*** Tropical rain forests are mainly found in central America, along Amazon and Orinoco rivers. South America, Congo river basin of Africa, Malagasy Republic and south east Asia including India.

Physical Characteristics. The biome occurs in equatorial or subequatorial regions where both rainfall and warmth are abundant. Rainfall is above 140 cm/yr usually between 200-500 cm/yr. It may be upto 1000 cm. Rain occurs through major parts of the year. Therefore, humidity is good. Plant growth is luxuriant. The forest is thick and almost impenetrable. As a result it is called jungle. The forests occupy about one-twelfth of the total land but possess more than half of the flora and fauna of the world. Diversity of life is so high that a hectare of the forest may have as many as 200 species of trees, 70-80% of all insects and 80-85% of all birds are known from tropical forests.

Productivity of the biome is very high (12000 kcal/m^2/yr as compared to 3000 kcal/m^2/yr for temperate deciduous, 2000 kcal for taiga and only 200 kcal/m^2/yr for tundra).

Flora. The vegetation shows stratification. Stratification is the grouping of plants in a forest into two or more well defined layers depending upon their height like tall trees, medium sized trees, small trees, bushes, herbs, etc. The different layers are called strata or storeys.

Tropical rain forest is multistorey (Fig. 3.10) and mainly contains broad-leaved evergreen plants. There are a minimum of Five storeys or strata or vegetation. The upper storey is occupied by very tall emergent trees (50 m or more). The canopy is open. The second storey is constituted by tall trees (35-40 m) which form a dense and closed canopy. There is an understorey or intermediate layer of small trees, a shrub layer and a ground layer of ferns, mosses and herbs. The forest floor receives very little sunlight. Epiphytic growth is rich due to humidity. It includes orchids, lichens, mosses and ferns. Vines and lianans are abundant especially on the edges of the forests. The forest trees have buttressed trunks. Cauliflory or formation of flowers on stems and branches is common. The leaves of tall trees are leathery with drip tips for the flow of rain water.

The important plants of tropical rain forest are Rosewood, Mahogany, Ebony, Rubber tree, Figs., Artocarpus, Nutmeg, Cinnamon, some palms and bamboos. Variations of tropical rain forest are semi-evergreen and *deciduous tropical* forests. The variations are produced by human interference, less rainfall, presence of dry season and shedding of leaves by some or most of the trees, *e.g., Dalbergia, Bombox, Butea, Shorea,* etc.

Fig. 3.10 : Stratification in a tropical forest

Fauna. Each storey or stratum has different fauna. Upper storeys have birds, insects, bats, monkeys, lemurs, tree frogs, lizards and anteaters. Ground fauna includes many snakes, some lizards, deer, forest goat, antelope, tapir, elephant, leopard, jaguar, etc.

5. **Chapparal.** (=Chapparal). *Location.* The biome occurs in Mediterranean area (hence Mediterranean scrub forest), pacific coast of north America, Chile, south Africa and south Australia.

Physical Characteristics. It is a broad-leaved evergreen *shrub forest* of hard and thick-leaved small trees and shrubs which usually contain resin but are resistant to fires. The area has frequent bush, fires during 'dry' summer. It receives humid air from nearby oceans which also keeps the temperature moderate. Rainfall is during winter only.

Flora and Fauna. Both plants and animals are adopted to frequent and long periods of drought. The common plants of chapparal are *Arctostaphylos* (Manzita), Sage, *Carnithus, Adenostema* (Cheemise), Oak and *Eucalyptus* (in Australia). Animals include rabbits, rats, chipmunks, deer, snakes, lizards, birds, tiger, etc.

6. **Tropical Savannah.** *Location.* The savannah is found in equatorial and subtropical regions of the world especially South America, Central Africa and Australia.

Physical Characteristics. It is a warm climate plain which contains coarse grasses with scattered trees and shrubs. It receives sufficient rain (100-150 cm/yr) but the same is seasonal. Wet seasons alternate with long dry seasons. The organisms of the biome are drought tolerant. The biome is also frequented by Fires otherwise it would be replaced by woodland. Productivity is about 3000 kcal/ m^2/yr.

Flora and Fauna. Tropical savannah of Central Africa is believed to be cradle of human evolution. A savannah does not have much species diversity. It is known after the dominant scattered tree, *e.g., Acacia savannah. Phoenix* savannah. *Eucalyptus* savannah. Hoofed herbivorous animals are quite common. Common animals are zebra, giraffe, antelope, gazelle, rhinoceros, goat, rabbit, mice, elephant, fox, wolf, tiger, lion etc. Kangaroo occurs in the savannah of Australia. Due to its easy accessibility of the tropical savannah is ideal for hunting.

7. **Grassland.** *Location.* A grassland possesses different types of grasses, non-graminaceous herbs and a few scattered bushes or an occasional trees. Depending upon the types of grasses and non-graminaceous flora, grasslands have been differentiated into (a) *Prairies* of Canada and U.S.A. (b) *Pampas* of South America (c) *Steppes* of Eurasia (d) *Iussocks* of New Zealand and (e) *Veldts* of South Africa. Prairies are further distinguished into long grass prairie, mixed grass prairie and short grass prairie or plain.

Physical Characteristics. It occupies a large chunk of the land which has flat or gently rolling topography. A large part of the grassland has been converted into cropland. Climate is 'continental' with cool to cold winters and hot summers. Rainfall is 25-75 cm. Dry periods are long. Fires occur occasionally. They help in maintaining the grassland.

Flora and Fauna. Productivity is 2500 kcal/m^2/yr. Dominant plants are short and tall grasses. Animal community of a grassland includes a number of insects, grain eating birds, herbivores and their consumer animals like jack, rabbits, mice, bison, elk, mule, deer, bighorn sheep, antelope (not a true antelope), wolves, prairie dogs, bears, budges, coyote, burrowing owl, etc.

8. **Desert.** *Location.* Deserts are found all over the world–in areas bordering cold oceanic currents, lacking cloud intercepting mountains, lying far away from cloud seeding regions or rain shadow. *Rain shadow* is the region of earth which does not receive rain either due to non-interception of clouds caused by lack of mountains or cutting off of all clouds because of the presence of high mountains ahead of it, e.g., Tibet north of Himalayas, Thar desert. Major deserts occur in Asia (Tibet, Gobi, Thar and West Asia), Central Western Australia, North Africa (Sahara), South-Western U.S.A., Mexico, coastal areas of Chile and Peru.

Physical Characteristics. Common features of all deserts is low annual rainfall of less than 25 cm. Depending upon the temperature, deserts are of two types, *cold (e.g.,* Gobi) and *hot (e.g.,* Thar, Sahara). Hot deserts have high evaporation rate, sand dunes, sand storms or rocky areas. Nights are cold but days are very hot with temperature reaching 50°-60°C. Whatever rainfall occurs, the same passes down into deeper strata or gets evaporated rapidly so that there is shortage of water even during rains.

Flora and Fauna. Flora and fauna are scanty. Some ephemeral or short-lived plants and insects occur during the rainy season. For example, *Boerhaavia repens* of Sahara desert has a life span of 10-14 days. A few plants live for 2-6 months because they are able to cut down water loss, *e.g.,* hardy grasses, *Echinops, Solarium suraltense.* Succulents survive desert conditions by storing water and opening their stomata only during night, *e.g.,* cacti, euphorbias. A few non-succulent perennial plants also grow in deserts due to having long spreading root system, thick bark and wooly covering, *e.g..* Date *(Phoenix)* Acacia, Prosopis, Tamarix. Most of the animals live in burrows and other protected parts during the day. Example of desert animals include ants, locusts, wasps, scorpions, spiders, lizards, rattles snakes, insectivorous birds like swifts and swallows, seed eating quails, doves, desert rats, hares, foxes, jackals and cats. Amongst higher animals only camel can survive desert conditions since its body cells can endure a loss of 40% of water while in man 15% loss causes death. In camel the feet are spreading and insulated. Another exceptional mammal of the desert is Desertor Kangaroo Rat. It does not lake water. Rather, it feeds on dry seeds. Its only source of water is the one produced in respiration and the meagre amount present in diet.

Altitudinal Biomes. On the basis of broad climatic regimes, only four types of terrestrial biomes are present—tropical rain forest, temperate deciduous, taiga and tundra. Other types are variations of these biomes. All these latitudinal biomes can be observed on the high mountain ranges found in tropical areas like Himalyas, Andes and Rockies. The four latitudinal terrestrial biomes are telescoped into altitudinal biomes each extending a few hundred metres in height from base to below the snow line. The basal part is called terai in India. It possesses tropical forest. The same is succeeded by deciduous forest, coniferous forest and tundra. High mountains growing in warm temperate areas do not have tropical forests at the base. Instead they begin with temperate deciduous forest. Low altitude mountains devoid of snow caps do not have tundra vegetation. Therefore, the number of biomes found on mountain depends upon its latitude and height. Taiga-like forests occur in the temperate area of the hills but in India such a forest is often of mixed nature having both conifers *(Pinus, Cedrus, Abies, Taxus, Picea)* and broad leaved trees *(Quercus, Betual Acer, Rhododendron, Ulmus, Aesculus).*

Alpine tundra is the highest which occurs near the top of very high mountains having permanent snow, e.g., Himalayas. It is a tree less region and lies above the timber-line. Trees of lower region become tiny shrubs in this area (e.g., *Rhododendron, Juniperus, Abies*). Other constituents of alpine tundra are lichens, mosses, grasses, herbs and small shrubs like *Artemesia Arenaria, Primula* and *Anemone.* The plants usually possess spreading or cushion habit and often grows in protected areas. Common animals include mountain goat, yak, wolves, snow leopard, snow bear, rabbit, willow grouse and some migratory birds. Alpine tundra differs from arctic tundra in being slopy, well drain with little peat or bog, more herbaceous flowering plants and dwarfed trees.

AQUATIC BIOMES

Some workers prefer to call aquatic biomes as aquatic ecosystems because their distinctive features are different from those of the terrestrial biomes. Aquatic biomes are of four types–(i) oceanic, (ii) lakes and ponds, (iii) marshes, (iv) *streams-rivers.*

1. ***Oceanic Biome (Marine Biome).*** It occupies more than two thirds of the earth's surface. It has high concentration of salts with an average of 3.5%. The most abundant ions are sodium and chloride. Other common minerals are sulphur, magnesium and calcium. The salinity does not vary much except for nearly closed areas (0.5% in red sea), near river mouths (0.5%, 1% in baltic sea) or poles. Despite high salt content of ocean, there is deficiency of some essential nutrients in the open sea so that productivity of oceanic biome (except coastal region) is 1000 kcal/m^2/yr, less than most terrestrial biomes.

Temperature of the ocean surface is about 28°C in the equatorial regions but less than 0°C near the poles. However, variations in temperature are less in deeper waters. Hot and cold currents occur at places.

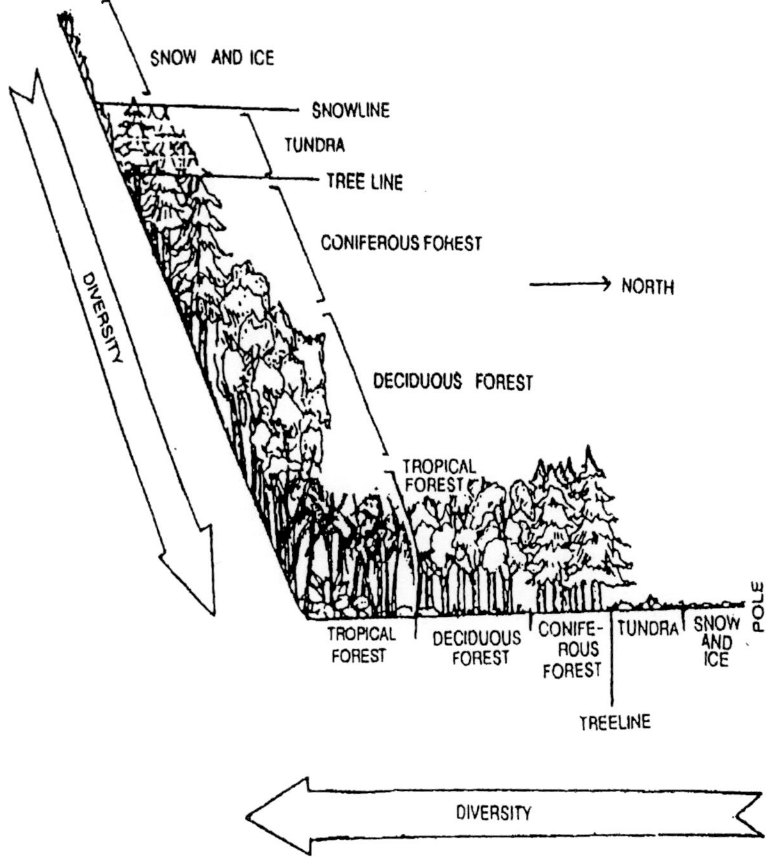

Fig. 3.11 : Latitude and altiudinal biomes

Rainfall and geographical location has little effect on oceanic habitat. The factors which determine the marine ecosystems are distance from shore, drainage of rivers or glaciers, and depth of water. Vertical zones occur in ocean depending upon the availability of light for photosynthesis. The upper 200 m layer of ocean floor receives good amount of sunlight. It is called *photic* or *euphotic* zone. Some light penetrates deeper though it is insufficient for photosynthesis. Its depth lies between 200 – 2000 m and is called *aphotic zone*. There is perpetual darkness below 2000 m upto the ocean floor (6000—10000 m). The dark zone is known as *abyssal zone*.

Ocean basin is like a wash basin or inverted hat. It is differentiated into three parts—continental shelf, continental slope and ocean floor, (i) *Continental shelf*. It is a gradually sloping area of sea shore that extends from coastline to about 160 km in the sea and has a depth of 8 – 200 m. The slope is imperceptible, hardly 0.1°. It has high productivity, (ii) *Continental slope*. It is a slanting area which connects the edge of continental slope to ocean floor. The slope is 3°—6°. Continental slope possesses ridges, trenches and basins of mud and sand brought by rivers, (iii) *Ocean Floor*. It is the bottom area of the open sea. Ocean floor is nearly horizontal with deep trenches at places. Its normal depth is about 6000 m but can extend to 10,000 m or more.

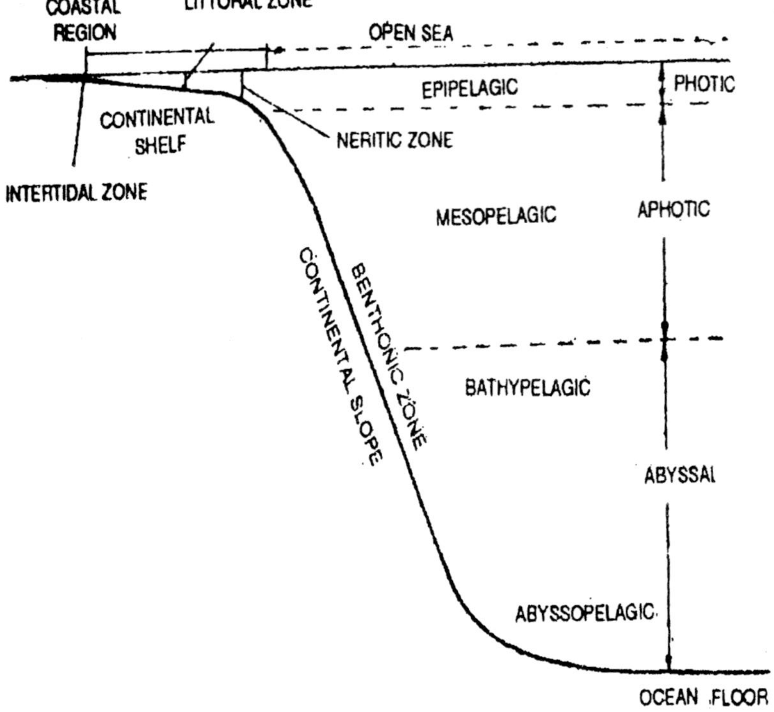

Fig. 3.12 : Ocean Basin

Three main types of environmental zones are recognised in the ocean basin – littoral, benthonic and pelagic, (a) *Littoral Zone*. Sea floor in the region of continental shelf, (b) *Banthonic Zone*. Sea floor along continental slope, aphotic and abyssal zones, (c) *Pelagic Zone*. Water of the ocean Pelagic zone is further differentiated into epipelagic, mesopelagic, bathypelagic and abyssopelagic.

Marine organisms are classified into three types depending upon the region of their activity-plankton, necton and benthon. (i) *Plankton*. They are organisms living in the surface layers of the aquatic habitat which float or drift passively alongwith the water currents due to either absence of locomotory organs or presence of weak locomotory organs. Certain workers restrict the term plankton for microscopic organisms. They use the term *neuston* for larger organisms. Plankton is of two types-*phytoplankton* (photosynthetic plankton) and *zooplankton* (phagotrophic plankton), (ii) *Necton* (== Nekton). They are organisms having well developed locomotory organs. Necton actively swim in surface or deep waters. The organisms are phagotrophic and feed on either plankton or smaller necton. (iii) *Benthon*. They are organisms of the bottom which are usually sessile, creeping or crawling. Scavengers and decomposers are components of benthon.

Oceanic biome is divided into three major ecosystems or smaller biomes – open sea, coastal region and estuary.

(a) Open Sea. It is the area of the sea beyond continental shelf. On the basis of light penetration, the open sea is divisible into photic, aphotic and abyssal zones. Otherwise it has two parts, pelagic and benthonic. The pelagic or water body is further differentiated into epipelagic mesopelagic, bathypelagic and abyssopelagic.

The photic region (= epipelagic) contains both producers and consumers. Aphotic part (= mesopelagic) possesses fewer producers but more of consumers. Producers are absent in the abyssal zone. The latter has consumers, scavengers and decomposers.

Producers of the open sea are phytoplankton. The latter include diatoms, dinoflaggellates and several other types of unicellular and multicellular algae. Some of the flagellates are luminescent. Large sized phytoplankton or photosynthetic neuston include *Sargassum*. Phytoplankton constitute the food of zooplankton. The latter include protozoans, several types of crustaceans, larvae, etc. Large sized zooplankton or phagotrophic neuston include jelly fishes, *Physalia,* some crabs and gastropods. Zooplankton become the food of several animals including the whales. The necton or free swimming forms include squids, fishes, snakes, turtles, etc. Marine birds may also visit the area.

The abyssal and benthonic zones of the open sea are devoid of producers. The abyssal zone contains only carnivores or predator animals. Most of them have luminescent organs for attracting other animals *(e.g.,* Devil Fish, Hatchet Fish), special baits *(e.g..* Angler Fish), telescopic eyes, large heads *(e.g.,* Granadier, Fish, Gulper Eel) with enormous jaws and sharp teeth, and brown or grey colour. The benthonic region has mainly scavengers and decomposers. The bottom also possesses oozoes of siliceous and calcareous shells.

(b) Coastal Region. It is the area of the continental shelf. Coastal region is usually divided into three parts–intertidal zone, littoral and neritic. Intertidal zone is alternately exposed. Littoral zone represents the floor or bottom of the continental shelf while neritic zone comprises the main water body of the area.

Coastal region has good productivity (2000-6000 kcal/m^2/yr as compared to 1000 kcal/m^2/yr of the open sea) with the exception of sandy beaches. Beaches belong to the *intertidal zone*. Very few plants occur on the sandy beaches. Crabs and a few burrowing animals like copepods are found. Rocky beaches and muddy flats have abundant biota. The area experiences alternate submergence and exposure. There are also variations in temperature, light and salinity. The plants of the area have abundant mucilage covering for protection against desiccation when water recedes. The important producers of the area are phytoplankton (brought by incoming tides), attached algae like *Cladophora, Enteromorpha, Polysiphonia, Dictyota, Ulva, Fucus,* etc. A seed plant called *Zostera* also occurs. Consumers include bivalves, snails, spring tails, hermit crab, other crabs, barnacles, sponges, bryozoans, tunicates, *Nereis,* star fish, sea urchin, etc.

In the littoral zone there is a strong wave action. A large number of brown and red algae occur attached. They include

Laminaria, Aland Macrocystic, Nerocytsis, Gelidium, coralline red algae (e.g., corallina, Lithothamnion) etc. Consumers consists of sponges, sea anemones, corals, polychaetes, tunicates, etc. Necton of the neritic zone also visit the area for feeding as it is not very deep. Decomposers are abundant.

The neritic zone contains phytoplankton (diatoms, dinoflagellates, blue-green and green algae) and free floating larger algae. Consumers include zooplankton like protozoans, copepods, small jelly fishes, salpa, etc. Necton consist of squids, fishes, turtles, larger crustaceans, seals, and marine birds.

A highly productive part of the coastal region is *coral reef*. It is formed of foraminferans, calcareous algae and calcareous skeletons of coelentrates especially corals. Maximum diversity of biota occurs in the reefs. Every group of marine algae and every animal phylum is represented here. The more common types, besides the reef forming individuals, are crustaceans, molluscs, echinoderms, sponges and Fishes.

(c) ***Estuary***. It is the name of coastal bay which is formed of tidal mouth of a river. Due to mixing of fresh and sea water at different levels during the day, there is a wide fluctuation of salinity from 0.5—3.5%. Fluctuations of temperature also occur. The region has the maximum stress due to turbulence. It is also one of the most productive ecosystems due to rapid circulation of nutrients, quick removal of waste products and year round succession of organisms. Further, both fresh water and marine biota occur in the area. The important phytoplankton consists of diatoms, green algae, blue-green algae and dinoflagellates. Zooplankton include protozoans, crustaceans, rotifers, etc. The necton are made up of Fishes and some crustaceans. The benthon have attached algae, snails, clams, small prawns and Fish larvae.

2. ***Lakes and Ponds***. They are stationary fresh water bodies which occur on land in almost all types of biomes. Ponds are small and shallow with a size of less than one hectare and a depth of less than 2 m. Lakes are larger and deeper. They may have a size of several hundred hectares and depth of over 100 m. Ponds are natural or man-made depressions which get filled with rain or run-off water. Water may be clear or muddy. In temporary ponds biota consists of phytoplankton and zooplankton with a few more animal types which can shift to other places. Permanent ponds, however, have biota of several types of organisms. It is similar to that of lakes in many respects.

Lakes develop in nature due to three reasons — (i) Result of glaciation (ii) Natural or man-made depressions getting filled with water, (iii) Ox-bow lakes which develop from the main stream of a river. Nutritional status and physical factors of a lake are dependent not only on its origin but also source of water, latitude, altitude, location and surrounding biome. Depending upon their productivity, lakes are of two types– eutrophic and oligotrophic. *Oligotrophic* lakes are usually deep, have rocky steep sides with little circulation of nutrients, especially phosphates.

They occur in arid areas, *e.g.*, Sambar Lake of Rajasthan. *Eutrophic* lakes are rich in biota. They have quick circulation of nutrients. These lakes are generally shallower, *e.g.*, Dal Lake of Kashmir.

Permanent ponds and shallow lakes usually possess similar types of biota. The *producers* include (i) Microscopic phytoplankton made of desmids, diatoms, *Chlamydomonas, Chlorella, Spirogyra, Oedogonium, Zygnema* and a number of other algae, (ii) Free-floating macroscopic plants like *Lemna, Azolla, Pistia,* and *Eichhomia* can also become rooted. It often covers up completely small ponds, (iii) Suspended macrophytes like *Utricularia, Ceratophyllum* and *Hydrilla.* (iv) Anchored submerged plants like *Vallisneria.* (v) Floating leaved anchored plants such as *Nelumbo* and *Nymphaea.* (vi) Emerged plants which are rooted at the bottom but part of them come out of water, *e.g., Typha, Scirpus, Sagittaria* and *Phragmites.*

The *consumers* are zooplankton, necton, benthon and water birds. Zooplankton is represented by ciliated, flagellates, rotifers, small crustaceans *(e.g., Daphnia, Cyclops, Cypris)* and larvae of several insects. Necton is constituted by water boatman, water spiders, diving beetles, amphibians, turtles, and fishes. Common water birds are ducks, herons, cranes and kingfishers. Benthon of the ecosystem is represented by snails, clams, mussels, crabs, prawns and decomposers.

3. ***Marshes***. They are wet lands which may possess turbid water body of a few centimetres in depth. Water is received from rains and drains. Marshes are quite common on the sides of roads and railway tracks where depressions are present due to removal of soil and mud. They are also found along embankments of rivers, streams and inside forest. Marshes have little plankton due to turbidity and temporary nature of water body. Typical flora of marshes consists of *amphibians* or emerged hydrophytes *(e.g., Typha, Phragmites)*. Some are floating plants capable of getting rooted in mud *(e.g., Eichhomia)*. Snails, mosquito larvae, some insects, frogs, etc. occur in marshes.
4. ***Streams and Rivers (Lotic Ecosystem)***. They are flowing fresh waters which differ in physical and chemical conditions, oxygen content, temperature, speed and volume of water. A perennial river originating from glaciers or high mountains has low temperature and high speed in the upper reaches, loses silt and attains moderate temperature as well as slower speed in the lower reaches. Streams, rivers and brooks differ from lakes and ponds in having (a) Speed or current (b) Higher land-water interchange and (c) Higher oxygen tension which is liable to decrease with pollution. Due to high speed and low temperature, plankton community is very small in the higher reaches. Detritus from land constitutes an important food for consumer communities. In some cases detrivores may function as 'primary producers' of ecosystem. Plankton growth occurs lower down due to slowing down of water current and warming up of water. River beds having sand are less productive as compared to ones having mud and stones. Besides phytoplankton important producer plant growing in flowing waters are (a) attached algae, e.g. cladopharo (b) Encrusting diatoms (c) Aquatic mosses (e.g. Fontinalis) (d) submerged seed plants (e) Reeds, water grasses and other amphibious plants growing on shallow banks.

The consumer animals include active swimmers with stream lined body like fishes otter mink muskart, crocodile etc. Other animals have flattened body so that the can find shelter in crevices and under stones e.g. flatworm nymphy or stomefly and maytly. Some animals posses possess hooks suchers and sticky under surfaces for attachment e.g. snails and many larvae. Many birds obtain their food from river ecosystem.

Man-made Ecosystems

Man made ecosystem are articles ecosystem which depend upon human efforts for their sustenances because they do not possess self regulatory machine. They have little diversity simple food webs, little recycling of nutrients and depend upon human efforts for protection and provisions of inputs. Some of the man made ecosystems are (i) Villages (ii) Towns and cities (iii) Orchards (iv) Plantations (v) Graden (vi) Parks (vii) Dams and reservoirs (viii) Lakes (ix) Canals (x) Fishery tanks (xi) Aquaria (xii) Agriculture.

Agriculture. It comprise animal husbandry and crop production Agriculture was the first man made modification of ecosystem which occurred due to compulsion of increasing human population and difficult in procuring food. At that time human being had learnt the use of fire. They domesticated some animals and therefore created an artificial ecosystem for them. They raised crops and created an artificial *agroecosystem*. In older days agroecosystem were temporary as they were raised by temporary as they were raised by cutting down forests cultivating the land for a couple of years and then abandoning the site. Now-a-day agro-ecosystem are of permanent nature though they cannot survive without our support. Pasture or pastoral land is similarly prepared for feeding domestic animals. Both crop lands and pastoral lands are essentially grasslands with choice plants since they have been mostly created by clearing grasslands. Forests have also been removed for this purpose at several places.

The study of relationships between agriculture crops and their surrounding animate as well as inanimate environment is called *agroecology*, agroecosystem are different at different places depending upon latitude, edaphic factors topography and climate factors. The crops mainly belong differs from area to area, season to season depending upon climate and preference of the farmers. The biotic community is the richest when the crop is in the field. It consists of some weeds pests earthworms, nematodes, insects rodents, many birds, pollinators, domestic animals and decomposers. All agroecosystem have some common features. The important ones are :

1. They are man-powered and artificial.
2. The ecosystems do not have self-regulatory mechanisms.
3. There is little natural circulation of biogenetic nutrient.
4. Stability of an ecosystem depends upon diversity. Croplands are mostly monocultures and lack diversity.
5. They are vulnerable to destruction due to drought, floods, disease pathogens and pests. Irish famine and Bengal famine are two such examples.
6. They are artificially irrigated, aerated and provided with required biogenetic nutrients in the form of fertilizers. The latter cause pollution of ground water as well as nearby above ground water bodies.
7. They are protected against weeds, pathogens and pests through chemicals. The chemicals are toxic to other living beings including human beings.
8. Technology is being increasingly used in raising and reaping crops.
9. Modern day crop plants have been modified genetically to give very high yields, several times those of natural ecosystems.
10. Biomass is not allowed to accumulate due to reaping. The stubbles are burnt these days.
11. Manures are added to the soils to keep them healthy.

BOUNDARIES OF ECOSYSTEMS

Ecosytem do not occur in isolation. They are in contact with other adjacent ecosystem. In the region of contact the boundaries of the different ecosystem are not sharply defined except in a few cases like pond and terrestrial ecosystem. Even in such cases some organisms are common (e.g., birds) and there is exchange of inorganic nutrients. For example Siberian cranes regularly visit Bharatpur in Rajasthan during winter. Similarly, phosphorus of sea food chain is brought to terrestrial areas by sea birds or soil from Himalayas is brought to plains by rivers like Ganga, Yamuna, Brahmaputra etc. In most cases the boundaries of ecosystems are overlapping. The overlapping area is often called transition zone (= ecotone). Transition or overlapping zone is quite wide between biomes like taiga and temperate deciduous forests, and evergreen chapparal, tropical savannah and tropical rain forest or grasslands and savannah. The overlapping areas usually possess plants and animals of both the ecosystems.

REVIEW QUESTIONS

1. Name the term used for
   (a) Woody climbers found on the edge of some forests
   (b) A grassland with scattered trees.
   (c) Arrangement of plant crowns in horizontal layers in a forest.
   (d) Gently sloping land mass submerged in sea.
2. Define : (a) Producers (b) Consumers (c) Decomposers (d) Food Chain (e) Food Web (f) Biomass (g) Ecosystem (h) Biome.
3. Explain the term ecological pyramids. Give their types.
4. Differentiate habitat, microhabitat and niche.
5. What do you mean by (a) Plankton (b) Nekton (c) Benthon (d) Neritic (e) Continental Shelf.
6. Distinguish between autotrophs and heterotrophs. What role do they play in energy flow of an ecosystem?
7. Name the two fundamental trophic levels and describe the general make-up of each.
8. Explain the concept of food chain.
9. Depict digrammatically a food web in any ecosystem. How many food chains are there in that food web?
10. Why is the concept of a food web more real ecotogically than the concept of a simple food chain?
11. Name the functional aspects of an ecosystem.
12. Briefly discuss the biotic components found in an ecosystem.
13. How do different abiotic constituents influence an ecosystem ?
14. Describe an ecological pyramid.
15. Why is atmosphere considered as a treasure-house of vital resources?
16. What are the basic differences of community, ecosystem and biome?
17. Can more than one species occupy the same niche? What would be the consequence of it, if they do?
18. Define the term 'biome' and list the factors that determine the characteristics of a biome.
19. List the different terrestrial biomes. Describe one of them.
20. Name the various aquatic ecosystems. Discuss one of them.
21. Name some man-made ecosystems. Describe one of them.
22. Why is it difficult to draw sharp boundaries between ecosystems?

Chapter 4

Biodiversity and its Conservation

BIODIVERSITY

The biosphere comprises of a complex collection of innumerable organisms, known as the biodiversity, which constitutes the vital life support for the survival of human race. As defined in the convention on Biological Diversity signed at Rio de Jeneiro (Brazil) in 1992 by 154 countries, the Biological Diversity is defined as the variability among living organism from all sources including "inter alia" terrestrial, marine and other aquatic ecosystems and the ecological complexes of which they are part this includes diversity within species between species and of ecosystems". Biological diversity abbreviated as Biodiversity represents the sum total of various life forms such as unicellular fungi protozoa, bacteria and multicellular organism such as plants, fishes and mammals at various biological levels including gene species, habitats and of ecosystems.

Nature has developed exceeding by complex spectrum of life forms over 600 million years. It is impossible to accurately assess the number of species of organisms present on earth. It is believed that anywhere between 10 to 80 million species exist on earth. However, only about 1.5 million species could be assessed and enlisted so far.

The existence of human race depends on the well-being of the other life forms present in the biosphere. However, we have been losing this accumulated heritage of millions of years at an alarmingly fast rate over the past 400 years by our own activities, thereby undermining the very basis of our own existence on this planet. Our biological systems are constantly impoverished by human activities. All over the world, human beings only have been mostly responsible for the destruction of habitats through intensive agricultural development, over-exploitation of resources, urbanization, industrialisation, overpopulation, deforestation, pollution, environmental degradation and ethical degradation.

ASSESSMENT OF BIOLOGICAL DIVERSITY

Biological diversity is analysed or assessed generally at the following four levels:

(i) ***Diversity of composition of a species within a community*** : The biotic component of an ecosystem may be composed of a number of species of plants, microbes and animals which interact with each other on one hand and interact with the abiotic factors of the environment on the other. The richness of species in an ecosystem is called *"species diversity"*.

(ii) ***Diversity of Genetic Organisation within a species*** : Within any given species, we can find several varieties or strains or races which slightly differ from each other in one or more characteristics such as size, shape, quality of their respective product, resistance against pests, insects, diseases, etc., and resilience to survive under adverse environmental conditions. Such a diversity in the genetic make-up of a species is called *"genetic diversity"*. A species having large number of varieties, strains or races is considered as rich and more diverse in its genetic organization.

(iii) ***Diversity of Biotic communities and Ecosystems*** : An ecosystem develops its own characteristic community of living organisms depending upon the availability of abiotic resources, environmental conditions and other relevant factors. Different types of ponds, lakes, rivers, wetlands, meadows, grass-lands, forests etc. represent diverse ecosystems with their own characteristic biotic community. For example, a pond may possess different sets of flora and fauna as compared to another ecosystem such as river.

(iv) ***Diversity within a land-scape*** : This refers to the size and distribution of several ecosystems and their interactions across a given land surface.

IMPORTANCE OF BIODIVERSITY

Biodiversity is of both aesthetic as well as practical importance.

(1) Biodiversity provides us valuable natural resources to satisfy the subtle needs of mankind. Our homes, livestock, fruits, vegetables, grains, grams, etc. are all derived from the products of diverse and healthy ecosystems. Our food, clothings, shelter, and a host of other useful products are derived from a variety of living organisms.

(2) Diverse communities of plants, animals and micro-organisms also provide us valuable and indispensable ecological services. They recycle wastes, maintain the chemical composition of the atmosphere, and play a major role in determining the climate of the different parts of the world. The "ecosystem services" provided to us by nature in "gratis" also include supply of fresh water, generating soils, supply of plant pollinators, and maintaining a huge "genetic library".

(3) In many cultures, maintenance of mountains and other diverse land-forms are of religious significance.

(4) A diversity of abodes of biological communities such as parks, gardens, natural animal habitats, forests, mountains, seashores, etc. are useful for picnicking and other recreational activities.

(5) In spite of our intensive scientific efforts, we know only fraction of the multitude of the species thriving on our planet, particularly in tropical ecosystems. Every year, some species are lost even before we had a chance to identify them or study their useful properties.

We have been cultivating hardly about 2500 plant species. This number may be double if we take into account the animals, wild plants, microbes and other organisms. However, all these species together account for only a fraction of the total number of plant species recorded, which is about 1.5 millions. Even this, in turn, is a minute fraction of the suspected or expected number of species on the earth, which may be 10 to 80 millions. Therefore, even today, we know very little about the biological world in which live. Out of about 2.5 lakh species, hardly about 25,000 species have been broadly assessed and only about 2,500 species have been studied in details. As far as the animal species are concerned, the situation is more inadequate. Thus it is clear that a huge wealth of biological resources are yet to be tapped by the mankind. The price we might have been paying for this neglect may be illustrated by the fact that if *"penicillium"* or *"cinchona"* would have been extinct before their curative properties were discovered, what would have been the plight of mankind.

(6) Biological diversity represents a valuable genetic resource for the mankind. The genetic organisation of wild plants and animals, which has been continually evolving for millions of years survived the trials, tribulations and vagaries of nature. While we are still cultivating some of those old varieties, we are also using some of those genes to improve the strains we are presently cultivating. Indeed, most of the improved varieties under cultivation today are synthetic or hybrid types by incorporating useful genes to produce better quality of the produce with longer shelf life or heaving better resistance to pests. Such breeding techniques are unlimited in scope. But, however, for getting better strains in future, it is essential to build up a gene-pool containing an assortment of traits.

Further, the quality, yield, and resistance to pests, diseases and adverse climatic vagaries mostly depend on genetic factors and combination of genes which may be different in different strains or varieties of species. Improved varieties containing higher number of useful traits and desirable gene combinations are always preferred which are more productive and have better adoptability. One of the reasons for this is that repeated and prolonged cultivation of a particular variety gradually loses its vigour, vitality, quality, and productivity, and increases its susceptibility for disease or attack by pests. There are several examples to illustrate how genetic modification helped in improved quality and pest-resistance of the product.

(a) A wild variety of rice grown in U.P. saved millions of hectares of paddy crop from "Grossy Stunt" virus.
(b) The Kans grass known as "Saccharum Spontaneum" from Indonesia provided genes for resistance to red rot disease of sugar cane.
(c) The genes from a wild melon grown in U.P. helped in imparting resistance to powdery mildew in musk melons grown in California.
(d) About 20 cultivars of rice grown in several countries were benefited from the useful genes identified from wild varieties from Kerala.

(7) The vast pool of genetic diversity contained within wild populations of plants and animals is of enormous value for the continuing research and development of agriculture, industry and medicine. For instance, the chances of survival of patients suffering from leukaemia have dramatically and appreciably increased in recent years because of synthesis of a drug containing active chemicals derived from a tropical tree from Madagaskar, called' "Catharanthus roseus" (rosy periwinkle).

(8) Thousands of species which are edible and demonstrably superior than those which are currently used, are waiting in the wings for their useful exploitation. For example, a plant found in west Africa, called as Katemfe (Thaumatococcus danielli), produces proteins that are 1600 times sweeter than Sucrose. Similarly, several plant species containing anti-cançer substances of varying potency, and some others having other potential medicinal applications, are yet to be screened and intensively studied for their applications. It should be remembered that about one third of the modern description drugs contain chemical compounds that plants have evolved to defend themselves from their enemies.

(9) The vast insect fauna contain large number of species that are potentially superior crop pollinators, weed-control agents and are parasites of insect pests.

(10) Biological diversity helps in maintaining a stable and healthy ecosystem. In a simple ecosystem, loss of even one or a few species could be disastrous because of the lack of alternatives. However, in a complicated ecosystem having several trophic levels, each of which, in its turn, is composed of several species, loss of one or more species do not cause any serious problems because, the alternatives available can maintain the functionality of the system. Thus, biological diversity helps in maintaining a stable ecosystem.

(11) Many species and creatures considered to be useless or undesirable have also their own roles to play in a natural ecosystem.

(a) Earthworms and termites help in aerating our soils. But this vital function is stopped by killing them using pesticides.

(b) Predators which were facing extinction can no longer control rodents that damage our crops.

(c) Mangroves which protect our coastline from erosion are being destroyed for use as firewood.

(d) Oyesters which were present at one time in large numbers in chespeak bay, (USA) could filter the entire water in a few days have declined by 99%, thereby leaving the bay water muddy and oxygen-deficient.

(e) An adult frog is supposed to consume its own weight of insect every day.

THE VALUE OF BIODIVERSITY

The value of the earth's biological resources can broadly be classified into the following two categories.

Direct values. One of the most important values of biological resource is in providing the food. Originally, plants were consumed directly from the wild. In due course of time, the wild species became the foundation for agriculture which began sometime between 5,000 to 10,000 years ago, more or less simultaneously in various parts of the world. Man has since utilized genetic diversity, through conscious or unconscious selection pressures. In traditional medical practices like Ayurveda– plants or their extracts are directly consumed or applied as medicines. In modern medicine too, around 199 pure chemical substances extracted from 90 species of plants are used in medicines. A host of microbial, antiviral, cardioactive and neurophysiological substances have been derived from marine fauna. Domesticated animals have given us hormones, enzymes, and food products while the fungi and microbes provide life saving drugs such as antibiotics. The possibilities of medicines from biological resources are immense and the scope also is wide. The biological resources are primary sources of wood, firewood and several industrial products. Ornamental plant's are a lucrative commodity today.

Indirect values. At the other end of spectrum, biological resources provide values which are not immediately seen but have far-reaching impact on our living conditions. Indirect benefits of biodiversity can be enumerated as follows:

- Carbon fixation through photosynthesis
- Pollination, gene flow, etc.
- Maintaining water cycles, recharging ground water, protecting
- watersheds and buffering from extreme conditions such as flood and drought

- Soil formation and protection from erosion
- Maintaining essential nutrient cycles
- Absorbing and decomposing pollutants
- Regulating climate at both macro and microlevel
- Preserving recreational, aesthetic, socio-cultural, scientific educational, ethical and historical value of natural environments.

Biodiversity and Ecosystem Function

Man depends on natural ecosystems for essential ecological services like maintenance of atmosphere, cycling of nutrients, soil fertility, control of past outbreaks and maintenance of genetic library etc. Would the loss of biological diversity affect an ecosystem's ability to carry out these functions? Is there a threshold of biodiversity below which the ecosystems lose their stability?

These two questions have been explained through two contradictory hypotheses.

(i) ***Rivet popper hypothesis*** (Ehrlich & Ehrlich, 1981). According to this hypothesis species are like the rivets on an airplane where each species plays a small but significant role in the maintenance of whole ecosystem.

(ii) ***Redundancy hypothesis*** (Walker, 1991). According to this hypothesis, most species are superfluous. The biomass of primary producers, consumers and decomposers is important. Only few species are needed to keep the system moving.

Several studies support the principle: *biodiversity begets productivity.* Species richness increases productive efficiency of the system. Reduced biodiversity may alter the function of other ecosystems. Similarly, there is another ecological tenet: *diversity begets stability.* Another set of experiments postulates that soil fertility is maintained and preserved by biodiversity. Addition or deletions of species provide direct evidence of the role of biodiversity in maintenance of ecosystems.

Removal of one species frequently leads to changes in the abundance and performance of other species. The impact may be so great that the very nature of ecosystem may change. Thus several studies of these key issues lent support to 'rivet' hypothesis. An extinction increases the probability that the next one will unravel the whole ecosystem. This rich biodiversity is needed to maintain an ecosystem.

Biodiversity—Hot Spots

Hot spots are the areas that are severely threatened by human activities. At the same time they contain outstanding examples of evolutionary processes of speciation and extinction. All those areas that support rich biodiversity, because of geologic formations and endemic flora and fauna, and exhibit exceptional scientific interest are called as hot spots.

Mittermeir and Werner (1990) using the criteria of species richness introduced the concept of Megadiversity centres. Following their concept 12 countries/regions namely Mexico, Columbia, Ecuador, Peru, Brazil, Zaire, Madagascar, China, India, Malaysia, Indonesia and Australia have been classified as Megabiodiversity centres. At the global level Myers (1988) has identified 10 hot spots of tropical forests to which he later added 8 areas. These areas collectively include 49,955 endemic species which constitute 20 per cent of the world's total plant species. Two regions from India namely the Eastern Himalaya and the Western Ghats are included as *hot spots*.

Broadly, the botanical hot spots lie in the (i) Western Ghats, (ii) North-east India, (iii) Himalayas and (iv) Andaman and Nicobar Islands.

The Western Ghats particularly the southern Western Ghat known as Malabar is the major genetic estate with an enormous biodiversity of ancient lineage. Out of five locations identified by IUCN as threatened, three areas Periyar Agastyamalai Hills, Silent valley and Periyar National Park are found in this region.

Northeast India represents the transition zone between the Indian, Indo-Malayan, Indo-Chinese biogeographic regions as well as meeting place of Himalayan mountains with that of peninsular India. Some of the hot spots in this zone are Namdapha, Tirap, Sirohi, Dzuko, Tura, Buxa valley, Senchal and Tale valley.

Western Himalaya is quite remarkable for its alpine *flora*. Gymnosperms are quite abundant. About 5000 flowering plants occur here of which 800 species are endemic. Some major hot spots of this region are Valley of Flowers, Pithoragarh, Gori valley, Mandal-Chopta valley, Karakorum and Ladakh.

The Andaman and Nicobar islands have a rich flora under littoral and inland types. About 2500 species of flowering plants are

The Global Hot Spots and the Number of Endemic Species

Region	*Number of plants*
Cape Region (South Africa)	6,000
Upland Western Amazonia	5,000
Madagascar	4,900
Philippines	3,700
Borneo (North)	3,500
Eastern Himalaya (India)	3,500
SW Australia	2,830
Western Ecuador	2,500
Colombian Choco	2,500
Peninsular Malaysia	2,400
Californian Floristic Province	2,140
Western Ghats (India)	1,600
Central Chile	1,450
New Caledonia	535
Eastern Arc Mts (Tanzania)	535
SW Sri Lanka	500
SW Coted Tvoire	200
	49,995

found here of which 250 species are endemic. Some of the hot spots of this region are North Andaman, Table islands. South reef island, Spike island an little and Great Nicobars.

Sacred forests. These are the real hot spots of biodiversity. They are protected by tribals due to some religious sanctity attached to them. These sacred forests contain a number of rare, endangered, endemic or interesting biota of the country. They are known as 'Devaskadu' in Karnataka, 'Devarahati' in Maharashtra, and 'La-Kyntok' in Meghalaya.

Mangroves. Mangroves are the salt tolerant forest ecosystems found mainly in the tropical and subtropical inter-tidal regions. The total area of mangroves in India is estimated to be 6740 km^2, of which the mangroves of Sunderbans itself contribute 4200 km^2, next being the Andaman and Nicobar islands. The quality and quantity of mangroves in India have greatly declined. The following mangrove areas must be considered as hot spots for conservation purposes.

Andaman. North Andaman, Prolobjig, Bomlungta, Havelock island and Wrafter creek (South Andaman).

Nicobar. Komorta, Noncowry islands, Bolloo Chengappa Bay and a few sheltered bays on Great Nicobar islands.

East coast. Sunderbans; Bhitarkanika; Mangroves and Godavari and Krishna Deltaic mangroves, Coringa Bay.

West coast. The Gulf of Kutch and Gulf of Kahambt, Vembanad in Kerala.

Wetlands and Swamps

India has rich variety of wetlands. Some major wetlands of India are Kolleru (AP), Wullar (J&K), Chilka (Orissa), Loktak (Manipur), Bhoj (MP), Sambhar (Rajasthan), Pichola (Raj), Ashtamudi (Kerala), Harike (Punjab), Ujni (Maharashtra), Sukhna (Chandigarh), Renuka (HP), Kadar (Bihar), Nalsarovar (Guj) and Kanjli (Punjab).

BIO-WEALTH

Biodiversity is not distributed uniformly across the globe. It is substantially greater in some areas than the others. Generally, species diversity increases from the poles towards the tropics. Tropical moist forests cover only 5–7 per cent of the land but possess 50 per cent of the world species. On the other hand, certain regions though may not have a high diversity, display a high degree of endemism. However, a less diverse system does not mean that the region is not important.

Bio-wealth of India

Biodiversity is indeed one of India's important strengths and is the bedrock of all bio-industrial developments in the unusually large rural sector (576,000 villages with 76 per cent of country's population) of the country.

India has over 108,276 species of bacteria, fungi, plants and animals already identified and described (Table 4.1). Out of these 84 per cent species constitute fungi (21.2 per cent), flowering plants (13.9 per cent) and insect (49.3 per cent). In terms of the number of species, the insecta alone constitute nearly half of the biodiversity in India (Fig. 4.1).

Known and Estimated Diversity of Life on Earth

Form of life	*Known species*	*Estimated total species*
Insects and other arthropods	874,161	30 million species, extrapolated from surveys in forest canopy in Panama; most believed unique to tropical forests.
Higher plants	248,400	Estimates range from 275,000 to 400,000; at least 10–15 per cent species believed undiscovered.
Invertebrates (excludes arthropods)	116,873	Ture invertebrates may number millions of species, Nematodes, eelworms and roundworms may each comprise more than one million species.
Lower plants (fungi and algae)	73,900	Not available
Micro-organisms	36,600	Not available
Fish	19,056	21,000, assuming that 10 per cent fish remain undiscovered; the Amazon and Orisnoco rivers alone may account for 2,000 additional species.
Birds	9,040	Known species probably account for 98 per cent of all birds.
Reptiles and Amphibians	8,962	Known species probably account for over 95 per cent of all reptiles and amphibians.
Mammals	4,000	Known species account for over 95 per cent of all mammals.
Total	13,90,982	10 million species are considered a conservative estimate. If insect estimates are accurate, total exceeds 30 million.

Table 4.1 : Number of Species of Bacteria, Fungi, Plants and Animals in India

Taxon	*Number of species*	*Percentage*
Bacteria	850	0.8
Fungi	23,000	21.2
Algae	2,500	2.3
Bryophyta	2,564	2.4
Pteridophyta	1,022	0.9
Gymnosperms	64	0.1
Angiosperms	15,000	13.9
Insects	53,430	49.3
Mollusca	5,050	4.7
Pisces	2,546	2.4
Amphibia	204	0.2
Reptilia	446	0.4
Aves	1,228	1.1
Mammalia	372	0.3
Total	1,08,276	100.00

These species occur on land, fresh and marine waters, or occur as symbionts in mutualistic or parasitic state with other organisms. In the world as a whole, 16,04,000 species of Monera, Protista, Fungi, Plantae and Animalia have been described so far. However, it is estimated that at least 1,79,80,000 species exist in the world, but as a working figure 1,22,50,000 species are considered to be near reality (WCMC, 1993).

Based on this data and the already described species, India is 10th among the plant rich countries of the world, fourth among the Asian countries, eleventh according to the number of endemic species of higher vertebrates (amphibia, birds and mammals), and tenth in the world as far as richness in mammals is concerned. Out of the 18 'Hot spots' identified in the world, India has four. These are Eastern Himalaya, North-east India, Western Ghats and Andaman & Nicobar Islands. The two areas–Eastern Himalayas and Western Ghats contain 5,332 endemic species of higher plants, mammals, reptiles, amphibia and butter flies (WCMC, 1993), to name a few. The following crops which first grew in India and

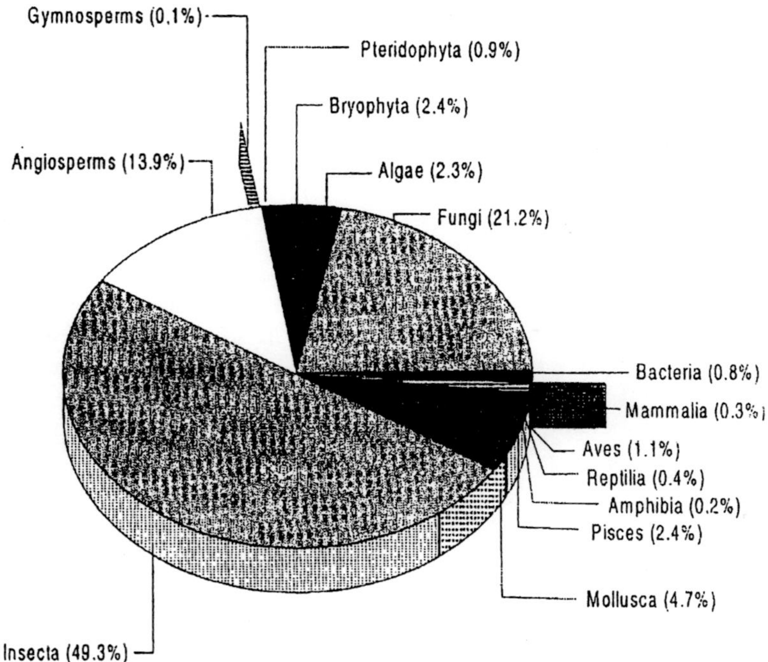

Fig. 4.1 : Diagram showing percentage of different biota in India.

Source: BSI and ZSI, 1994 (see also Khoshoo, 1990)

spread throughout the world include rice, sugarcane, Asiatic vignas, jute, mango, citrus, banana, several species of millets, spices, medicinals, aromatics, and ornamentals. India ranks sixth among the centres of diversity and origin as far as agro-biodiversity is concerned (Khoshoo, 1990).

BIOTIC IMPOVERISHMENT

Species are now dying out as fast as it did since the mass extinction at the end of cretaceous period, some 65 million years ago (Reid *et al.* 1992). Tropical deforestation is the cause of today's extinction crisis. About 17 million hectares of tropical forests are cleared every year. If this trend continues, at least 5-10 per cent of tropical forest species will face extinction in the next 30 years (Fig. 4.2).

The list of species at risk is swelling. Some 633 species are listed as being in danger of extinction in the United States alone, and more than 3,000 other species have been proposed for listing. In the past several hundred years, 80 species are believed to have become extinct in the continental United States and Hawaii, and a further 210 species are likely to be extinct. One-third of North America's fresh water fish are rare, threatened or endangered. Worldwide, more than 700 extinctions of vertebrates, invertebrates and vascular plants have been recorded since 1600 (Reid *et al.*, 1989). In Indonesia, 1,500 local rice varieties have become extinct in the past 15 years. Worldwide, some 492 genetically distinct populations of tree species are endangered. In north-western United States, 169 genetically distinct populations of anadromous fish are facing the risk of extinction.

Though biological extinction has been a natural phenomenon in geological history, the rate of extinction was perhaps one species per 1,000 years. However, between 1600 and 1950, the rate increased to one every ten years and currently it may be one every year. According to an estimate of Forest Survey of India, only 640,164 sq. km. of forest is left now out of a 4,017,009 km. Faunal losses have been mainly due to over exploitation, habitat alteration and pollution. The other possible reasons for the loss of species could be improper use of agro-chemicals and pesticides, a rapidly growing human population, inequitable land distribution and economic and political policies and constraints.

However, extinction is not caused by a single reason but by the cumulative effect of a host of causes. The earth's biodiversity being one big interlink, the loss of one species in most cases triggers off the downward slide and consequent loss of many other dependent species.

Decline in genetic diversity in agriculture is costly as well. The Irish potato famine in 1846, the Soviet wheat crop loss in 1972, citrus canker in Brazil in 1991, all have been the result of loss of genetic diversity.

We pay high cost for the loss of a species. The water we drink, the air we breathe, our fertile soils, and our productive seas are all products of healthy biological systems. The loss of ecosystems for instance, wetlands that provide critical services such as flood control, fish production, and pollutant assimilation is a direct economic cost

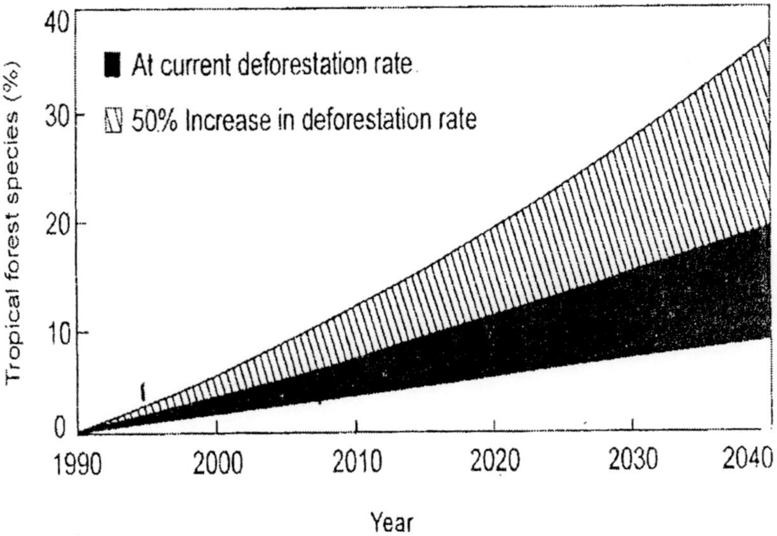

Fig. 4.2 : Percentage of tropical forest species likely to become extinct.

as well as threat to species survival. Loss of biodiversity also destroys opportunities of recreation and tourism business–now worth as much as $12 billion a year worldwide.

FACTORS RESPONSIBLE FOR LOSS OF BIODIVERSITY

The sum total of all varieties of genes present in a population or species constitutes a gene-pool. Loss of biodiversity means the genetic erosion of the loss of genes from a gene pool. Deforestation and the consequent soil erosion resulted in the categorisation of over 1,500 species of animals, plants and micro-organisms as "endangered species" in India alone. In fact, any disturbance in the ecosystem tends to reduce biodiversity.

The estimates of endangered wildlife grossly understates the real magnitude of the loss of biodiversity because it generally deals with higher life-forms commonly known to us. However, all amphibians, reptiles, birds, fishes and mammals put together constitute only about 2-3% of the known species of living organisms.

More than a million species of invertebrates and one third of a million species of recorded green plants have been reportedly existing on our planet, quite apart from the enormous number of species in nature which have not yet been assessed and enlisted. The gravity of the situation should be assessed while considering the entire life-spectrum on our planet.

Over the past few decades, the rate of global loss of biodiversity and the consequent biotic impoverishment has been increasing alarmingly. Exponential growth in human population and the consequent growth in consumption of the world's natural resources have led to the accelerated loss of species and habitats. The fragmentation of populations by destruction of habitats has also been contributing to the extinction of species and loss of genetic diversity. In some cases, the introduction of alien species (known as exotic species) has led to the loss of native plant and animal species. For instance, the native flora of *"fynbos biome"* in South Africa (Mediterranean type climate) is under threat, in parts, from the invasion of accacias and pines. Island communities, which have evolved in isolation, are particularly vulnerable to the introduction of exotic species.

Serious problems are faced in the tropical countries where biodiversity is highest and loss of species as well as ecosystems is relatively rapid. The issues are more serious in developing countries because of huge population, poverty, and lack of awareness.

The destruction of forests in the Third World Countries is estimated to be more than 11 million hectares every year. Recent tropical deforestation is associated with a pervasive cycle of initial timber extraction, followed by shifting cultivation, land acquisition, and subsequent conversion to pasture which leads to loss of forest resources, reduction in biodiversity and impoverishment or rural masses.

Almost about two thirds of the world's plants are present in tropics. Over 1.6 lac species known to us occur in the so-called megadiversity countries (i.e. tropical regions of the developing countries) of the world. It is feared that by 2020, about 25% of the tropical plants may be eliminated, if no efforts are made to protect them. In our country, about 15,000 species of flowering plants are known to occur. The "Red data book" compiled by the Botanical Survey of India enlists about 1000 species as rare and threatened plants. About 16% of the flowering plant population in our country have dwindled to alarmingly low levels, which include about 35% of the endemic plants i.e. plants found nowhere else. Among the non-tropical countries, the arid and semi-arid regions of South Africa contains large collection of rare and endangered plants which include large number of endemic species. About 12% of the 300 plant species of the USA are reported to be dangerously rare.

It is interesting to note that some of our ancient civilization (e.g., Olmec and Mayo civilizations) existed with high population densities integrated within tropical forest ecosystems. The knowledge of management techniques practised by such ancient civilizations could help in reverting current processes of landscape degradation in the tropics.

In some ecosystems, richness of species may be very high, as in the trophical rain forests and coral reefs. Apart from these, the proportion of endemic species (i.e. those species which are confined to a particular area) is also very significant. The conservation of areas of high endemism is particularly important for the conservation of genetic diversity.

The loss of genetic diversity through the mono-culture of crops is another serious cause of concern. The natural ecosystem, with its high number of species is replaced by a much simplified ecosystem, namely the agroecosystem, which is typified by only a few species. The world is dependent only on a tiny percentage of the 2.5 lac species of flowering plants for the provision of its staple food crops. The conservation of the genetic diversity found in related wild strains is thus vital to the continuing development of crop plant cultivars, in terms of increased productivity and resistance against disease and pests.

Other adverse effects of decrease in biodiversity apart from loss of valuable species are degradation of land, carbon dioxide build-up, acid rain, depletion of stratospheric ozone layer, etc.

The preservation of biodiversity at all levels, is a global concern now. Biodiversity is our most valuable natural resource and it is an insurance for our food and ecological security. Hence its components and genetic resources should be utilized only to support our socio-economic need in a sustainable manner.

The major causes responsible for reduction in biodiversity may be summarised as follows:

(1) Destruction of natural ecosystems to meet the requirements of food, space, shelter, etc. for the ever-expanding human settlements.

The species in the Boreal coniferous forests at the northern most belt skirting the Arctic Sea, which is thinly populated, are getting further reduced due to harsh and unfavourable climatic conditions. The biodiversity in several temperate and sub-tropical regions of the world (e.g., the zone between coniferous forests of north and the tropical zone in the south in the northern hemisphere, which support extensive agriculture and cattle renching) is already under threat due to encroachments.

The tropical regions of the world colonized by developing and under-developed countries (which are called mega-diversity countries) are endowed with rich and diverse assortment of fauna and flora. These occupy about 1/4th of total area of the world and contain about 3/4th of the world population. In order to feed and sustain millions of poor people, these countries are forced to expand agriculture and industries, which led to over-exploitation of resources and large-scale deforestation. Since tropical forests contain about 50% of the total number of species on the earth, destruction of natural habitats are proving disastrous for the entire biosphere.

Mangrooves and other coastal wetlands provide a buffer zone between the sea and land, which protects the coastline from erosion from the oceanic waves and at the same time it clears the silt, sediments, pollutants, etc. before the water flows into the sea. Mangrooves also are rich in fauna and flora. Loss of mangrove ecosystems in Asia, West Africa and Latin America are causing extensive damage to biological damage and fish productivity apart from erosion of coast-line.

Coral-reefs are marine equivalents of tropical forests and they represent highly productive and richly diverse ecosystems. Loss of mangrooves result in destruction of coral-reef ecosystems. The destruction of mangrooves in Costa-rica and Philippines is already a matter of grave concern for their rich and diverse biota.

(2) Unfavourable changes in the biotic and abiotic factors result from

  (a) *Natural Upheavals :* Natural calamities such as earthquakes, floods, volcanic eruptions, forest-fires, epidemics, droughts, etc., cause extensive damage to plant and animal life.

  (b) *Environmental Pollution :* Pollutants such as pesticides, chlorinated hydrocarbons, and toxic heavy metals destroy the weak and susceptible species, and damage the living organisms. Over-fertilization leads to entrophication which promotes the growth of some species while suppressing others. Industries emitting SO_2 and NO_x cause acid rains which destroy plants and fishes.

  (c) *Invasion by exotic species :* Any species which is not a natural inhabitant of the local habitat but is introduced into the system accidentally or deliberately, is called *"exotic species"*. Several instances are known when a natural biotic community of the ecosystem suffered extensive damage because of the invasion by exotic species. A classical example is the contamination of wheat seeds imported by us from USA under PL-480 scheme by *"parthenium hysterophorus"* (known as congress grass) and *"Agrosteamma githago"* (known as corn cockle). Both the species grew and spread all over the country as pernicious weeds in wheat fields. Parthenium is an aggressive plant which matures rapidly and generates thousands of seeds. Because of these plants, indigenous herbs and grasses as well as cultivated crops suffered as it depletes the nutrients from the soil. Moreover, the enormous quantity of pollens produced by parthenium cause allelophatic effects on cultivated plants like tomato, chillies and brinjals and inhibited their growth. Similarly, myrtaceae (eucalyptus) and casurinaceae (casurina) are the species introduced form Australia and Tropical America into our country. Their extremely fast growth made them a valuable source of rough timber. However, they tend to suppress the native inhabitants and hence are ecologically harmful. Another striking example is the introduction of Nile perch from north in Lake Victoria in Africa, because of which about 50% of the 400 original fish species of the lake were driven to near extinction.

(d) *Over-exploitation of selected species:* About ten thousand years ago, man used over 5000 species and varieties of edible plants for his sustenance. Now, several traditional varieties are neglected and only about 150 species of plants are cultivated on large scale and out of these only about 15 types provide nearly 90% of the world's food supply. Most of these important cultivars used today are synthetic varieties in which desirable traits from a number of varieties are pooled together to obtain better yields, tastes and flavours. This trend of neglecting traditional varieties and substituting them with a few selected strains for some immediate benefits has been causing considerable damage to the biological diversity.

Over-exploitation of certainty types of species is leading to dwindling of some plants of scientific, medical, decorative and other values. Medicinal plants such as podophyllum sp., are fastly disappearing due to huge collection. Similarly timber-producing trees of economic value such as Dysoxylon malabaricum, santalum album and pterocarpus santalum, and orchids producing decorative flowers are diminishing due to over-exploitation.

Further, only limited varieties of rice, wheat, corn, apples, potatoes etc. are widely cultivated today, neglecting totally the traditional varieties. Hybrid varieties have replaced old strains of wheat and rice to obtain higher yields. Such a reduction of genetic diversity among the cultivated species and disappearance of their wild relatives, may restrict the creation of new cultivars in future, which doesn't augur well for the posterity.

Elephants, tigers, rhinoceros, snakes, crocodile whales, minks and some select groups of animals-and birds are mercilessly hunted for food, recreation, and monetary benefits from their horns, tusks, hides skins, etc. This has resulted in rendering certain types of selected group of organisms as endangered category.

(3) Increasing concentration of green-house gases in our atmosphere which threatens our planet with rise in mean global temperature. This may cause extensive changes in environmental conditions and precipitation patterns, which may

endanger the plant species as a whole. As the climatic changes occur, large-scale migrations of species take place to exist in favourable climate. Such species which are incapable of migration will be adversely affected. Further, global warming will inevitably lead to expansion of tropical belt, displacement of sub-tropical and temperate zones northward and southward away from equator, while the temperate and polar regions may reduce in size. Analysis of fossil pollens indicate that under such a scenario, the forests will have to migrate 20 kms to 200 kms per century. Many species that cannot migrate at such a faster rate may perish. Even species that migrate will have to overcome barriers like oceans, mountain ranges and large human establishments and under unfavourable conditions, many of these species may become extinct. Further, since different species respond to changing climatic patterns in different ways, the entire global biotic spectrum will change and in such a sweeping upheaval, the biological diversity is likely to be drastically reduced.

CONCEPT OF THREATENED SPECIES

The International Union of Conservation of Nature and Natural Resources (IUCN) has classified the threatened species of plants and animals into three categories for the purpose of their conservation. The classification is based on

(i) the present and past distribution,
(ii) the decline in the number of population in the course of time,
(iii) abundance and quality of natural habitat and
(iv) the biology and potential value of the species.

According to the degree of danger to the threatened species, the latter are classified as endangered, vulnerable and rare species.

(i) ***Endangered (E) Species.*** These species are in danger of extinction and they are not likely to survive if the factors threatening their extinction continue. Their numbers have been reduced to a critical level or their habitats have been so drastically reduced that they are in immediate danger of extinction. Great Indian bustard, pink headed duck, lion, tiger, musk deer, kashmir stag and some other animals (to be mentioned ahead) are facing extinction in India.

(ii) Vulnerable (V) Species. These species are likely to be in danger of extinction in the near future if the factors threatening their extinction continue. Their populations have been greatly reduced and their survival is not assured. This category of species also includes those species whose populations are still abundant but are somehow threatened throughout their range.

(iii) Rare (R) Species. These species have small population in the world. They are usually confined to limited areas or are thinly scattered over a more wide area.

Threatened (T) Species. The term "threatened species" is used for any one of the above categories (endangered, vulnerable or rare species) in the context of conservation of wild life.

Conservation of Wildlife

Meaning. Wildlife conservation means the management of human use of the biosphere so that it may give maximum benefit to present generation while maintaining its potential to meet the requirements of the future generations.

Objectives. Conservation of wildlife has three specific objects:

(i) To maintain essential ecological processes and life supporting systems — air, water and soil.

(ii) To preserve the diversity of species or the range of genetic material found in the world's organisms.

(iii) To ensure a continuous use of species and ecosystems which support rural communities and major industries.

Thus conservation of living resources is concerned with living organisms (plants, animals and micro-organisms) and with those non-living elements of the environment which support them.

Conservation strategies. Scientists of 100 countries of the world have evolved a comprehensive ***world conservation strategy*** for the proper use of resources. Some of the steps proposed to protect the wildlife are given below:

(i) The wildlife protection requires proper planning and management of land and water uses. The wildlife should be protected both in their natural ***habitats*** and in zoological and botanical gardens.

(ii) The species that are threatened should be given ***preference*** over others in their conservation. The species that are only representative of their family or genus should receive special attention. An endangered species should be given priority over a vulnerable one, a vulnerable species over a rare one and a rare species over other categories.

(iii) All varieties of useful plants and animals and their wild relatives and micro-organisms should be preserved. Preference should be given to those varieties that are most threatened and most needed for national and international ***breeding programme.***

(iv) The ***critical habitats*** the feeding, breeding, nursery and resting areas) of the species should be preserved in protected areas.

(v) The habitats of ***wild relatives*** of useful plants and animals should be safeguarded.

(vi) Protected areas should be established to preserve the habitats of ***migratory animals.***

(vii) *Bilateral* or multilateral ***agreements*** should be made to set up the required network for a species that migrates from one national jurisdiction to another.

(viii) The ***unique ecosystems*** should be protected on priority basis.

(ix) The species and ecosystems should not be exploited beyond their ***productive capacities.*** Industries, communities and countries that are over-exploiting the living resources, should be convinced that they would be better if they utilize the resources at a sustainable level.

(x) International trade in wild plants and animals and their products should be regulated by appropriate ***laws.***

(xi) Only licensed persons should be allowed to shoot animals. ***Hunting*** should not be permitted during the breeding seasons of the animals. Killing of young growing animals should be prohibited.

(xii) ***National parks*** and ***Sanctuaries*** should be set up to protect wildlife.

(xiii) People must be ***educated*** regarding the preservation of wildlife. This is very essential for wildlife conservation.

Protected Areas in India

India has rich flora and fauna because of its diverse climate and physical conditions, ranging from cold deserts of Ladakh and

Spiti to the hot deserts of the Thar; the temperate forests in the Himalaya to the tropical rain forests of lowlands. India is also gifted with large fresh water bodies (like the Wular and Manasbal lakes in Kashmir, the Chilka in Orissa and the Kolleru lake in Andhra Pradesh) and rich ocean resources of Deccan.

In 1952, the government of India constituted an advisory body on country's wildlife called the *Central Board for Wildlife.* Later, it was named as the *Indian Board for Wildlife* (IBWL). Many Indian states have formed their own *Wildlife Advisory Board.* The government of India has already enacted a *Wildlife Protection Act, 1972.* The killing and capturing of wildlife without prior permission from suitable authority have become punishable under the law. National parks and sanctuaries could be created under this law. This Act has been adopted by all the states except Jammu and Kashmir which has its own similar Act. Creation of biosphere reserves has also been put into practice since 1986. Zoological Survey of India (ZSI) and Botanical Survey of India also have wildlife protection programmes. Certain societies like *The Bombay Natural History Society* (B.N.H.S.) and *Wild Preservation Society of India* (W.P.S.I) are also doing useful work for the conservation of wildlife. Protected areas in India include National parks, sanctuaries and biosphere reserves.

(i) ***National Parks.*** A national park is an area which is strictly reserved for the welfare of the wildlife and where activities such as forestry, grazing or cultivation are not allowed. Private ownership, rights and habitat, manipulation are not permitted in a national park. There are at present 66 national parks in India covering an area of 33,988,14 square kilometres (about 1% of India's total area).

(ii) ***Sanctuaries.*** A sanctuary is an area which is reserved for the conservation of animals only and operations such as harvesting of timber, collection of minor forest products and private ownership rights are allowed so long as they do not affect the animals adversely. There are at present 368 sanctuaries in India covering about 1,07,310,13 square kilometres (about 3.2% of India's total area). Some important national parks and sanctuaries of India are mentioned in the Table 4.2. Out of the 434 national parks and sanctuaries, 17 have been selected for "Project Tiger." To save the tiger from extinction in India/

Project Tiger" was launched in 1972. This project planned to create tiger reserves in selected areas of India. Due to this effect considerable improvement has been observed in tiger population.

(iii) ***Biosphere Reserves***. Recently the concept of biosphere reserves has been evolved by the Man and Biosphere (MAB) programme of the UNESCO.

Zones of Biosphere Reserve : In a biosphere reserve, multiple land use is permitted by dividing into different zones These zones are as follows:

(a) ***The Core Zone.*** No human activity is permitted in this zone.
(b) ***The Buffer Zone.*** Limited human activity is permitted in this zone.
(c) ***The Manipulation Zone.*** Several human activities can occur in this zone.

Role of Biosphere Reserve. A biosphere reserve preserves:-

(a) Wild population (b) traditional life styles of tribals (c) genetic resources of varied domesticated plants and animals.

Biosphere Reserves of India. 14 areas have been declared as Biosphere reserves in India. Their names and locations are given below.

Table 4.2 : Some Protected Indian Wild Animals

Biosphere Reservers	*States/Union Territory*
1. Niligiri (First Biosphere Reserve established in 1986)	Kerala, Karnataka and Tamil Nadu.
2. Namdapha	Arunachal Pradesh
3. Nanda Devi	Uttaranchal
4. Uttarakhand (Valley of flowers)	Uttaranchal
5. North Islands of Andamans	Andaman and Nicobar islands
6. Great Nicobar	Andaman and Nicobar
7. Gulf of Mannar	Tamil Nadu
8. Kaziranga	Assam
9. Manas	Assam
10. Sunderbans	West Bangal
11. Thar Desert	Rajasthan
12. Kanha	Madhya Pradesh
13. Nokrek	Meghalaya
14. Little Rann of Kutch	Gujarat

Mammals

1. Black buck *(Antelope cervicapra)*
2. Capped lungur *(Presbytis pileatus)*
3. Chital *(Axis axis)*
4. Clouded *(leopard Neofelis nebulosa)*
5. Crab-eating macaque *(Macaca irus umbrosa)*
6. Fishing cat *(Felis viverrina)*
7. Flying squirrels *(Petaurista, Eupetaurus Belomys, Hyloptes. All species)*
8. Four-horned antelopes *(Tetraceros quacs quadriconis)*
9. Gangetic dolphin *(Platanista gangetica)*
10. Giant squirrels *(Ratufa macroura, R. indicai, R. biocolor)*
11. Golden cat *(Felis temmincki)*
12: Golden langur *(Presbytis eei)*
13. Himalayan black bear *(Selenarctos thibetanus)*
14. Himalayan brown bear *(ursus arctos)*
15. Himalayan crestless porcupine *(Hystrix hodgsoni)*
16. Hispid hare *(Caprolagus hispidus)*
17. Hoolock or gibbon *(Hylobates hoolock)*
18. Hyaena *(Hyaena hyaena)*
19. Indian elephant *(Elephas maximus)*
20. Indian lion *(Panthera leo persica)*
21. Indian pangolin *(Monis crassicaudata)*
22. Indian wild ass *(Equus hemionus khur)*
23. Indian wolf *(Canis lupus)*
24. Kashmir stag or hangul *(Cervus elaphus hanglu)*
25. Leopard or panther *(Panthera pardus)*
26. Leopard cat *(Felis bengalensis)*
27. Lesser or red panda *(Ailurns fulgens)*
28. Lion-tailed macaque *(Macaco silenus)*
29. Loris *(Loris tardipadus)*
30. Musk deer *(Moschus moschifeeru)*
31. Nilgai *(Boselaphus tragocamelus)*
32. Nilgiri langur *(Presbytisjohni)*
33. Otters *(Lutra lutra, L. perspicillata, Aonyx cinerea)*
34. Pig-tailed macaque *(Macaca nemestrina)*
35. Pigmy hog *(Sussulvanius)*

36. Redfox *(Mpesvulpes)*
37. Rhinoceros *(Rhinoceros Unicornis)*
38. Sambhar *(Cervus unicolor)*
39. Sloth bear *(Melursus rsinus)*
40. Slow loris *(Nycticebus coucang)*
41. Snow leopard *(Panthera uncia)*
42. Swamp deer or gond *(Cervus duvauceli, all species)*
43. Tibetan antelope or chiru *(Panthelope hogsoni)*
44. Tibetan fox *(vulpes ferrilatus)*
45. Tibetan wild ass *(Equus heminonus king)*
46. Tiger *(Panthera tigris)*
47. Wild buffalo *(Bubalus bubalis)*
48. Wild dog or dhole *(Cuon alpinus)*
49. Wild pig *(Sus scrofa)*
50. Wild yak *(Bos grunniens)*

Reptiles

51. Estuarine crocodile *(Crocodilus porosus)*
52. Gharial *(Gavia lisangeticus)*
53. Leathery turtle *(Demochelys coriacea)*
54. Marsh crocodile *(Crocodilus palustris)*
55. Monitor lizards *(Varanus griseus, V. bengalensis, V. flavescens, V. salvator, K. nebulosus)*
56. Pythons *(Python molurus, P. reficulatus)*

Birds

57. Cheer pheasant *(Catreus wallichi)*
58. Great Indian bustard *(Choriotis nigriceps)*
59. Great Indian hombill *(Buceros bicoaris)*
60. Large falcons *(Falco peregrinus, F. biarmicus, F. chicquera)*
61. Mountain Quail *(Oppassiasuperciliosa)*
62. Peafowl *(Pavo cristatus)*
63. Pink headed duck *(Rhodonessa caryophyllacea)*
64. Siberian white crane *(Grus leucogeranus)*

Amphibians

1. Himalayan newt *(Tylototriton verrucosus)*
2. Viviparous toad *(Nectophryne tuberculosa)*

Table 4.3 : Some National Parks and Wildlife Sanctuaries of India

Name and Location	*Area in sq. km*	*Important Animals found*
1. Kaziranga National Park. District Sibsagar, Assam	430	Rhinoceros, elephant, wild buffalo, bison, tiger, leopard, sloth bear, sambhar, swan deer, barking deer, wild boar, gibbon, python, and birds like pelican, stork and ring-tailed fishing eagles. This is a famous National park for famous one-horned rhinoceros of India.
2. Sundarbans (Tiger Reserve) 24-Parguna, West Bengal.	2585	Tiger, wild boar, deer, gangetic dolphin, estuarine crocodile.
3. Jaldapara Sanctuary Madarihat, West Bengal	1155	Rhino elephant, tiger, leopard, gaur, deer, sambhar, different kinds of birds.
4. Hazaribagh National Park, Hazaribagh, Bihar	186	Tiger, leopard, hyaena, wild boar, gaur, sambhar, nilgai, chital, sloth bear, peafowl.
5. Corbett National Park, District Nainital-Uttaranchal	525	Tiger, elephant, panther, sloth bear, wild boar, nilgai, sambhar, chital, crocodile, python, king cobra, peafowl, partridge. This is the first National Park of India which is famous for tigers.
6. Gir National Park District Junagarh, Gujarat	1412	Asiatic lion, panther, striped hyaena, sambhar, nilgai, chital, 4-horned antelope, chin-

Name and Location	*Area in sq. km*	*Important Animals found*
		kara, wild boar, langur, python, crocodile, green pigeon, partridge. This national park is famous for the Asiatic lions.
7. Keoladeo Ghana Bird Sanctuary, Bharatpur, Rajasthan	29	Siberian crane, storks egrets, herons, spoon bill, etc. Drier parts of this marshy sanctuary have spotted deer, black buck, sambhar, wild boar, blue bull, phython. This sanctuary is famous for birds.
8. Desert National Park, Jaisalmer, Rajasthan	3000	Great Indian bustard, black buck, chinkara.
9. Sultanpur Lake Bird sanctuary, District Gurgaon, Haryana	12	Crane, sarus, spot bill, duck, drake, green pigeon, wild boar crocodile, python.
10. Bir Moti Bagh Wildlife sanctuary, District Patiala, Punjab	8.3	Nilgai, wild boar, hog deer, black buck, blue bull, jackal, peafowl, partridge, sparrow, myna, pigeon, dove.
11. Shikari Devi Sanctuary, Mandi, Himachal Pradesh	213	Black bear, snow leopard, flying fox, barking deer, musk deer, chakor, partridge.
12. Dachigam Sanctuary, Srinagar, Jammu & Kashmir.	89	Hangul or Kashmir stag, musk deer, snow leopard, black bear, brown bear.
13. Kanha National Park. Mandla and Balaghat, Madhya Pradesh.	940	Tiger, panther, chital, chinkara, barking deer, blue bull, four horned deer, langur, wild boar, black buck, nilgai, wild dog, sloth bear,

Name and Location	*Area in sq. km*	*Important Animals found*
		sambhar, crocodile, grey horn bill, egret, peafowl.
14. Tandoba National Park, Chandrapur, Maharashtra.	116	Tiger, sambhar, sloth bear, bison, chital, chinkara, barking deer, blue bull, four horned deer, langur, peafowl, Crocodile.
15. Bandipur National Park, District Mysore, Karnataka,	874	Elephant tiger, leopard, sloth bear, wild dog, chital, panther, barking deer, langur, porcupine, gaur, sambhar, malabar squirrel, green pigeon.
16. Annamalai Sanctuary, District Coimbatore, Tamil Nadu.	958	Elephant, tiger, panther, gaur, sambhar, spotted deer, sloth bear, wild dog, barking deer, langur, porcupine.
17. Mudumalai Wildlife Sanctuary, District Nilgiri, Tamil Nadu.	321	Elephant, gaur, sambhar, chital, barking deer, mouse deer, four horned antelope, langur, giant squirrel, flying squirrel, wild dog, jackal, wild cat, civet, sloth bear, porcupine, python, rat snake, monitor lizard, flying lizard.
18. Nundanthurai Sanctuary. Distt. Tirunelveli, Tamil Nadu.	520	Panther, tiger, sambhar, chital.
19. Nagarjuna Sagar (Ikshawaka sanctuary) Guntur, Prakasham, Karnool, Mahbubnagar and Nalgonda, Andhra Pradesh.	3568	Tiger, panther, wild bear, chital, nilgai, sambhar, black buck, fox, jackal, wolf, crocodile.

Name and Location	*Area In sq. Km*	*Important Animals found*
20. Periyar Sanctuary, District Idukki, Kerala	777	Elephant, tiger, panther, gaur, leopard, sloth bear, sambhar, barking deer, wild dog, wild boar, black Nilgiri langur, hornbill, egret.
21. Chilka Lake Bird sanctuary, Balagaon, Orissa.	900	Water fowls, ducks, cranes, ospreys, golden plovers, sandpipers, stone curlews, flamingoes.

CONSERVATION OF BIOLOGICAL DIVERSITY

Biotic resources represent the very essence of life on the earth and everything that is possible should be done to preserve the biological diversity. There is an urgent need to educate people to adopt environment friendly practices and re-orient our activities in such a way that our development is harmonious with other life-forms and is sustainable.

The following steps will go a long way in achieving these objectives:

(1) *Biodiversity Inventories:* Current scientific understanding of ecological processes has to be strengthened by adequate research efforts aimed at improving methodologies, distributional and status information. Strategies based on sound information will ultimately provide the basis for pragmatic policies and management decisions. Better inventories and assessments are needed of current conditions, abundances, distributions and management direction for genetic resources, species populations, biological communities and ecological systems. For this purpose, the following activities should be supported.

   (a) Rapid assessment programmes to provide a "Snap-Shot" of richness of species in an area.
   (b) Surveys, to map out the distribution of earth's ecosystems.
   (c) Broad-based inventory efforts, focused on poorly known habitats, degraded and multiple-use habitats as well as

on vertebrates and vascular plants that can be used as "benchmarks" for habitat quality.

(d) Intensive inventories, to determine all the species present, from microbes to vertebrates, should be prepared at a few selected sites around the world. Although this requires intensive effort over a long time-frame, that will be highly useful to decipher processes such as how the biodiversity at species level affect diversity at the ecosystem and community levels.

Such an effort will lead to significant improvements in inventory processes, methodologies and technologies such as the following :

(a) Provide estimates of resources in specific geographical units and evaluate their reliability.
(b) Enable spatial display of the estimates and units.
(c) Enable the best use of existing information and new technologies, such as remote sensing and geographic information system (GIS).
(d) Provide a base-line for monitoring changes in the extent and condition of the resource.
(e) Eliminate redundant data collection, develop common terminology and promote data sharing through corporate data bases.
(f) Provide up-to-date data bases using modelling techniques, accounting procedures, and re-inventories.
(g) Ensure optimum utilisation of information management systems to provide maximum flexibility for data integration, manipulation, and sharing.

2. ***Conserving Biodiversity in protected habitats:*** Wildlife conservation efforts are mostly concentrated on protecting animal and plant life in zoos, sanctuaries, gardens, biosphere reserves, etc. The two basic approaches to the wildlife conservation in protected habitats are as follows :

(a) *"Ex-situ" conservation :* It means the wildlife conservation in captivity under human care. In this, the endangered plants and animals are collected and bred under controlled conditions in gardens, zoos, sanctuaries etc. wildlife management in captivity have the following advantages :

(i) The organisms are assured of food, water, shelter and security and hence can have longer life span and should be motivated to conserve resources, to avoid extravagance, and educated properly regarding ecological issues. All efforts should be made to respect and conserve indigenous knowledge, traditions and environment-friendly practices, longer span of breeding activity, thereby increasing the possibility of having more number of offsprings.

(ii) The chances of survival of endangered species increase because of human care under secure conditions.

(iii) This offers, the possibility of using genetic techniques to improve the species concerned. However, there are some disadvantages and limitations of wild-life management in captivity :

  (i) Since maintenance and breeding of plants and animals under captivity is very expensive, it can be adopted only for a few selected species.

  (ii) Wildlife captivity only under a set of favourable environmental conditions deprives the organisms the opportunity to adopt to ever-changing natural environment. Therefore, new life forms cannot evolve and thus the gene-pool gets stagnant.

(b) *"In-situ" conservation :* This involves setting aside large portions of earth's surface for wildlife. However, many protected habitats are used for logging, tourism and other profitable activities, thereby pushing wildlife conservation to a less important objective. Further, many protected habitats are not large enough, not maintained properly for want of adequate staff, and are not properly protected from environmental pollution. These factors, to some extent reduce the laudable advantages of "In-situ" conservation. Large pockets of protected zones are essential for not only conserving vast number of species of living organisms but also provide opportunities to evolve. Otherwise, man-made habitats may end up with static gene-pool.

Monitoring the effects of land management and conservation programmes is also very important.

(c) *"Ex-situ" conservation:* It means the wildlife conservation in captivity under human care. In this, the endangered plants and animals are collected and bred under controlled conditions in gardens, zoos, sanctuaries etc. wildlife management in captivity have the following advances :

(3) *Conserving Biodiversity in Seed-Banks and Gene-Banks :* Most of the plant species form seeds with variable periods of dormancy after which they can be germinated to yield daughter plants. The seeds can be stored in seed-banks or gene-banks or germ-plasm banks. The germplasm of a plant is any of its parts from which new plants can be generated.

The International Board for Plant Genetic Resources was constituted in 1974 with its Headquarters in Rome. By 1985, a chain of 43 gene banks were set up in different parts of the world.

Recent techniques of preservation have greatly improved the use of germ plasm banks. Seeds with low moisture content when cooled at -196°C under liquid nitrogen temperature can be preserved for almost indefinite periods, say for several centuries. By application of tissue culture and cryo-preservation techniques, most of the plants can be maintained in viable state for very long periods of time. Meristematic cells of plants or small parts of simple vegetative tissues to be preserved are incorporated into suitable synthetic medium under aseptic conditions and are allowed to grow into small clumps, known as *"Callus"*. These calluses can be preserved under liquid nitrogen for almost unlimited periods of time. They can be multiplied as and when desired.

There is a dire need to establish more and more regional germ-plasm banks particularly in the mega diversity countries like India.

(4) *Restoration of Biodiversity:* Restoration of ecosystems and biological communities is an important means for maintaining biodiversity or at least of retarding its net loss. Biodiversity is threatened not only by reduction of habitat area and of connections between habitats, but also by degradation of quality of the remaining habitats. By restoring both the extent and quality of important habitats, restoration programmes provide refuge for species and genetic resources that might be lost otherwise.

Several techniques are used to restore ecosystems, depending on the nature of the ecosystem and the impact type being addressed. These techniques include vegetation planting to control erosion, fertilization of existing vegetation to encourage growth, removal of contaminated soils, fencing to prevent cattle, reintroduction of extirpated species, restoration of hydrologic connections to wetlands, etc.

Rapid reforestation to reinstate green plant cover on barren lands, hill slopes, highways, roads, etc. taking care to ensure that the species planted are ecologically compatible to the region and acceptable and useful to the local people, will go a long way in restoring biodiversity.

However, we should remember that restoration is not a substitute for the preservation or good management and it is both expensive and time consuming.

(5) *Imparting Environmental Education :* Educating people from all walks of life regarding eco-friendly practices goes a long way in conservation of plant and genetic biodiversity. People at large should be motivated to conserve resources, to avoid extravagance, and educated properly regarding ecological issues. All efforts should be made to respect and conserve indigenous knowledge, traditions and environment-friendly practices.

(6) *Enacting Strengthening and Enforcing Environmental Legislations :* Existing environmental laws against ecologically unsound practices should be strengthened and enforced ruthlessly. Simultaneously, voluntary organizations should be motivated to include protection of biodiversity as a major and priority issue in their agenda.

(7) *Population Control :* Effective population control measures have to be taken as a top-priority issue in the national agenda by involving people of all political parties, religious faiths and social organisations. Suitable incentives and disincentives should be in-built into the strategies specially formulated for this purpose.

(8) *Reviewing the Agricultural Practices :* We should refrain from temptation of high yields and making a fast buck at the cost of sustainable development. We should try to infuse diversity in our agricultural practices by restoring to mixed cropping polyculture and tolerance to wild plants and other life forms around our agricultural fields. A healthy soil should be maintained by using more of organic manures and less of synthetic fertilizers and pesticides. All efforts should be made to maintain a balanced prey-predator relationship in our agro-ecosystems.

(9) *Controlling Urbanization :* Ever-increasing urbanisation and expansion of urban settlements should be controlled. Biological diversity should be infused into the urban localities.

CONCLUSION

Maintaining global biological diversity should go much beyond crisis-management of increasing number of endangered species. Long-term maintenance of species and their genetic management requires cooperative efforts across entire landscapes. Biodiversity should be dealt with at the scale of habitats or ecosystems rather than at species level.

REVIEW QUESTIONS

1. Define biodiversity. Explain the importance of biodiversity?
2. What are the major factors that are responsible for the loss of plant and genetic biodiversity? What steps are needed to prevent them.
3. Explain briefly :
   (i) In-side conservation of biodiversity.
   (ii) Ex-side conservation of biodiversity.

Chapter 5

ENVIRONMENTAL POLLUTION

Definition. Pollution may be defined as an undesirable change in the physical, chemical or biological characteristics of our air, water and land that may or will harmfully affect human life, the lives of the desirable species, our industrial processes, living conditions and cultured assets, or that may or will waste or deteriorate our raw material resources.

Pollution is different from *contamination* which is the presence of harmful organisms or their products causing disease or discomfort. Pollution can be *natural* or man made. Natural pollution comes from volcanic eruptions, emission of natural gas, soil erosion, ultraviolet rays, cosmic rays, etc. Most of the pollution is man-made.

Pollutants. They are substances *(e.g.,* smoke), chemicals *(e.g.,* sulphur dioxide) or factors (e.g., heat) which cause a potential or actual adverse effect on the natural quality of any constituent of the environment. Pollutants are generally waste products or by-products. Sometimes, pollutant is a constituent in a wrong proportion. For example, addition of nitrate and phosphate is a must for soil fertility. They, however, cause water pollution. Similarly, excess of carbon dioxide or low concentration of oxygen in the atmosphere are also pollutants. Pollutants are classified into different ways:

1. According to the form in which the pollutants persist after release into the environment, they may be primary or secondary.
   (i) ***Primary Pollutants.*** Pollutants persisting in the environment in the form they are passed into it, e.g., DDT.
   (ii) ***Secondary Pollutants.*** Pollutants which are formed by reaction amongst the primary pollutants. For example, peroxyacyl nitrates (PAN), are formed through reaction between nitrogen oxides and hydrocarbons in the presence of sunlight.

2. According to their existence in nature, the pollutants may be quantitative or qualitative.
   (i) ***Qualitative Pollutants.*** They are pollutants which do not normally occur in the environment but are passed into it through human activity, *e.g.*, DDT and other pesticides, fungicides, herbicides, etc.
   (ii) ***Quantitative Pollutants.*** They become pollutants only when their concentration reaches beyond a threshold value in the environment, *e.g.*, CO, CO_2, nitrogen oxides.

3. According to their natural disposal, the pollutants may be biodegradable or non-degradable.
   (i) ***Biodegradable (= Degradable) Pollutants.*** They are actually waste products which are slowly degraded by microbial action. Pollution results when their production exceeds the capacity of the environment to degrade them.
   (ii) ***Non-degradable (= Non-biodegradable) Pollutants.*** They are pollutants which do not get easily decomposed. They include wastes *(e.g.,* phenolics, plastics, glass or metallic containers) or poisons *(e.g.,* pesticides like DDT, salts of heavy metals, radioactive substance).

Types of Pollution. Pollution is of five types : air pollution, water pollution, soil pollution, radioactive pollution and noise pollution.

I. AIR POLLUTION/ATMOSPHERIC POLLUTION

The World Health Organization defines air pollution as "the presence of materials in the air in such concentration which are harmful to man and his environment." In fact air pollution is the occurrence or addition of foreign particles, gases and other pollutants into the air which have an adverse effect on human beings, animals, vegetation, buildings, etc.

Causes/Sources of Air Pollution

The various *causes of air pollution* are (i) combustion of natural gas, petroleum, coal and wood in industries, automobiles, aircrafts, railways, thermal plants, agricultural burning, kitchens, etc. (soot, flyash, CO_2, CO, nitrogen oxides, sulphur oxides), (ii) metallurgical processing (mineral dust, fumes containing fluorides, sulphides and metallic pollutants like lead, chromium, nickel, beryllium, arsenic, vanadium, cadmium, zinc, mercury), (iii) chemical industries

Table 5.1 : Showing Important Air Polliition Sources And Emmission

Category	*Examples*	*Important Pollutants*
Chemical plant	Petroleum refineries, fertilisers, cement, paper mills, ceramic, clay products and glass manufacture.	Hydrogen sulphide, sulphur oxide, fluorides, organic vapours and dusts.
Crop spraying	Pest and weed control	Organophosphates, chlorinated hydrocarbons, lead, arsenic.
Fuel burning	Domestic burning, thermal power plants	Sulphur and nitrogen oxides
Metallurgical plants	Aluminium refineries, steel plants	Metal fumes (Lead and Zinc) fluorides and particulates
Nuclear device testing	Bomb explosions	Radioactive fallout, Sr-90, Cs-137, C-14, etc.
Ore preparation	Crushing, grinding and screening	Uranium and Beryllium dust, other particulates, Argon-41 Iodine-131.
Spray painting, solvent extractions, inks, solvent cleaning	Furniture and appliances dyeing, printing and chemical separations, dry cleaning.	Hydrocarbons and other oreanic vapours
Transportation	Cars, trucks, aeroplanes and railway	Carbon monoxide, nitrogen oxides, lead. smoke, organic vapours, etc.
Waste recovery	Scrap metal yards, rendering plants	Smoke, soot, odours, organic vapours, metal fumes.

including pesticides, fertilizers, weedicides, fungicides, (iv) cosmetics, (v) processing industries like cotton textiles, wheat flour mills, asbestos, (vi) welding, stone crushing, gem grinding. Natural air pollutants include (a) pollen, spores, (b) marsh gas ,(c) volcanic gases and (d) synthesis of harmful chemicals by electric storms and solar flares. The major cause of pollution in the *urban areas* are automobiles which inefficiently burn petroleum, release 75% of noise and 80% of air pollutants. Concentration of industries in one area is another major cause of air pollution, *e.g.*, cotton dust in Ahmedabad, Surat and nearby areas.

Thus sources of air pollution can be divided into the following groups.

(a) *Stationary Combustion Sources.* These include products of burning of fuels by people.
(b) *Mobile Combustion Sources.* These include locomotives, automobiles, aircrafts, etc.
(c) *Industrial Processing and Other Sources.* These include chemical industries and operations such as crushing, blasting, drilling, dyeing, mixing and grinding.

EFFECT OF AIR POLLUTANTS

Air pollutants are broadly classified into particulate and gaseous. The particulate substances include solid and liquid particles. The gaseous include substances that are in the gaseous state at normal temperature and pressure. *The air pollutants have adverse effect on human beings, animals, vegetation, buildings.* Air pollutants also change earth's *climate. Aesthetic sense* is also influenced by air pollutants. The different air pollutants and their effects are as follows.

1. ***Particulate Matter.*** It is of two types–*settleable* and *suspended.*

The settleable dusts have a particle longer than 10μm. The smaller particles are able to remain suspended for long periods in the air. They are further distinguished into two types (i) less than 1 μm—*aerosol* (ii) more than 1 μm–*dust* (if solid) and *mist* (of liquid). The important effects of particulate matter are:

(i) Dust and smoke particles cause irritation of the respiratory tract and produce bronchitis, asthma and lung diseases. In the area of Bhatinda Thermal Plant in Punjab, bronchitis and asthma have afflicted almost every second person.
(ii) Smog is a dark or opaque fog which is formed by the dust and smoke articles causing condensation of water vapours around them as well as attracting chemicals like SO_2, H_2S,

NO_2, etc. Smog harms plant life through silvering, glazing and necrosis besides reduced availability of light. In human beings and animals it produces respiratory troubles, asthma, allergies, etc. Because of reduced visibility, smog causes a number of accidents. A five day smog in London in 1952, killed about 4000 persons. Many more persons suffered from illness. A large scale destruction of plant life also took place (Royal Meteorological Society, 1954).

(iii) Particulate matter suspended in air, scatters and partly absorbs light. In industrial and urban areas, sunlight is reduced to 1/3 in summer and 2/3 in winter. It is hazardous to vehicles. Widespread decrease in transmittance of solar radiations can occur by volcanic eruptions which sprout huge quantities of particulate matter into the stratosphere.

(iv) Pandit *et al.* (1972) have found that concentration of cotton dust of less than 2mm size is 176mg/100m^3 in fine mills, 274mg/100m^3 in coarse mills and 763 mg/100m^3 in ginning presses. At a concentration above 150 mg/100m^3, cotton dust produces *pneumoconiosis* or lung fibrosis called *byssinosis*. Lung fibrosis produced in other industries includes asbestosis (in asbestos industry), sillicosis (stone grinders), siderosis (iron mill), coal miners' pneumoconiosis, flour mill pneumoconiosis, etc. Manganese poisoning has been reported in welders. A survey by Industrial Toxicology Research Centre, Lucknow has reported 4% incidence of pneumoconiosis in flour mill workers and upto 63.5% in gem grinders.

(v) Smoke and other dust particles settling over the vegetation not only clog stomata but also form a thin layer on the leaves to reduce light absorption and hence photosynthesis.

2. ***Carbon Monoxide.*** Carbon monoxide accounts for 50% of the total atmospheric pollutants. It is formed by incomplete combustion of carbon fuels in various industries, motor vehicles, hearths, kitchens, cigarette smoke, etc. A good amount of CO is also produced naturally by plants and animals. It is estimated that plants alone give out 10^8 tonnes of carbon monoxide out of a total production of 2.5×10^9 tonnes from various sources. Carbon monoxide combines with haemoglobin of blood and impairs its oxygen carrying capacity. At 100 ppm concentration, CO produces giddiness and headache within less than an hour.

Such concentration is quite common at the time of traffic jams on the busy roads of a city. Other symptoms include laziness, exhaustion, psychomotor disturbance, cardiovascular malfunction, and decreased vision. At higher concentration, carbon monoxide proves lethal.

3. ***Sulphur Oxides.*** They occur mostly in the form of sulphur dioxide. It is produced in large quantity during smelting of metallic ores *(e.g.,* iron, copper, lead, zinc, nickel, etc.) and burning of petroleum and coal in industries, thermal plants, home and motor vehicles. In the air, SO_2 combines with water to form sulphurous acid H_2SO_3. It is the cause of *acid rain.* The dreadful British smog of 1952 contained SO_2. The different effects of sulphur dioxide include :

   (i) It causes chlorosis and necrosis of vegetation in as low concentration as 0.032 ppm. Lichen vegetation *(e.g., Paramelia, Usnea, Cladonia)* and mosses are completely destroyed. Garden pea is another SO_2 pollution indicator. Besides chlorosis, SO_2 causes membrane damage, metabolic inhibition, growth and yield reduction. Chlorosis results from destruction of chlorophyll which is changed to phaeophytin (Roy and Le Blanc, 1966, 1967).
   The leaves often assume water-soaked appearance. Monocotyledons are more sensitive. Therefore, cereal crops are damaged in the area around smelters and industrial belts. Coniferous forests, apple and mango orchards are also destroyed.

   (ii) Sulphur dioxide, above 1 ppm, affects human beings. It causes irritation to eyes and injury to respiratory tract (asthma, bronchitis, emphysema). Hicky (1971) believes that SO_2 pollution is related to higher death rate in aged persons. It kills fish and other animal life. Brehm (1976) has recorded the destruction of fish from thousands of lakes and rivers of Scandinavia due to SO_2.

   (iii) It results in discoloration and deterioration of buildings, sculptures, painted surfaces, fabrics, paper, leather, etc. The reported threat to Taj Mahal of Agra from nearby refinery of Mathura is on account of it.

   (iv) SO_2 corrodes metals like iron and zinc. Therefore, it impairs electrical and other metallic equipment.

   (v) The compound has mutagenic properties.

4. ***Nitrogen Oxides***. They are produced naturally through biological and non biological activities from nitrates, nitrites, electric storms, high energy radiations and solar flares. Human activity forms nitrogen oxides in combustion process of industries, automobiles, incinerators and nitrogen fertilizers. The important effects of nitrogen oxides (N_2O, NO, NO_2, N_2O_4, N_2O_5) are as follows:
   (i) They act on unsaturated hydrocarbons to form peroxyacylnitrates or PAN.
   (ii) Nitrogen oxides give rise to photochemical smog.
   (iii) In the presence of moisture, nitrogen oxides have a corrosive effect on metals.
   (iv) They cause fading and deterioration of different types of textiles.
   (v) Nitrogen oxides produce lesions, necrosis, defoliation, die back and death of many plants.
   (vi) They cause eye irritation, respiratory troubles, lung edema, blood congestion and dilation of arteries. At a concentration of 15–50 ppm, nitrogen oxides are known to bring about injury to lungs, liver and kidneys. They are suspected to produce cancer.
   (vii) The oxides possess mutagenic properties.
   (viii) Nitrogen oxides (NO_x) are formed in the stratosphere due to solar flares. Large flares shall form enough oxides to act upon and destroy the protective ozone layer.

5. ***Carbon Dioxide***. Due to excessive combustion activity, the content of CO_2 has been steadily rising. For example, a single transatlantic flight by a jet plane releases about 70 tonnes of CO_2. CO_2 is opaque to infra-red waves. Therefore, it allows the sunlight to fall on earth but checks the loss of heat during night. As carbon dioxide accumulates in the atmosphere it absorbs more and more of the reflected infra-red radiation. This could cause an increase in temperature referred to as the *green house effect*. Melting polar ice caps and glaciers could cause sea levels to rise, flooding most of the major population centres and fertile lands.

6. ***Flourides***. They are emitted during refinement of aluminium, rock phosphate and other fluoride containing minerals or processes using fluorides. Gaseous fluorides cause leaf chlorosis, necrosis of margins and tips and abscission of leaves, fruits, etc. Maize is highly sensitive to them.

7. ***Benzpyrene.*** It is a carcinogen produced in Tobacco smoke, automobile exhausts and industrial effluences.
8. ***Phosgene and Methyl Isocyanate.*** Phosgene ($COCl_2$) is a poisonous and suffocating volatile liquid which is employed in dye industry and synthesis of organic compounds. Release of phosgene and methyl isocyanate in industrial accident of Bhopal (December. 4, 1984) killed over 2500 and maimed several thousand persons.
9. ***Aerosols.*** They are chemicals which are passed into the air in the form of vapours or fine mist. Aerosols are widely used as disinfectants. Another source are jet plane emissions which contain chlorofluorocarbons. Chlorofluorocarbons (*e.g.*, CCl_3F_1, CCl_2F_2) are also used in refrigeration and formation of certain types of solid plastic foams. Carbon tetrachloride (CCl_4) is produced naturally as well as due to human activity. Burning of plastics produces polychlorinated biphenyls (PCB). The latter are persistent and pass into the food chain whence they become concentrated in the body of animals of higher trophic levels, *e.g.*, 1400 mg/kg of pectoral muscle in Sea Eagle and 3.5 mg/kg of fat in mother's milk. In animals they impair reproductive ability while in human beings they damage liver, central nervous system, vision and change in pigmentation.

   Chlorofluorocarbons released by jet aeroplanes flying at higher altitude react with ozone layers of stratosphere and hence deplete the same. Carbon tetrachloride also does the same.
10. ***Photochemical Oxidants.*** Hydrocarbons (*e.g.*, methane, ethylene) are released into the atmosphere as part of natural gas, incomplete combustion of organic fuels and decomposition of organic debris under water logged conditions. Many of the hydrocarbons have carcinogen (cancer-causing) properties. Some of the hydrocarbons are also harmful to plants because they cause senescence and abscission. In the presence of sunlight, hydrocarbons react with nitrogen oxides to produce ozone, peroxyacyl nitrates, aldehydes and other compounds.

    Ozone. Ozone in the stratosphere protects the earth from high energy ultra-violet radiations. Ozone found in the troposphere has a warming effect. It has also toxic effects which are as follows:

    (i) At concentration of only 0.02 ppm. ozone destroys chlorenchyma cells and produces necrotic areas. There is a positive correlation between ozone concentration and destruction of foliage in a large number of plants (Moore, 1975).

(ii) Ozone hardens rubber.

(iii) It discolours and damages textiles.

(iv) At less than I ppm., ozone injures mucous membrane.

Peroxy-Acyl Nitrates. The common ones are peroxy-acytyl *nitrate (PAN),* peroxypropionyl nitrate *(PPN) and peroxy-buteryl nitrate (PBN).* Peroxyacyl nitrates are a major constituent of air pollution. They cause eye irritation and respiratory distress. In plants they impair photosynthesis, respiration, enzyme activity and anabolic processes, reducing their yield. The effects on foliage include silvering, glazing, banding, bronzing and necrosis at different places depending upon the species and age of the leaves. Spongy parenchyma and newly differentiated tissues are the prime targets of the pollutants.

Aldehydes. They produce irritation in the gastro-intestinal and respiratory tracts.

Phenols cause damage to spleen, kidneys, liver and lungs.

11. ***Automobile Exhausts.*** They are one of the major sources of air pollution. The important pollutants are (i) Carbon monoxide (ii) Benzpyrene (iii) Lead (iv) Nitrogen oxides (v) Sulphur compounds (vi) Ammonia. 1000 litres of petrol produce on combustion about 320 kg of CO, 20—40 kg of organic vapours, 2-7.5 kg of nitrogen oxides, 1.8 kg of aldehyde. 1.7 kg of sulphur compounds, 0.6 kg of lead, 0.2 kg of organic acids, 0.2 kg of ammonia and 0.03 kg of solid carbons. 90% of lead pollution is caused by gasoline combustion which contains tetra-ethyl or tetra methyl lead as antinock agent (non-lead antinock agents like olefinic or aromatic chemicals are costlier). Lead is emitted as aerosol and dust particles. The aerosol parts pass into the lungs and hence blood. Bombay residents have a lead content 16—18/μg/100ml of blood while lead begins to show toxic effect beyond 2μ/100 ml. Traffic policeman and shop owners along the busy city roads receive a higher lead content from the air. The larger lead particles settle down on the soil and pass into food chain.

12. ***Pollen and Microbes.*** They are normal constituents of air. Excess of them are produced in certain seasons. Microbes directly damage the vegetation, food articles and cause diseases in plants, animals as well human beings. Excess of pollen cause

allergic reactions in several human beings. The common reactions are also collectively called *hay fever*. The important allergic pollen belong to *Amaranthus spinosus, Chenopodium album, Cynodon dactylon, Ricinus communis, Sorghum vulgare, Prosopis chilensis.*

Control of Air Pollution

1. Industrial estates should be established at a distance from residential areas. 2. Use of tall chimneys shall reduce the air pollution in the surroundings. 3. Compulsory use of filters and electrostatic precipitators in the chimneys. 4. Removal of poisonous gases by passing the fumes through water tower scrubber or spray collector. 5. Use of high temperature incinerators for reduction in particulate ash production- 6. Desulphurisation of fuel or removal of sulphur from gas after combustion. 7. Development and employment of non-combustive sources of energy, e.g., nuclear power, geothermal power, solar power, tidal power, wind power. 8. Use of non-lead antiknock agents in gasoline. 9. Complete electrification of railway track. 10. Attempt should be made to develop pollution free fuels for automobiles, *e.g.,* alcohol, hydrogen, battery power. 11. Automobiles should be fitted with exhaust emission control. 12. Industrial plants and refineries should be fitted with equipment for removal and recycling of wastes. 13. Switching over from coal to gas fuel when atmosphere is calm. 14. Growing plants capable of fixing carbon monoxide, *e.g., Phaseolus vulgaris, Coteus blurnei, Daucos carota, Fici variegata)* (Bidwell and Bebee, 1974). 15. Growing plants capable of metabolising nitrogen oxides and other gaseous pollutants, *e.g., Vitis Pinus, Juniperus, Quercus, Pyrus, Robinia pseudo-acacia, Viburnun Crataegus, Ribes, Rhamnus.* 16. Afforestation of the mining area on priority basis.

WATER POLLUTION

Water pollution is defined as the addition of some substance (organic inorganic, biological, radiological) or factor *(e.g.,* heat) which degrades the quality of water so that it either becomes health hazard or unfit for use. Surface water is never pure. It contains small quantities of suspended particles, organic and inorganic substances plus a number of organisms (bacteria, fungi, algae, viruses,

Table 5.2 : Showing Effects Of Air Pollution on Man, Vegetation and Other Materials

Pollutant	*Effects on Man Vegetation, and other Materials*	
Carinogenic hydrocaibon	*On man*	Cancer
Carbon *monoxide*	*On man*	Poisning, increased accident liability
Dust	*On man*	Respiratory, diseases, diseases like silicosis (cough, cold, sneezing, allergic diseases, etc.), asbestosis, byssinosis, poisoning from metallic dust.
Hydrogen sulphide	*On man*	Irritation of respiratory passages, danger of respiratory paralysis and asphyxiation.
	On materials	Darkening of painted surfaces, corrosion.
Hydrogen fluoride	*On man*	Irritation, diseases of bone (fluorosis), mottling of teeth, respiratory diseases.
	On vegetation	Destruction of crops.
Heavy metals	*On man*	Specific poisoning, retardation of activities of brain, interference in enzyme activities in liver and kidney.
Nitrogen dioxic	*On man*	Irritation, bronchitis, oedema of lungs.
Photochemical smog (oxidants)	*On man*	Lung irritation, asthma, bronchitis, etc.
	On vegetation	Destruction of vegetation.
	On material	Deterioration of rubber products such as tyres and insulating wires.
Sulphur dioxide	*On man*	Suffocation, irritation of throat and eyes respiratory diseases.
	On vegetation	Destruction of sensitive crops and reduced yield.
	On material	Corrosion.

protozoans, higher organisms). When concentration of these organisms increases, the water becomes polluted and hence unfit for use. This type of pollution is dependent upon concentration of substances. Other substances pollute the water even in low concentration. They include poisons, toxic chemicals and pathogens. Most of water pollution is man made. Natural water pollution consists of silt, clay (through soil erosion), deposition of animal wastes, leaves, minerals, etc.

Sources of Water Pollution and Effect of Water Pollutants

Water pollution is a serious health hazard in India, especially in villages. It is estimated that 50—60% of Indian population suffers from diseases caused by it. 30—40% of all deaths are believed to be due to it. Pollutants produce five types of effects : (i) addition of suspended particles (ii) addition of non-toxic salts (iii) water deoxygenation (iv) addition of poisonous substances (v) heating. A pollutant may cause one or more of these effects. The principal sources of water pollution and effect of water pollutants are as follows:

1. ***Domestic Wastes and Sewage***. Treated or untreated sewage (municipalities, boats, ships, etc. is poured into water bodies. Many organic wastes are similarly added from tanneries, slaughter houses, and canning industries. The villagers often wash their animals, clothes, and take bath in the same pond. Such waters get contaminated with infectious agents for cholera, typhoid, dysentery, jaundice and skin diseases. Municipal sewers also contain a lot of industrial wastes and effluents. The effects of pollution by organic wastes are:

   (i) Micro-organisms causing degradation or decomposition of sewage take up most of the oxygen present dissolved in water.

   (ii) Raw sewage contaminates water with pathogens. Number of intestinal bacteria *(e.g.,* coliforms, enterococci) is an indication of pollution caused by raw sewage. Table 5.3 gives the degree of contamination of river Yamuna (= Jamuna) in Delhi, Agra and Mathura.

   (iii) Sewage stimulates the activity of several types of decomposers collectively called sewage fungus. It is made up of bacteria *(e.g., Spherotitus, Beggiotoa, Escherichia,* fungi *(e.g., Leptomitus, Mucor, Fusarium)* and algae (diatoms,

Table 5.3 : Contamination Characteristics of river Yamuna

	Delhi	*Mathura*	*Agra*
Coliforms mpn/100 ml	10	84000	240000
Enterococci mpn/100 ml	21	46000	150000
BOD (mgm/litre)	2000	9000	12000

Oscillatoria and other blue green algae, *Chlamydomonas, Cladophora* and other green algae). Algae provide oxygen to sewage flora and fauna. Other sewage tolerant plants do the same, *e.g.,* moss *Fontinalis antipyretica,* angiospermic *potamogeton.* Sewage rich water also contains a number of bacteria eating protozoans *(e.g., Paramecium, Glaucoma, Colpidium)* and protozoan eating animals. Since growth of algal component is essential for sewage fungus, decrease in it reduces bacterial and fungal activity. Animal population declines except for bottom dwelling blood worms and sludge Worms (chironomidae and tubifidae).

(iv) Sewage produces foul odour and makes the water brownish and oily.

(v) Organic waste gives rise to scum and sludge that makes the water unfit for recreational and industrial use.

(vi) It induces the growth of some algal blooms that add to the depletion of oxygen, addition of more organic matter and fouling of water.

(vii) Modern day detergents used for brighter and quicker washing are very slow to get degraded. They, therefore, accumulate and render the water unfit for human and animal use. The phosphates present in detergents (upto 40% in some cases) further stimulate algal growth that add to the organic loading of water.

(viii) Anaerobiosis gives rise to many secondary pollutants like H_2S, NH_3, organic sulphides, methane (CH_4), etc. H_2S combines with metallic ions to form a blackish or brownish precipitate that floats on and inside water.

2. ***Surface Run-off.*** The pollutants present on the surface of land and fertilizers added to the soils for increased crop yield are washed down into water reservoirs and water courses during rains.

Flow of fertilizer rich water into streams and lakes gives rise to eutrophication, that is, increased growth of aquatic plants or blooms. Bloom producing plants generally occur on the surface. They cut off light from submerged plants. The latter, therefore, do not produce oxygen. Oxygen, given out by bloom causing plants diffuses into the atmosphere from the surface. Night time respiration of plants and animals further depletes the oxygen content of water. On their death, the plants increase the organic loading of the water body. This results in further oxygen depletion and hence death of aquatic animals. The water also becomes unfit for drinking, industrial and recreational use. Therefore, addition of excess fertilizers to the fields be avoided. Surface or agricultural run off can be minimised by judicious use of small doses of fertilizers and manure. Fertilizer Association of India has recommended use of ammonium sulphate and urea for submerged paddy. International Rice Research Institute in Philippines has recommended that urea and other fertilizers be mixed in mud to form mud balls. The latter are placed 2 — 3 cm below the soil surface. Ammonium sulphate is also recommended instead of nitrates for tea plantations and saline soils.

3. ***Silt.*** Wrong agricultural and forestry practices cause soil erosion or removal of top fertile soil during rain. This makes the water muddy. The latter does not support much plant growth because of the shutting of light. Animal population correspondingly gets depleted. The water ways get choked due to deposition of mud and silt.

4. ***Industrial Effluents.*** They are industrial wastes which are allowed to pass into water bodies. The important toxic chemicals present in them are:

   (i) *Mercury*. It is released during combustion of coal, smelting of metallic ores, chlor-alkali, paper and paint industries. Mercury is persistent. In water it gets changed into water soluble dimethyl form $[(CH_3)_2Hg]$ and enters the food chain accompanied by biological or ecological amplification. It kills fish and poisons the remaining fauna. Human beings feeding on such poisoned animals develop a crippling deformity called *minamata disease* (numbness of lips and limbs, impairment of tactile, sense, speech and hearing, narrowing of vision, meningitis and death). Mercury also causes genetic changes (Ramel, 1974).

(ii) *Lead.* The source of lead pollution are smelters, battery industry, paint, chemical and pesticide industries, automobiles' exhausts, etc. It is mutagenic and causes anaemia, headache, irritability, colic, loss of muscle power, bluish lines round the gums.

(iii) *Cadmium.* The pollutant is given out by metal industries, electroplating, pesticide and phosphate industries. Cadmium shows biological amplification and accumulates inside kidneys, liver, pancreas and spleen. It causes renal damage, emphysema, hypertension, anaemia, testicular necrosis and damage to placenta. Cadmium enters the food chain through soil and crops like wheat and rice (Nath, 1986).

(iv) *Other metals.* Copper, Zinc, nickel, titanium, etc. cause toxaemia and change in enzyme functioning.

(v) *Liquid Effluents.* Several types of liquid effluents containing toxic chemicals, acids and bases, are added to the rivers and other water bodies. They kill fish and other aquatic life besides being toxic to human, beings. Some examples of large scale effluent addition into rivers are Yamuna (near Okhla, Delhi), Gomti (near Lucknow), Ganga (near Kanpur) and Hoogli (near Calcutta).

Table 5.4 : Effects of Heavy Metal Water Pollutants on Man

Metal	*Toxic effects on man*
Cadmium	Diarrhoea, growth retardation, bone deformation, kidney damage, testicular atrophy, hypertension, tumour formation hepatic injury, central nervous system injury, anaemia.
Copper	Hypertension, sporadic fever, uremia, coma, etc.
Barium	Excessive salivation, colic, vomiting, diarrhoea, muscular paralysis of nervous system.
Zinc	Vomiting, renal damage, cramps.
Mercury	Abdominal pain, headache, diarrhoea haemolysis, chest pain.
Lead	Anaemia, brain damage, vomiting, loss of appetite convulsions, liver damage, kidney damage.

Metal	*Toxic effects on man*
Arsenic	Disturbances of peripheral circulation mental disturbances, liver cirrhosis, hyper-keratosis, kidney damage, ulcers in gastrointestinal tract, lung cancer.
Selenium	Fever, nervousness, vomiting, dental carles, fall in blood pressure, and even death. Liver kidney and spleen damage, blindness.
Chromium (hexavalent)	Cancer, nephritis, gastro-intestinal ulceration, diseases in central nervous system.
Cobalt	Paralysis, diarrhoea, low blood pressure lung irritation and bone defects lung irritation and bone defects.

5. ***Oil.*** At many places petroleum is extracted from the area of continental shelf. Further, it is transported from one country to another through sea. After unloading, the tankers are washed in sea before being filled with water. Therefore, oil slicks or spills are quite common in the sea, especially near the ports and shore-lines. Accidents during transport and extraction spread oil over many hundred kilometres. A total of over 10 million tonnes of oil is spilled into the ocean annually. Refineries also discharge a lot of oil present in their effluents into rivers. Barauni refinery has destroyed aquatic life of Ganges for many kilometres in 1968 and 1974. The effects of oil pollution are as follows:

   (i) Oil spreading on the surface of water prevents its oxygenation. Rather, it depletes the small quantity of oxygen present in water for its own degradation.
   (ii) Oil pollution inhibits plankton growth and photosynthetic activity of other aquatic plants.
   (iii) Animal life is destroyed due to reduced availability of oxygen, food and toxic effects of oil. .
   (iv) Oil spilled over the water surface may catch fire and hence kill all organic life.
   (v) Detergents used to clean oil spill are equally harmful. This happened during Torrey Canyon accident on British Coast (1969).
   (vi) Sea birds smeared with oil fall sick and die.

6. ***Thermal Pollution.*** It is caused by addition of hot effluents and hot water bodies. Warmer water contains less oxygen (14 ppm at 0°C, 1 ppm at 20°C). Therefore, there is decrease

in the rate of decomposition of organic matter. Green algae are replaced by less desirable blue-green algae. Many animals fail to multiply. Trout eggs fail to hatch while Salmon does not spawn at higher temperature.

MARINE POLLUTION

Coastal industries dump industrial wastes straight into the sea, however, inland industries do it through rivers. Inland waters of certain areas are full of pesticides and fertilizers used in farming. Sewage of coastal cities and distant places also reach the sea. Oil, grease, petroleum products, garbage, sewage and detergents from ships also pollute sea water.

Twenty-five per cent of the people who reside by the coastal areas in South Asia make their living from polluted seas and its contaminated products. South Pacific Islanders, who live largely on sea food, suffer frequently from cholera, gastroenteritis and infective hepatitis.. Several people of Nigeria's capital Lagos use their beaches and lagoons as toilets. This makes the population especially susceptible to water borne bacteria, viruses and parasites.

Sometimes the oxygen levels in many coastal waters have been so reduced to such levels that their capacity to support any form of life within them has been severely impaired.

Biological Magnification

Chlorinated hydrocarbons, such as DDT, are not very soluble in water. It (water) generally contains only 1 to 2 parts of pesticide per billion parts of water. However, these compounds are very soluble in fats. Aquatic micro-organisms absorb them in fat and oils, where they accumulate to form concentrations many times greater than in water. Zooplankton that feed on countless contaminated phytoplankton cells concentrate the pesticide still further in their tissues. Their levels may be as high as 1 part pesticide per million parts of fat, 1,000 times more concentrated than in water. Fish feeding on zooplankton further concentrates the pesticide to levels as high as 3 to 8 parts per million. Birds feeding on fish concentrate the compounds further upto 3,000 parts per million. The increased accumulation of toxic substances in the food pyramids is called *biological magnification* (Fig. 5.1). Many species of predatory birds like eagles, hawks have shown serious adverse effects from this accumulation. It interferes with egg shell production in many birds.

The shells are thin and are easily broken by the bird's weight during incubation. It adversely affects the developing embryos. As a result, the populations of many birds have decreased. Pesticides on vegetation can also be concentrated by herbivorous animals such as cattle. By eating meat and drinking milk humans concentrate pesticides in their tissues.

Degree of Water Impurity

Different aquatic organisms have different level of tolerance to change in temperature, *pH,* oxygen content, calcium content, phosphate level, organic matter, poisons and other pollutants. For example, *Daphnia* and most of the fishes are sensitive to poisons and metals. From the study of aquatic life the degree and type of water pollution can be known.

Water pollution by organic wastes is measured in terms of biochemical oxygen demand or BOD. BOD is the amount of oxygen taken up by the micro-organisms present in water. It is oxygen required in milligrams for five days to metabolise waste present in one litre of water at 20°C. A weak organic waste will have BOD below 1500 mg/litre, medium between 1500 -1400 mg/litre while a strong waste above it. Since BOD is limited to organic wastes, it is not a reliable method of measuring water pollution. Another slightly better mode is COD or chemical oxygen demand. It measures all oxygen consuming pollutant materials present in water.

Treatment of Waste Water

A sewage treatment system often involves three stages (Fig. 5.2):

1. ***Primary Treatment Stage***. In this stage, sewage is processed through screens and settling tanks to remove solid matter, such as large lumps of organic matter and coarse materials such as sand or grit. Primary treatment does not remove the population of pathogens and minerals.
2. ***Secondary Treatment Stage***. In this stage, the sewage is held in aerated tanks where micro-organisms decompose the organic matter into simpler compounds. The effluent is filtered and then chlorinated to kill any pathogens and minerals or the threat of eutrophication (rich growth of micro-organisms consumes much of dissolved oxygen, tending to deprive other

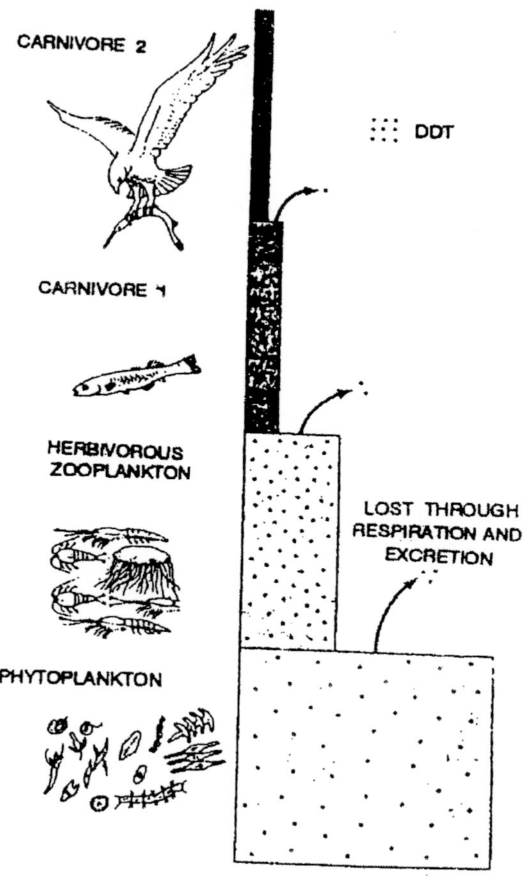

Fig. 5.1 : Biological magnification of the pesticide DDT in the tissues of organisms in an aquatic food pyramid.

organisms) in rivers and lakes. Most sewage plants then return this relatively clean liquid to the environment. This secondary effluent still carries a considerable load of nitrates, phosphates and other minerals.

3. ***Tertiary Treatment Stage.*** In this stage, the sewage undergoes additional filtering and chemical treatment to remove its mineral load. The effluent from this treatment is pure enough to drink. But few communities provide tertiary treatment, because of the cost, however, it is necessary to get drinking water.

Another method of tertiary treatment is *reverse osmosis*. The method was developed to demineralise brackish water. Water is rid of its salts by pumping it through a semipermeable membrane under strong pressure.

Control of Water Pollution

Water pollution can be controlled to a large extent on the principle, "the solution to pollution is dilution."

The various methods for the control of water pollution are discussed below:

1. The sewage pollutants are subject to chemical treatment to change them into non-toxic substances or make them less toxic.
2. Water pollution due to organic insecticides can be reduced by the use of very specific and less stable chemicals in the manufacture of insecticides.
3. Oxidation ponds can be useful in removing low level of radioactive wastes.

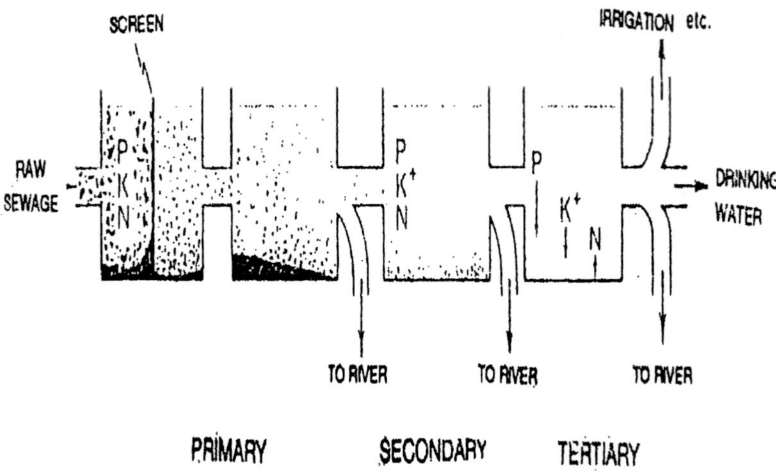

Fig.5.2 : The three stages of sewage treatment. The primary stage removes solid material. In the second stage, organic wastes are decomposed by micro-organisms. Tertiary treatment removes minerals that may cause eutrophiction.

4. Thermal pollution can be reduced by employing techniques– through cooling, cooling ponds, evaporative or wet cooling towers are dry cooling towers. The purpose is that the waters in the rivers and streams should not get hot.
5. Domestic and industrial wastes should be stored in large but shallow ponds for some days. Due to the sun-light and the organic nutrients present in the waste there will be mass scale growth of those bacteria which will digest the harmful waste matter.
6. Polluted water can be reclaimed by proper sewage treatment plants and the same water can be reused in factories and even irrigation. Such treated water being rich in phosphorous, potassium and nitrogen can make good fertiliser.
7. Suitable strict legislation should be enacted to make it obligatory for the industries to treat the waste water before being discharged into rivers or seas.
8. Water hyacinth can purify water polluted by biological and chemical wastes. It can also filter out heavy metals like cadmium, mercury, lead and nickel as well as other toxic substances found in industrial waste waters.

SOIL POLLUTION AND LAND DEGRADATION

Unfavourable alteration of soil by addition or removal of substances and factors which decrease soil productivity, quality of plants and ground water is called soil pollution. It is of two main types: negative and positive.

Negative Soil Pollution

It includes overuse of soil and erosion. Erosion of soil is caused by two factors, water and air. Water erosion of soil is found near the hills where high speed rivulets and flooding removes top soil. Floods are quite common in India. The major cause of their recurrence is the destruction of forests in the catchment area of rivers by excessive felling and overgrazing.

Soil erosion also occurs by high speed winds which bring sand particles from dry desert. Today about 40% of the land has turned into deserts. The Thar desert of India (Rajasthan) was a fertile land only 5000 years back. Overgrazing and felling of trees for timber allowed the sand of Gujarat coast to settle on this area and produce a desert. It is spreading at a speed of 9 km/year.

Fertile land is also being converted into barren areas by unplanned urbanisation, building of roads, houses or industrial complexes. Rubbish, empty cans, garbage, broken furniture, empty bottles, building material, sludge, ash, etc. are all dumped outside the towns on vacant lands which not only become barren but also make the nearby lands so. It is sometimes also called *third pollution or landscape pollution.* The best way is to perform sanitary land fill which involves the hurrying of waste in low lying areas. Some rubbish can be recycled while garbage is burnt or converted into compost.

Positive Soil Pollution

It is a pollution caused by (i) pesticides, herbicides, and fumigants (ii) chemical fertilizers and (iii) air pollutants washed down from atmosphere through rain.

1. ***Pesticides and Weedicides***. A number of chemicals have been developed to kill insects (insecticides), fungi (fungicides), algal blooms (algicides), rodents (rodenticides), weeds (weedicides or herbicides) in order to improve agriculture, forestry, horticulture and water reservoirs. The most widely used among them are insecticides. They are sprayed over the plants in the form of fine mist or powder. Most of these insecticides or pesticides are broad-spectrum and affect other animals, man and even plants. They are, hence, also called biocides.

   (i) *Chlorinated Hydrocarbons*. They include DDT (dichlorodiphenyl-trichloroethane), DDE, Chlordane, Aldrin, Dieldrin, Endrin, Heptachlor, BHC (benzene hexa-chloride), etc. Chlorinated hydrocarbons are toxic. Dieldrin is 5 times more toxic than DDT when ingested and 40 times more poisonous when absorbed. Endrin is the most toxic amongst chlorinated hydrocarbons. Besides, being toxic these pesticides are both persistent and mobile in the ecosystem (over dust particles in air, over organic matter in water). The chlorinated hydrocarbons are fat soluble and, therefore, tend to accumulate inside living organisms. Their concentration per unit weight of the organisms also rises with the rise in trophic level due to the phenomenon of ***biological amplification*** or DDT and other chlorinated hydrocarbons affect central nervous system, cause softening of brain, cerebral haemorrhage, Cirrhosis of liver, hypertension, cancer, thinning of egg shells in birds, malformation of sex hormones, etc. Ecological amplification of chlorohydrocarbons, therefore, proves fatal to higher trophic level animals, especially Fish and birds. Over 100,000 ducks from Kerala died recently in six weeks due to pesticide poisoning. The population of certain birds has declined because of these biocides, e.g., Bald Eagle. Excessive spray of hard biocides sometimes causes an imbalance in prey– predator population. For example, in Australia the population of predator Lady bird beetle *(Novius cardinalis)* declined while that of its prey, scale insect *(Icerya purchasi)* increased due to DDT use (Rudd, 1971). The balance was restored only after DDT spraying was stopped. DDT also affects the photosynthetic activity of plants, especially phytoplankton.

(ii) *Organo-Pesticides.* They include organo-phosphorous compounds *(e.g.,* malathion, parathion, diazonin, triothion, ethion, tetraethyl pyrophosphate or TEPP) and carbamates. Organo-pesticides are degradable but being poisonous they influence the workers handling them causing sweating, salivation, nausea, vomiting, diarrhoea and muscular tremors.

(iii) *Inorganic Pesticides.* The pesticides usually contain arsenic and sulphur. Their continued use is poisonous to both plants and animal life since the pesticides are of persistent nature.

(iv) *Weedicides (Herbicides).* The chemicals are used in clearing area of forests for building new residential or industrial colonies, highways, rail-road, weed control in agriculture, horticulture and in forest management. The weedicides or herbicides are usually metabolic inhibitors which stop photosynthesis and other metabolic activities and hence kill the plants. Some weedicides cause death due to proliferation of phloem cells so as to block transport of organic food. Weeds of Aswan dam in Egypt were controlled by weedicides. It not only affected the agricultural fields irrigated by that water but also marine fish production in the sea where this water was discharged. The phenomenon of producing adverse ecological effects of substances or actions in later period is called ecological *bomerang* or *backlash.*

2. ***Industrial Wastes***. Both solid and liquid wastes of the industry are dumped over the soil. The wastes contain a number of toxic chemicals like mercury, copper, zinc, lead cadmium, cyanides, thiocyanates, chromates, acids, alkalies, organic solvents, etc.
3. ***Mine Dust.*** It is a major source of pollution in mining areas. Mine dust not only spreads with wind but also splits over a large area during transportation to purification plants. The waste of the purification plants also adds upto the pollution. The pollution destroys vegetation in the area by deposition of particulate matter over the foliage, poisonous gases and the toxic mineral components. It also produces numerous types of-deformities and diseases in animals and human beings.
4. ***Fertilizers.*** Chemical fertilizers added to the soils enter the crop plants as well as leach down into water table to become part of underground water. Nitrogen fertilization produces toxic concentration of nitrate or nitrite in the leaves and fruits (Schuphan, 1976), *e.g.*. Spinach, Mustard, Lettuce. Nitrate containing canned food causes corrosion of tin lining of the can, increases tin content of food and produces nitrous oxide (N_2O) gas. The toxicity increases if the drinking water also possesses sufficient nitrates. In the alimentary canal, the activity of bacteria changes nitrates into nitrites. The latter enter the blood and combine with haemoglobin to form ***met-haemoglobin***. As a result oxygen transport is reduced. It gives rise to disease known as ***methaemoglobinaemia***. In infants it produces cyanosis (blue babies due to bluish tint of skin). In adults it produces breathlessness. In infants nitrate poisoning can be fatal unless and until methylene blue is injected in time. Excessive use of chemical fertilizers causes soil deterioration through the decrease in natural bacterial population (nitrogen fixing, nitrifying, sulphofying) and destruction of crumb structure. The salt content of the soil is also bound to increase with continuous use of fertilizers.
5. ***Other Soil Pollutants.*** Air pollutants and many water pollutants ultimately become part of the soil. The soil also receives toxic chemicals during the weathering of certain rocks. A major part of lead given out in the automobile exhausts settles down on

the roadside areas and becomes part of it. Khandekar (1984) has reported a lead concentration of 41 μg/gm in the surface soil of Bombay due to automobiles. The same enters the food chain. Fluorides similarly pass both into pumped water and food chain. In plants fluorides combine chemically with Mg^{2+} of chlorophyll and hence inhibit photosynthesis, cause abscission of leaf and fruit, and hence destroy vegetation. Maize is a sensitive indicator in fluoride pollution. In human beings the typical symptoms of excess fluorine or fluorosis is the mottling of teeth. Later on bone fluorosis follows. The latter consists of weak bones, boat shaped posture and knocking of knees. Animals grazing over fluoride rich foliage show ill health, weak teeth, weak bones and swelling of knee bones.

Salination of Soil

Increase in the concentration of soluble salts in the soil is called salination. Origin or development of saline soil depends upon following factors:

(i) *Poor Drainage of Soil.* Salts dissolved in irrigation water accumulate on the soil surface due to inadequate drainage especially during flood.

(ii) *Quality of Irrigation Water.* The ground water of arid (dry, barren having not enough rainfall to support vegetation) regions are generally saline in nature. The irrigation water may be itself rich in soluble water and add to salinity of soils.

(iii) *Capillary action.* Salts from the lower layers move up by capillary action during summer season and are deposited on the surface of the soil.

(iv) *Excessive use of basic fertilizers.* Excessive use of alkaline fertilizers like sodium nitrate, basic slag, etc. may develop alkalinity in soil.

(v) *Salts blown by wind.* In arid regions near the sea, lot of salt is blown by wind and get deposited on the lands.

(vi) *Saline nature of parent rock materials.* If soil develops from saline nature of parent rock materials, soil would be saline.

India has about six million hectares of saline land. About 6,000-8,000 hectares of farm land becomes unfit for agriculture every year in Punjab alone.

Control of Soil Pollution

(i) Use of pesticides should be minimised. Biological control should be known and implemented. (ii) Use of fertilisers should be highly judicious. (iii) Cropping techniques should be improved so that weeds are unable to take foot hold in the fields. This would automatically reduce the use of weedicides or herbicides. (iv) Special pits or low lying areas be selected for dumping of industrial wastes. (v) Improvement in mining techniques and transport of extracted materials so that spread of mine dust should be minimised. The area should not be left barren and dry. Instead, afforestation should be carried out as soon as it becomes feasible. (vi) Controlled grazing. (vii) Proper forest management. (viii) Wind breaks and wind shields in areas exposed to wind erosion. (ix) Planting of soil binding grasses and other perennials along the banks and slopes prone to rapid erosion. (x) Afforestation and reforestation.

Recycling of Wastes

1. ***Paper***. Waste paper can be recycled. Old books, newspapers, old answer books, magazines, etc. are now regularly purchased by mills for manufacturing new paper.
2. ***Agricultural Wastes***. Coconut waste, jute waste, cotton stalks, bagasse of sugarcane, stem of rice and other cereals can be changed into paper and hard board. Paddy husk is now used in the preparation of livestock feed and hard board. Forestry wastes, corn cobs and other crop refuse can now be profitably utilised.
3. ***Food Processing and Cannery Wastes***. About 25000 tonnes of peeling and other wastes of vegetables and fruits worth about 27.5 crore rupees are allowed to go waste annually in India. They can be allowed to undergo fermentation. It produces organic acids which have preservative action. The unpalatable treated waste is used as manure while the palatable one is given to cattle as fodder. Gas released during fermentation is employed as a source of energy.
4. ***Seeds***. Seeds of forest *(e.g., Sal or Shorea robusta)* and roadside trees *(e.g.,* ahua or *Madhaca,* Neem or *Azadirachta indica)* are often allowed to go waste and add to the pollution problem. These seeds contain sufficient quantity of oil which can be extracted. Oil cakes may be used both as fodder and manure.

5. ***Composting.*** All types of organic wastes of a town are used to prepare a manure called compost. In composting the sludge obtained after primary treatment of sewage alongwith other wastes are allowed to decompose in an open space. In 4-6 months compost is ready for use as a manure. In some cases minerals are also added to the organic matter to enrich the compost.
6. ***Gobar Gas Plants.*** Cowdung and other organic wastes of farm houses can now be profitably placed in gobar gas plants which not only enrich manure but also provide gas for domestic use.
7. ***Sludge Burning.*** Where electricity is obtained from a thermal plant, sludge from sewage treatment can be mixed with coal to form fuel for power generation.

Land Degradation

Besides pollution, land and soil face many other problems such as deforestation, erosion, flooding, water logging, salination, desertification and urbanisation. If these problems are not checked about one-third may be degraded by the end of this century.

(i) *Soil Erosion.* Soil erosion is world spread phenomenon but it is more extensive in Australia, Central Africa, India, Nepal, China, Spain, U.S.A. and erstwhile Soviet Union. It is estimated that over 40,000 hectares of land are affected by wind and water erosion every year. The quantity of top soil lost is maximum in our country. It is about 18.5 of the total global loss. India has largest livestock in the world. Therefore, the pressure of grazing causes soil erosion. Cause and prevention have been discussed in Chapter 25.

(ii) *Shifting cultivation.* A practice of slash and burn agriculture is common among many tribal communities of tropical and subtropical regions of Africa, Asia and Oceania. It consists of cutting down trees and burning the felled trees and raising crops on the ash formed. This practice is called "Jhuming" in the Jhum forests of north-east India. It is not harmful if the jhuming is not done frequently. However, when jhuming is done in less than ten years, it destroys forests and causes soil erosion.

(iii) *Desertification.* It is the change of a normal soil into desert. Desertification is the result of deforestation, soil erosion, over grazing and shifting of sand dunes by wind. Many deserts in the world are man made.

(iv) *Developmental Activities.* Various developmental activities such as rapid urbanisation, mining, construction of dams, canals, roads, railways, airports, playgrounds and industries have played major role in land degradation.

Control of Land Degradation. Land degradation can be checked by the following measures:

(i) Reforestation and plantation of grass can check soil erosion, floods and water logging.
(ii) Crop rotation and mixed cropping improve fertility. It would increase production, which would support large population.
(iii) Salinity of the soil can be checked by providing adequate drainage. Salinated lands can be recovered by leaching them with more water, especially were the ground water table is not high.
(iv) Desertification can be checked by artificial bunds or covering the area with suitable vegetation.

Land and Water Management in India

Total land area of India is 305 million hectares (ha). However, nearly half of this is considered as waste land. The Himalaya is the major water shed in India. It has very high sedimentation rate. The use of its slopes has created serious problems in the plains of India. These problems are recurring floods, damage to water survivors and irrigation system. The following steps should be taken for the proper management of land and water.

(i) *Maintenance of Catchment Area.* It should begin from the topmost reaches of the catchment. The trees planted for land conservation should be economically and socially useful. Grasses chosen for binding the soil should be selected according to edaphic and climatic conditions and local needs.
(ii) *Proper drainage and desalination.* Water-logging and salination are main reasons for lower productivity from canal irrigated lands in India. Proper drainage and desalination techniques can over come these problems.
(iii) *Proper use and management of ground water.* The following are some advantages in using ground water.

(a) It can be tapped in a very short time.
(b) The cost of tube wells is low.

(c) It is economical as it does not involve storage and transport.
(d) It does not suffer evaporation and seepage losses.
(e) The ground-water helps in lowering the water tables in canal command areas where water level is high.
(f) It is also available in a large amount in desert areas.

Therefore, proper use and management of ground water should be undertaken.

Wasteland Development

Productive land is decreasing due to degradation. However, the human population is increasing. Therefore, the development of wasteland is essential. As explained earlier, wasteland forms about half of our country. Wasteland is of two types : culturable and unculturable. (i) The culturable waste-lands include ravinous land, waterlogged land, marsh and saline lands, forest land, degraded land, strip land, mining and industrial wastelands, (ii) Unculturable wastelands include barren rocky areas, steep slopes, snow-capped mountains and glaciers. Development of culturable wastelands can provide more land for agriculture. Reclamation of waste-land should be undertaken immediately.

RADIOACTIVE POLLUTION

Radioactive pollution is a special form of physical pollution of air, water and soil with radioactive materials.

What is Radioactivity ? Radioactivity is the property of certain elements (radium, thorium, uranium, etc.) to spontaneously emit protons (alpha particles), electrons (beta particles) and gamma rays (electromagnetic waves of very short wave length) by disintegration of their atomic nuclei. The elements that give radiation are called **radioactive elements.**

Sources of Environmental Radiation – Sources of environmental radiation are both natural and man made.

(i) *Natural (Background) Radiation.* This includes cosmic rays that reach the surface of the earth from space and terrestrial radiations from radioactive elements present in the earth's crust. Many radioactive elements such as radium 224, uranium 235, uranium 238, thorium 232, radon 222, potassium 40 and carbon 14 occur in rocks, soil and water.

(ii) *Man-made Radiation.* This includes mining and refining of plutonium and thorium, production and explosion of nuclear weapons nuclear power plants, nuclear fuels and preparation of radioactive isotopes.

(a) *Nuclear Weapons.* The First atomic bomb was exploded in Nagasaki and the second in Hiroshima in Japan in 1945. This caused large-scale destruction of human, animal and plant life. In spite of these heinous events, the nuclear race by the big powers is still continuing.

Production of nuclear weapons involves the tests of nuclear arms. These tests produce large amount radioactive elements into the environment and make other materials also radioactive. They include strontium 90, cesium 137, iodine 131 and some others.

The radioactive materials are transformed into gases and fine particles which are carried to distant places by wind. When rain drops, the radioactive particles fall on the ground, it is called *fall out.* From the soil radioactive substances are taken by plants, thence they reach humans and animals through food chains. Iodine 131 damages white blood corpuscles, bone marrow, spleen, lymph nodes, skin cancer, sterility and defective eye sight and may cause lung tumours. Strontium 90 accumulates in the bones and may cause bone cancer and tissue degeneration in most animals and man.

The radioactive materials are washed from land to water bodies where the aquatic organisms absorb them. From these organisms radioactive materials may reach man through food chains.

(b) *Atomic Reactors and Nuclear Fuels.* The operation of a nuclear power plant releases large amounts of energy. This energy is used in large turbines, which produce electricity. Both the fuel elements and coolants contribute to radiation pollution. Wastes from atomic reactors also contain radioactive materials. The biggest problem is the disposal of these radioactive wastes. If these wastes are not properly disposed of, can harm the living organisms wherever they may be dumped. If they escape, they can cause havoc. Inert gases and halogens escape as vapours and cause pollution as they settle on land or reach surface waters with rain.

(c) *Radio Isotopes.* Many radioactive isotopes such as ^{14}C, ^{125}I, ^{32}P and their compounds are used in scientific research. Waste waters containing these radioactive materials reach water sources like rivers, through the sewers. From water they enter human body through food chains.

(d) *X-rays and radiation therapy.* Human beings also voluntarily receive radiation from diagnostic X-rays and radiation therapy for cancers.

(e) People working in power plants, nuclear reactors, fuel processors or living nearby are vulnerable to radiation exposure.

Effects of Radiation Pollution

Harmful Effects. The effects of radiation were first noted in 1909 when it was found that uranium miners suffer from skin burn and cancer due to radiations from the radio-active mineral. Different organisms show different sensitivity to ionising radiations. For example, tests have shown that pine trees are killed at radiations in which oak trees continue to thrive comfortably. It has also been reported that high altitude plants have developed polyploidy (increase in the number of chromosome sets) as a protective mechanism against radiations. Parts of coastal areas in South India have a high degree of background radiation which was formerly considered to be quite harmful to human beings.

The cells which actively grow and divide are quickly damaged. This category includes the cells of skin, intestinal lining, bone marrow, gonads, and embryo. The cells which are less active in growth are not so easily damaged. These are the cells of bones, muscles and nervous tissue. Radiations have both immediate or short-range and delayed or long ranged effects.

(i) Short Range (Immediate) Effects. They appear within days or a few weeks after exposure. The effects include loss of hair, nails, subcutaneous bleeding, change in number and proportion of blood cells, changed metabolism, and proportion of blood cells, etc.

(ii) ***Long Range (Delayed) Effects.*** They appear several months or even years after the exposure. The effects are caused by development of genetic changes, mutations, shortening of life span, formation of tumour, cancers, etc. The effect of mutations can persist in the human race.

All organisms are affected by radiation pollution. Some organisms preferentially accumulate specific radioactive materials. For example, oysters accumulate ^{65}Zn, fish accumulate ^{55}Fe, marine animals accumulate ^{90}Sr.

Controls of Radioactive Pollution. The following preventive measures should be followed to control radioactive pollution.

(i) Leakage of radioactive materials from nuclear reactors, industries and laboratories using them should be totally stopped.
(ii) Radioactive wastes disposal must be safe. They should be changed into harmers for or stored in safe places so that they can decaying a harmless manner. Radioactive wastes only with very low radiation should be discharged into sewerage.
(iii) Preventive measures should be taken so that natural radiation level does not rise above the permissible limits.
(iv) Safety measures should be taken against accidents in nuclear power plants.
(v) Workers using radioactive materials should wear protective garments.

NOISE POLLUTION

The quality of our environment is judged, apart from other factors, by the amount of noise present. Noise has been well defined as "unwanted sounds" which is being "dumped" into the atmosphere to disturb the unwilling ears. It adversely affects our physiological and mental health.

Sources of Noise Pollution. Main sources of noise pollution are

(i) Various industries such as textile mills, printing presses, engineering establishments,
(ii) agricultural machines like tractors, harvesters, tubewells, lawn mowers,
(iii) defence equipment such as tanks, artillery, rocket launching, shooting practices, explosions,
(iv) entertaining equipment like radios, record players, television sets,
(v) domestic gadgets such as desert coolers, airconditioners, vacuum cleaners, exhaust fans, mixers, pressure cooker,
(vi) Public address systems like loud speakers,

(vii) transport vehicles like scooters, motor-cycles, car, buses, trucks, trains, jet planes,
(viii) dynamite blasting,
(ix) crackers used at occasions like marriages and festivals,
(x) bull dozing, stone crushing, construction work, etc.

Effects of Noise Pollution. Noise pollution affects on hearing and general health of man.

1. ***Effects on Hearing.*** Prolonged and continued high intensity noise not only cause some hearing loss but may cause a permanent loss of hearing. A sudden loud noise such as an explosion can damage the tympanic membrane (ear drum). Noise is measured in units called *decibels* (dB). Noise over 115 dB is regarded as highly avoidable. The World Health Organization (WHO) recommends an industrial noise limit of 75 dB. A type writer at work measures about 60 dB.

2. ***Effect on General Health***

   (a) The first effects of noise are anxiety and stress, however, in extreme cases it may lead to fright.
   (b) Noise causes headache by dilating blood vessels of the brain, eye strain by dilating the pupil, digestive spasms through anxiety and high blood pressure by increasing cholesterol level in the blood.
   (c) Noise pollution also causes increase in the rate of heart beat, constriction of blood vessels, decreased heart output, and defective night and colour vision.
   (d) A sudden high intensity of sound produces a startle reaction which may affect psychomotor performance.
   (e) Noise also causes emotional disturbances.
   (f) Noise can impair the development of nervous system of unborn babies which leads to abnormal behaviour in later life.
   (g) It has been reported that prolonged noise pollution causes damage to heart, brain and liver in animals.

Control of Noise Pollution. The control measures for noise pollution depend upon three factors, (i) to reduce the source of noise, (ii) to put checks in the path of its transmission, and (iii) to safeguard the receiver of the noise. Vehicular traffic should be diverted away from the residential colonies, educational institutions and hospitals. Redesigning of the machinery and proper maintenance of the vehicles can lessen the noise. Sound absorbing techniques like acoustical furnishing should be extensively employed. Measures should be taken to deflect the noise away from the receiver, *e.g.*, by mechanical devices the ear splitting jet exhaust noise can be diverted upwards instead of downwards. New industrial complexes, aerodromes, etc. should be built quite away from inhabited areas. There should be legal enforcement of restrictions on noise pollution. Above all a general awareness should be developed among the people that the noise is a public nuisance and poses a serious danger to our mental and physical health.

Pollution without Human Intervention

Pollution of the environment can occur without human intervention. It includes the following factors (i) Volcanic eruptions, (ii) Forest fires, (iii) Natural organic and inorganic decays, (iv) Electric storms, (v) Cyclones and, (vi) Floods.

To conclude we may observe that pollution, be it in air, water or in the form of noise, is posing a major health problem threatening the very existence of life. Therefore, immediate effective and decisive environmental control measures should be taken on a global scale. Otherwise there is a horrible spectacle of future world with poor natural resources, stinking surroundings, suffocating atmosphere, ill-nourished sick children and in brief suffering humanity is steadily, but surely, going to extinction.

UTILISATION OF URBAN REFUSES

Everything around us is changing foster than what we anticipate. World is changing and India is only a part of this changing world. What is not changing is the human soul and our own social thoughts.

Faster than the rural India it is our urban areas are changing. It is our education, developing technology and influence of westernisation are probably main reason for fastness of urban changes. We have our architects to design modern buildings, our town planners to plan modern townships, our doctors to give us modern treatments, our technocrats to give us modern gadgets for comforts and our bureaucrats and managers to give us modern administration. Yet with all the above we are bringing more problems to our society day by day than solutions. Our cities and towns are either already problem ridden or approaching fast towards decay. Impact of such decay is very prominent. For comfortable living a city needs minimum 25 per cent free space to provide space for roads, play ground, park and sufficient free air to breathe. If we look at our cities Like Calcutta and Bombay they have not more than 10 per cent free space, Kanpur is fast following them, Madras is the next. Probably Delhi and Bangalore may be having around 25 per cent but if precautions are not token, well the cancer will invade these two cities too. Due to certain modern amenities it may be noted that industrialisation is growing faster in our cities and towns. While giving us all important products to us, these industries are also creating a lot of problems whose impact on our society is very very deep. Due to lack of proper precautions, these industries are major cause of pollutions in air as well as water. It is already experienced that industrial townships are creator of many law and order problems. Industrial accidents has caused death of many innocents not only those who work in the industry but also otherwise. The impact of industrialisation has already surfaced in our cities. With growth of industrialisation and growth of population in cur urban belt demand for electricity has increased considerably. Electricity was only a luxury in the turn of this century has become one of our essential requirement. As generation of electricity could not match its demand due to growth rate, it becomes more a problem. Calcutta is already limping with this problem and other are just following the footstep. Besides above there are several other problems crippling our urban life like problem of conveyance, bad road condition, uneven distribution of essential commodities, insufficient medical facilities and so on. All the above problems are, however, well thought over and attempt are being made to solve the above problems.

One of the major urban problem which is given low priority is handling the urban refuse and garbages. With diversities in city life there are several types of urban refuse which needs immediate attention or else our urban life will definitely cripple down. It is a very common site now-a-days in cities like Calcutta and Bombay where garbages are dumped at road side for days together in such a way that we cannot but have to use hanky to cover our nose while walking on the road. Even after 37 years of independence garbages in our cities are just thrown on a particular area marked for the purpose. This way of handling the garbages has brought back malaria and other diseases to our society. After the area is saturated with garbages problems are being faced to search out another area for the purpose. This has become a chronic problem. But if we identify these urban refuses properly and treated properly we will not only solve the problem it will also solve many other problems while giving many useful product; finally giving us a happy disease free society.

There are several types of urban refuses. Some may be in gaseous form to pollute our air we breathe, some may be in liquid form to pollute water sources, some may be of solid to occupy space and chemically disintegrate to cause diseases. Considering all above, urban refuse may be categorised as follows :

(A) Industrial refuse

- (a) spent gases from industry
- (b) metallic wastes from industry
- (c) spent solids both metallic and non-metallic

(B) Domestic refuses

- (a) human excreta and cowdung
- (b) metallic wastes like broken utensils
- (c) vegetable wastages
- (d) cellulosic wastes like torn clothes and paper
- (e) liquid refuses like urine etc. (degradable)
- (f) gaseous refuses like smoke from oven
- (g) liquids (not degradable)
- (h) other miscellaneous refuses.

(C) Common refuses does not fall in above categories

- (a) dead animals on road side
- (b) smokes from automobile etc.
- (c) any other common refuses.

The above refuses flows out almost every hour of the day. Except taking study on pollution practically no attempt were made so far to handle the above whereas some of these refuses if processed properly may turn out to be gold mine.

Spent gases from industry may be of many types. In majority of the cases the gases from industrial chimney contains mostly carbon-dioxide with carbon, ash and dirt particles. It is needless to say if this type of gas can be channelised to absorb all floating particles and carbon-dioxide then purifies and put in gas cylinders, this may become a boon for many industries like aerated water manufacture etc. In fact one such industry manufacturing carbon-dioxide utilising flue gases of Coromandal Fertilizers Ltd., has come up successfully in Vishakhapatnam in Andhra Pradesh. The other type of spent gases comes out from industry may contain various other types of gases like sulphur-dioxide, ammonia and so on which can be utilised for manufacture of sulphuric acid, ammonium compounds, etc. If the same are not possible commercially at least the gases can be absorbed before release to atmosphere to avoid pollution thus avoiding future health hazard. Spent refuses from industries are of various types. They may be dealt with merit of each case either to produce by product or may be discarded after treatment so as to avoid pollution. Spent solids again are of several types. The metallic scraps can easily be segregated and can be remoulded into useful products. Same is the case with mouldable plastics and PVC. If possible cellulosic materials can be segregated and may be utilised for manufacture of handmade paper. Those materials which can perish easily and ferment may be passed through a digester to produce bio-gas which can be a source of energy while balance material may be sold as a manure. This applies to both solids and liquids as the case may be.

Coming to domestic refuse as categorised above they are of several types. Sum total of all the above or some of the above are termed as city garbage which is one of the major menace in our urban life. As in case of industrial refuses, metallic refuse from the garbage can be easily separated with the help of parators like magnetic separators. It is not very easy to separate out cellulosic waste like paper or cotton cloth, etc. But, however, if consciousness can be brought among citizens this type of materials can be thrown separately and handmade paper can be manufactured. This will not only solve part of our unemployment problem but it will be able to solve part of problem like paper famine. As regards gaseous refuse coming out from our homes are mostly smoke from our kitchen. It is not easy to utilise this as in case of industrial flue gases since it is not available in large quantity at one place. Yet it is one of the greatest nuisance in our city life. As an example kitchen smoke in conjusted North Calcutta area during winter causes suffocation at times. One can very well experience the same. This smoke contains mostly carbon-dioxide with carbon particles, moisture dirt and ash particles cloating with it. Our city planners can think of having a common pipe with domestic smoke from all the houses of a particular area can be channelised through it. If necessary slightly modernised chullah or oven like Maganwadi Chullah or Sarkar Chullah can be made compulsory for all houses giving some subsidy initially to popularise it. Once the smokes from all houses can be channelised at one place we will be able to produce carbon-dioxide as described before and carbon particle can be reused as fuel by converting it as briquette. Modern world has given us hazards too. From soap and soda as cleansing material we have slowly switched over to detergent powder for cleaning our clothes which is preferred due to high cleaning capacity. Today's detergent powder or synthetic detergent or syndet are manufactured from a petrochemical aromatic compound generally known as dodecyle benzene or dobane. They are of two varieties *i.e.,* hard variety or straight chain ones and soft variety or branched chain ones. Accordingly, depending on type of raw material there are two types of syndets *e.g.,* hard syndet and soft syndet. Physically there is no difference between the two. But there is lot of difference chemically hence there is difference considering pollution factor. The solution in water of hand syndet which goes to drain after we wash our cloth is not biodegradable hence remains as it is for years together. This goes to river as well as reaches our subsoil water sources. Total quantum of such solution comes out from our city homes probably much higher than polluted water flowing out from our industries. One can very well notice the changes around us after detergent was introduced in our country. One can very well notice there is no fish available in Ganges passing dose to Calcutta. Stomach ailment has become common disease for our city dwellers. It may be noted that in almost all developed countries use and manufacture of hard syndet was officially banned. But it is not so in one country. As a result hard variety dobane produced in petrochemical complexes in western countries, are easily. We make its way to our country. The western capitalists amasses their wealth at the expense of our health. Presently, however, Indian Petrochemical Corporation Ltd., in their factory at Vadodara has started producing soft dobane and it is proposed hard variety will be banned after we will be in a position to produce sufficient soft dobane.

Vegetable wastes from our kitchen comes out in bulk. The vegetable wastes are normally mixture of waste fruits and vegetables. They can be further classified in different categories according to process we adopt to produce and products for useful purposes. There is lot of seeds of fruits and vegetables comes out from kitchen like mango seed, orange seed, apple seed, sitaphal seed, bael (wood apple) seed, etc. It may be noted that these seeds contains lot of oil edible and non-edible. Some of these oil has capacity to prevent plant diseases. Presently these seeds are thrown alongwith other garbages causing substantial loss to our notion. If we are able to arrange these seeds being collected separately they can be either expelled or solvent extracted thereby giving us either edible oil back to kitchen or non-edible oil as raw material for our soap and glycerine industry. Oil cakes can be used either as cattle feed or as manure or even can be utilised for producing bio-gas as described in subsequent paras in addition to rich manure. Properties and fatty odd content of some of above oils are given below:

Properties	*Mango fat*	*Sitaphal oil*	*Bael fat*
Nature	*semisolid*	–	–
	bland fat	–	–
Colour	*Cream*		
sp gr	09139 at 30°C	0.9126 at 15°C	0.918 at 30°C
Acid value	0.28	0.8	-
Sop Value	194.8	188.3	193.6
Iodine Value	93.2	85.6	108.0
Unsop matter per cent	2.8	0.2	1.6
Fatty acid content	Per cent	Per cent	Per cent
Myristre	0.69	Linsleic 55.2	Linoleic 36.0
Palmitic	8.83	14.7	16.6
Stearic	33.96	10.7	8.8
Archidic	6.75	Creotic 0.9	Linoledo 8.1
Oleoc	49.78	18.1	30.5

Second type of vegetable waste that comes out of our kitchen are peal of some of our fruits which contains essential oil. One of the glaring example being orange. Peal of such fruits can be segregated and collected every day to extract essential oil needed for our perfumery Industry. Rest of the vegetable waste can be used with dung to produce biogas as described in next paragraph.

Most neglected waste from our urban society is human and animal dung. Its may be noted that it is now very difficult to handle the huge quantity of dung that comes out from urban society. It may be noted that the dung if fermented gives us mixture of methane gas and carbon-dioxide in the proportion of one in to four. The mixture gas is now popularly used as kitchen gas and can easily be used effectively for generation of electricity. Due to fermentation in digester the germ or bacteria it contains dies and odour less slurry that comes out from digester becomes better manure since NPK value rises. This will also help preventing spreading of communicable diseases. If we can segregate methane gas and carbon-dioxide separately, we may go of producing valuable ingredients. Carbon-dioxide thus segregated can be purified and filled in cylinder for industrial use as described earlier. Methane, can be purified after separation and can be used for manufacture of many organ chemicals. On chlorination methane produces chloro methane and carbon tetra chloride and hydrochloric acid. The Chemical reactions for above is below:

$CH_4 + Cl_2$	$CH_3CI+HCI$	
Methane	Chloro	Hydrochloric content
$CH_3Cl + Cl_2$	$CH_3Cl_2 + HC1$	
	Di Chloro methane	
$CH_2Cl_2 + Cl_2$	$CH\,Cl_2 + HCl$	
	Tri chloro methane	
$CH\,Cl_3 + Cl_2$	$CCl_4 + HCl$	
	Carbon tetra Chloride	

Carbon tetrachloride has good cleansing property while being extremely volatile. This can replace petrol in dry cleaning of our fabric and cleaning of machines. The other chloro methanes as above are important raw material for dye industry. Besides above methane con also be used for manufacture of many important chemicals by different methods like fuffonation, nitration, etc. It is not only dung, the vegetable wastes, oil cakes, garbages or any other perishable materials can be used alongwith dung for production of biogas. To utilise this gas property and commercially again we need better coordination and planning in each medical college of a city can have at least one or more biogas plant utilising dung of patients and students. This will solve the problems of Preventing spreading of communicable diseases which might come out from the dung of patients. The gas that comes out can be used in the kitchen of students hostel and preparation of food of patients.

Secondly in case of power failure methane lamp can be used in operation theatre for conducting emergency operations. For domestic dung the city can be divided in three or four zones depending on quantum of dung available. Each zone may have one community biogas Plant producing biogas in bulk. Part of this gas may be used for generation of electricity while rest may be used for manufacture of industrial product as described above. The slurry coming out from biogas plant after giving away the gas can be dried and sold as manure. The liquid refuses which are easily degradable can also be used with dung in biogas digester.

Lastly the refuses that does not fall in above two categories are mainly of two types. The animals like cow, buffalo, goat, sheep, etc. Which are of wandering types of dies due to accident or meets natural death on the road creates nuisance on the street as they rot. We may organise one big carcass utilisation unit in a city having sufficient conveyance so that those dead animals can immediately be removed. In the carcass unit hide can be pealed off and sent to tanning centre for using in leather industry. Rest of the body may be boiled and fat may be separated for utilising in manufacturing of soap.

Bone, blood and meat may be dried and crushed to produce bone meal, meat meal or blood meal to be utilised as either manure or cattle or poultry feed. Horn may be used for producing horn product while feet for neat foot oil. It may be noted that Khadi and village Industries. Commission is successfully running such a unit near Delhi. As regards smoke from automobile there is no fruitful method yet invented. It may be noted that automobile smokes causes lead poisoning resulting calcium deficiency. This is a chronic problem not only in our country but also all over the world. Scientists are still at work to solve this problem but atleast we can take care to see that our automobiles are maintained properly to give less smoke.

The urban refuses are one of the greatest menace in our society but unfortunately we are not giving due recognition to this problem. Of course, the treatment of the refuses as described above needs planning and investment which can only be implemented by an organisational set up. But at least if we improve our civic sense and sanitation habits we may be able to improve our hygiene upto certain extent. We are limitating western world many ways like style, dress, drinking, etc., but not their good habit like sanitation. Our Father of the Nation Mahatmaji rightly observed "The one thing which we can and must learn from West is the science of sanitation." We are just ritualists. We spin charkha as pleased by Mahatmaji but ignore economics behind it. That is why Mahatmaji concept of new India has failed. But in case of sanitation even if we follow as ritual things may change. We need this change for the sake of our children.

REVIEW QUESTIONS

1. Define Pollution and Pollutant.
2. Name the sources of air pollution in big cities. What type of pollutants do they produce?
3. What are the effects of SO_2 as pollutant on living organisms.?
4. What is negative soil pollution? Suggest the causes which have produced the Thar desert of India?
5. Describe the mechanism of sewage disposal.
6. (i) What is importance of recycling of wastes?
   (ii) Describe what is BOD and COD.
7. What are the main water and air pollutants?
8. What are the harmful effects of ultraviolet radiations? How can they be minimised?
9. What are natural and artificial sources of ionising radiation?
10. Write briefly how air pollution can be controlled.
11. Name some important constituents of effects. Describe the methods for prevention of water pollution.
12. Discuss briefly the pollution of environment by pesticides. How can this be minimised?
13. Describe the sources and effects of noise pollution.
14. Mention the steps which can check noise pollution.
15. (i) How recycling of water material can prevent soil pollution?
    (ii) Explain the effects of air pollution on vegetation?

Chapter 6

SOCIAL ISSUES AND THE ENVIRONMENT

I. SUSTAINABLE DEVELOPMENT

The world commission on Environment and Development in its Report *Our Common future,* described sustainable development as 'development that meets the needs of the present without compromising the ability of future generations to meet their own needs. So defined and from the terms of reference to the Commission, sustainable development can be said to involve three conceptual key elements. They are firstly, integration of environmental concerns into economic policies, secondly, appraisal of development from a proper perspective and thirdly, a commitment to equity—a concern for others.

INTEGRATION OF ENVIRONMENTAL CONCERNS INTO ECONOMIC DEVELOPMENT

The objective of sustainable development is certainly not zero growth because such an attitude is an insurmountable block 10 initiating any workable policy. Any policy which aims to be of operational value in a complex and contradictory human situation cannot afford to be solely value oriented leaving absolutely no choice to the participants. It has to be logical, pragmatic and one that offers a common meeting ground for conflicting interests. Thus, zero growth policy which views environmental protection and development as antagonistic hinders the promotion of the very objective it seeks to promote namely, the protection of the environment. It is not denied that the present human life-style involving appropriation of natural resources, indiscriminate energy consumption, exponential use of natural resources, indiscriminate energy consumption, exponential population growth and inappropriate technology etc., are too vulgar, irrational, unbalanced and unsustainable. If we go on being as destructive and as greedy as we have been, we will soon find ourselves freezing or stifling, starving in teaming warrens on a plundered and polluted planet. It is also true that if current forms of development were employed, a further live to ten fold boost in economic activity would be required over the next 50 years to meet the needs and aspirations of 10 billion people. Economic growth of this enormous magnitude will undoubtedly have far reaching environmental consequences by stretching the planet earth beyond the enduring capacity. But the question is whether 'no growth' is the only solution for restraining environmental decline. Is the relationship between economics, ecology and human well-being so simple so as to accommodate zero growth demand without any opposition'? Indeed, the relationship between economic growth and environmental protection is not so simple. They do interact but in a system which is both natural and man-made. The complexity of the system restrains anyone from arriving at the drastic conclusion that zero growth alone is what is required. The environmental impact of productive activities often involves a long chain of causation involving the confluence of a multiplicity of factors. Further, in a system predominantly occupied by poor states, a call for zero growth is impractical and unacceptable. It is inappropriate more so because of inadequate knowledge of ecological consequences. 'Moreover, advocating zero growth policy with total disregard for the complex system in which it has to be implemented, encourages an attitude or dismissing environmental concerns altogether, leading to the defeat of the very objective of environmental protection. This is perhaps the reason why economics has continued to violate the fundamental principles of environmental maintenance.

Traditionally, environmental protection and economic development are considered parallel goal that can never meet. On the contrary, sustainable development implies that environmental protection and economic growth are complementary and resource each other. The once separate issues of environment mid economic development are considered to be inextricably linked. Heavily influenced by the former West German Chancellor Willy Brandt's

North-South Report, the Bruntland Commission called for restructuring mankind's future so that economic growth would be 'based on policies that sustain but expand the environmental resource base. Thus the primary language of sustainable devèlopment is the integration environmental and economy of It seeks to define the goals of economic and social development in terms or ecologic sustainable. Therefore, for reaching modifications the present pattern or economic policies is the prerequisite of sustainable development.

The environmental debate till recently fòcused mainly on the adverse impacts of development on the environment the impacts of a degraded environment on prospects for development were largely ignored. The convergence or ideas about environment and development ultimately culminating in the concept of sustainable devèlopment must be reckoned as a historic advance in both the fields. Recently many on both sides have begun to relict that they need the insights of each other if the goals of either are to be met. It is relished that economy is not quit the production of wealth and ecology is not just about the protection of nature: Both are relevant for improving the lot of humankind. Indeed, the first world, wide public opinion survey on environment conducted by Louis Harris : and his Associate for the United Nations Environmental Programme which was released in May, 1989, provided striking evidence that people in developing countries as well as the industrial nations are concerned that environmental quality is deteriorating and that it requires urgent attention. This is because certain significant environmental threats have proved to be truly international. Sustainable development provides the magic formula which, while by passing the zero growth stance of conservative environmentalists endeavours to bring both economy and environmental concerns within a common framework.

The concept of sustainable development which speaks not of zero growth but of development with environment concurring provided till much needed respite for those who dreaded to consider zero growth is the one and only alternative to environmental protection. The maxim sustainable development is not limits to growth. It is growth of limits. Harmonising economic expansion with environmental protection requires the recognition that there are environmental benefits to growth, just as there are economic benefit flowing from healthy natural by Sustainable development is premised on the underacting that natural resources which are necessary for economic: development are finite that overstepping these limits could lead to disastrous consequences for present as well as future generations unless a determined effort is made to change the course of economic development.

A move towards sustainability is different from environmental movement in that it has a much broader orientation since it directs attention to systems rather than single issues. It insists on a type of development based on such forms and professes that do not undermine the integrity of the environment on which they depend. The phrase sustainable development in essence has integrative effect though it does not in itself provide conceptual integration. So, environment which was viewed largely as an add-out to development is now treated as an integral 'build-in' by the advocates of sustainability. It is indeed a paradox that while environment and economics and inextricably linked in reality, a politically divided world has required a sustained effort over a period of time to formulate the same relationship as a policy guide line sustainable development.

DEVELOPMENT: A PROPER PERSPECTIVE

The goal and essence of the concept of sustainable development is derived from the adjectival use of the key word 'development'. Defining development from the perspective of sustainability or vice versa is a critical task, yet, is very crucial to render direction to the policy of sustainable development. In order to understánd any global strategy which talks about development, it is essential explore the core meaning of the word 'development'. However, it should be understood that the concept of development cannot be defined with scientific precision. A jurist who deals with the concept of development in skating on thin ice.

Development depend upon a combination of factors but all of .them have a common factor, namely, the human factor. Taking care of people, or more accurately creating conditions within which people take care themselves–is what development is supposedly all about. Yet, in what has usually been called 'development' this primary goal has rarely been taken into account and what was actually been promoted is industrial growth and modernisation.

'Development is an evolutionary product' of the idea 'progress', while means change of the world. The idea progress was, conceived in the modern world as an antithesis to the sociology of the middle ages "which believed ill the status quo was introduced in the... era of the industrial Revolution. The economic conception of the notion of progress advocated by Lord Keynes and followed by other economics gave rise to the phenomenon of growth. It is believed that progress could be achieved by generating wealth through maximisation of the productivity of labour and capital. Attempts were made to prove that such an attitude was conductive to the maximum satisfaction of all and that it created the best of all conceivable world. This was typically the' attitude of economic science, represented by Adam Smith, and it continued 10 be the approach adopted by the: economists for a considerable part of the 20th century, The 'invisible hand' advocated by Adam Smith never worked in the interests of humanity Instead what has occurred unequal and inequitably economic growth. This has led to problem like environmental degradation, consequent health hazards, resource depletion, alienation and other assorted social pathologies—all of which have become global concerns.

To understand real development, it is essential to distinguish between 'growth' and 'development' 12th especially in view the fact that the two terms came to be used synonymously, 'Growth refers to explanation In the scale of physical dimension whereas 'development' signifies qualitative improvement in a non physical dimension. Growth defined in the above sense has both biophysical and ethico-social limits unlike the development. Our belief in technology made us continue to avoid making a distinction between growth, a quantitative change and development. A qualitative change and hence has made us assume that there are no limits to growth It needs to be noted that the Brundtland Commission has used the phrase 'sustainable development' but not 'sustainable growth. It has described sustainable development as 'development that meets.....,' thereby delineating its goal away from the narrow connotations of growth. So, any attempt to treat the two terms as synonymous is neither desirable nor acceptable.

Development is generally accepted to be process that improves the living conditions of the people. It includes both material and non- material requirements. Thus, development in its true sense is a much wider term and has connotations apart from those of economic growth. This does not, however, mean that economic growth is devoid of the development dimension. Economic growth to the extent it contributes to the quality of life of mankind in both material and non material dimensions can certainly be considered as development. In fact it is absolutely necessary for certain parts of the world. For a period of time at least to improve the basil: living standards of the poor. But it will have to be a different type of a growth targeted to the needs of people and sensitive to the needs of the environmental. What is undesirable is economic growth achieved at the expense of development, and is not enough in itself to ensure improvement of qualitative social conditions.

Suffixing 'development' to sustainability presupposes an entirely new perspective of economic policy from a variety of angles. When one recognises that development is a combination of factors, any missing factor can retard development. Economic growth as conceived presently and pursued vigorously is virtually deficient in its environmental dimension. Environment at present, is viewed as an object of utility, not as a precondition for the well being of humankind. Economic growth is revered without taking into account its negative consequences especially so far nature is concerned and consequently detrimental for human development. Too often, our vision of development whether capitalistic or Marxist tends to be a simple one of capital accumulation. In failing to value nature as capital, economic growth ceased to be developments in true sense.

Environment and development are inextricably linked together, The welfare of an individual depends upon those goods and services he consumes as well as many intangibles such as environmental quality. A close connection between environment and development is implicit in the very definitions of the environment (as the conditions and influences that interact with man) and of development (as a process to improve human welfare). Hence, separate approach to environmental and development problems is futile.

Environmentally ignorant growth is unsound and inherently unsustainable. Unfortunately, there are some who still argue that

what has to be sustained is not the environment but growth. According to Wolfgang Sachs, there are those who focus attention on the 'impact of ecological destruction upon economic prospects: not nature but growth has to be sustained. To them, everything should subserve economic growth and they feel betrayed by nature which like errant workers in a factory threatens to go on strike when growth is accelerated and the world is poised for new horizons in production and consumption'. The use of the term 'development' instead of 'growth' in the phrase 'sustainable development' should provide the necessary insight to discourage such arguments which refuse to take stark realities into consideration and so are ultimately fatal. 'Sustainability' has to be interpreted in conjunction with 'development' -'development' in its proper sense. So, any attempt either explicitly or impliedly, to substitute 'growth' for 'development' either due to either ignorance or over confidence in the prospect of growth, should be discouraged and rejected. Any reluctance to change the exclusively growth centred development vision would indeed be a great folly and will inevitably lead to disastrous consequences.

No discussion on development is complete without reference to human rights. In fact the influence of human rights activists in the conceptualizing of sustainable development by the Brundtland Commission is profound in many parts of its Report and in the very definition of the concept which speaks of meeting human needs. The Commission did not expressly cast its analysis in terms of human rights. But, 'sustainable development' in its inherent perception aims at improving the lot of mankind and thus includes a perspective of human rights as well. It involves tackling two variables, namely, environmental protection and economic development within the ambit of a broader variable -human rights. They interact with one another as well as with other related factors in different combinations to achieve or obstruct the goal of development. Sustainable development, about sustaining worldwide human development. Development must encourage local self reliance including relative self sufficiency. Such self reliance should be built from local knowledge, traditions, and skills, rather than the transfer of unsuited technology. It is a self reliance based on appropriate technology–appropriate to the environment and appropriate to the background and knowledge of people. It is a goal and a criterion for people's development with their participation. It is a concept that can mobilize broader political consensus and one on which the international community can and should build.

EQUITY

Sir Henry Maine defined equity as 'any body of rules existing by the side of the civil law, founded on distinct principles and claiming incidentally to supersede the civil law in virtue of superior sanctity inherent in these principles'. But the traditional definition which is formulated in the context of common law background has no application to the development process aimed at levelling the existing economic, social and political disparities. Equity, now, is a principle of international ethics which is invoked by those who seek to promote a process of change for evolving a new international order. In this sense, equity is a matter of paramount importance for achieving the goal of sustainable development which demands appropriate changes in the existing international order. The function of principles of equity is to guide international standard -setting and law making in the process of promoting a new social-political order of sustainable development. According to professor Edith Brown Weiss, two issues in environmental law which gave importance to consideration of equity are (a) the allocation of natural resources and (b) responsibility and liability for pollution.

The explicit reference in the Brundtland Commission's definition to 'needs of the present without compromising the ability of future generations to meet their own needs', introduces equity another conceptual dimension. Sustainable development reflects a choice of values for managing the planet earth in which equity matters–equity among people around the world today, equity between parents and their grandchildren. Perhaps more important than the professional vertigo caused by the Earth Summit preparations was the realization that at the core of the now common, and overused phrase 'sustainable development', there lies something profound–not just economics or pollution control but equity and justice. The clubbing of the needs of the present and future generations in the policy of sustainable development as formulated by the Commission indicates that questions of both intergenerational and intragenerational equity must be squarely faced and trickled simultaneously.

Intergenerational equity

Equity as a key component of sustainability implies that of development is to be the goal, it should mean something which applies not only to the present generations but also to the future generations. The worst effect of environmental degradation is that its adverse effects pile up for future generations. For many centuries it 'has been wrongly assumed that human progress is ever continuous and that each generation involved in rapid exponential growth activities leaves a richer legacy to those that follow. But the rapidly deteriorating environmental condition casts serious doubts about such overconfident and rather oversimplified assumption. As the scale of economic activity proceeded steadily, the variety and intensity of environmental problems triggered by the activity have transcended both geographic and generational boundaries. Whereas each generation of humans used to have the luxury of being able to satisfy its own needs without worrying about the needs of those generations to come, that may no longer be the case. The present signals are indicative of resource shortages, widespread inequity, poverty and starvation, famine, political instability and violent conflicts. So, the future generations while being rich monetarily would in fact lead a poor life style emanating from an environment which is severely degraded due to the short gain activities of present generations. Futurologists have little disagreement as to the major global needs in international future. They are, population stability, adequate supplies of energy fund raw materials, well-being, economic equity control of power and use of force and environmental protection. So sustainability as projected by World Commission on Environment a and Development has a profound concern that the environment be conserved and protected for the welfare of future generations as well.

Over the past two decades, several economists and environmentalists have debated the present-value criteria and indeed the role of cost-benefit analysis in decisions affecting the environment. A utility oriented economist believes in maximization of the benefit referred as 'well-being'. In this process, the utilitarian discounts and does not mind sacrificing the welfare of the vast majority of human beings. For him, allocation of natural resources is only a matter of expediency. Accordingly, if due to any economic activity the gains in the present are greater than the costs to the future, such an activity is deemed proper even if future generations are worse. Reacting off vigorously to such arguments, environmentalists point out that in such a case, economic decisions will always be in favour of present generations and always tend to be myopic about the future needs. In view of this, intergenerational equity of sustainable development requires that society's tendency to discount (i,e., to reduce the value of) future costs of benefits and burdens, especially in comparison to short-term benefits must be modified.

There are a number of ways in which the future generations will suffer from the scourge of environmental degradation. For example, in the case of non-renewable resources like oil, any use increases their scarcity which means future generations will have to do with completely or badly depleted stock of such resources. Even in the case of renewable resources, over consumption results in their scarcity as well. Similarly building up of pollution levels denies the future generations chances of healthy existence.

One of the objections raised against the orthodox approach of environmental protection was that in valuing the environment the interests of future generations were kept outside its purview. The approach of sustainable development obviously fills this gap by taking into account the needs of future generations as well Sustainable development implying the environmental capacity to last or to continue, is a policy for the future. It is different from most past social policies which were concerned exclusively with immediate problems and had not envisaged an intergenerational perspective.

Sustainable development is promised on the understanding that as major adverse effects became imminent for future generations due to practices of their ancestors, equity demands the accommodation of social relationships and sociological concerns that transcend time. It is very much essential to have a social relationship between two generations even if their life spans do not overlap because the actions of one generation can have long lasting effects long after its death, thus raising issue of fairness, equity and justice like.

Is it fair for the present generations to mindlessly destroy those .natural assets without any concern for disastrous consequence for future generations?

Is it just to indulge in and enjoy benefits from the over utilization of natural capital common to all generations and pass on the terrible costs to future generation who have had no say In the matter?

Is it not fair to asses the impact of our pleasant activities on the need and interests of future generations and drastically modify and change our course of action accordingly?

To help to resolve the question of fairness., to future generation, any ethical system, be it utilitarian, egalitarian or libertarian, has to address the issue of obligation of the present generations to future generations of people.

Conventional economic analyses by confining only to the so called well being of present people have the advantage of knowing the individual preferences and degree of the willingness of the existing generations to protect the environment. On the other hand, a future oriented, social-centric, strategic perspective of sustainability development undertaken in response to question of 'What portable will be?', 'What could be?' 'What should be', and 'How what should be could be?', results in visions of what is probable, possible, desirable and feasible. So, in the formulation of any future oriented policy, value assessment of future generations is an uphill task fraught with many uncertainties.

Therefore, proponents of intergeneration equity are confronted with the arguments that

(a) there are no circumstances for equity because future people are only possible people.
(b) there is a crisis of non identity. It is difficult to identify one's relationship with far too instant generations extending beyond great grandchildren.
(c) judging future people's preferences and their wants is very difficult.
(d) a great variety of alternate and better futures can be envisioned given the vast potential for human intervention.
(e) mutual cooperation and assistance becomes impossible when there is no overlap of lifestyles. Later generations can offer nothing in return for the sacrifices of their ancestors.
(f) there is no ombudsman figure to act on behalf of future generation in decisions of today.

However, the above arguments invite the following questions in turn.

(a) is it just that future people are denied a fair deal simply because they do not exist.
(b) should identity be extended to many unknown generations in order to protect the future people from environmental crisis?
(c) is it essential that individual preferences should be precisely judged?
(d) is the possibility of an alternative future an excuse for not planning available future right now?

All the question raised above deserve a negative answer on the following grounds:

Firstly, it is not fair to discount future generations simply because they do not yet exist. As long as some people' existence i: anticipated, equity demands that their interests be taken into consideration. When the activities of the existing generations are capable of carrying forward their devastating effect into the future, it is but fair to plan for a secure future. It is all the more essentiality plan in advance because future generations do not exist today to have any say in the present day lifestyles of over consumption and exploitation of natural resources.

Secondly, to exercise restraint it is not essential to extend the psychological bond of identity beyond a limited period of time say that of great grandchildren. So long as each generation looks after the next, each succeeding generation will be taken care of.

Thirdly, it not essential to be precisely certain about future people's preference. Whilst preferences of 'want' might vary, 'needs' remain substantially the same within and even across generations. So, reasonable speculation about the need-based interests of future generations is not impossible.

Fourthly, the possibility of alternative futures can never be a valid reason for not discharging our present obligation to future generations, at least of planning a sustainable future. It is possible that given the accelerating and increasing complexity of scientific advances and technological change, a better future might be in store for future generations. On the contrary, given the potential for human intervention, it is equally possible that a completely different and worse future might come into existence contrary to our prediction, plans or desires. Hence, risk factor is high to let things happen for better or for worse, especially when there are visible indicators if a negative future.

Fifthly, humankind need not always i.e. calculative of getting something in return. There is an element of ethical choice guided by socio-centric concerns. And for equity to be relevant, it is not essential that there should only be contemporary relationship

Edith Brown Weiss ha developed three principle which form the core of the concept of intergenerational equity. They are, conservation of option, conservation of quality and conservation of access. Conservation of options entails that 'each generation should be required to conserve the diversity of the natural and cultural resource base so that it does not unduly restrict the option available to future generations in solving their problems and satisfying their own values, and should be entitled to diversity comparable to previous generations'. Conservation of quality entails that 'each generations should be required to maintain the quality of the planet that it is passed on in no worse condition than the present generation received it, and should be entitled to a quality of the planet comparable to the one enjoyed by previous generations'. Conservation of access entails that 'each generation should provide all its members with equitable rights to access to the legacy from past generations and should conserve this access for future generations'.

The efforts of economists to define sustainable development in terms of mere utility results in a very weak version of sustainability since it discounts future people. Weak sustainability takes into account only the minimum expectations of the future generations which is that they will not face an environmental catastrophe. So, the present generations are expected to refrain only from those activities which have they potential of causing an environmental catastrophe, e.g., complete deforestation. Economists justify the discounting processes on the ground that future generations are going to be better off as they get higher incomes and facilities. But this argument is fallacious because their vision and perspective is limited and they value environment only in terms of material gains. The utilitarian approach ignores the fact that what can be passed to future generation is only capitally—either man-made or natural, but never utility. Moreover, the utility approach bypasses the physical approach of the ecologist.

A much better version of sustainability is the one which is concerned about the transfer of aggregate capital which includes both man-made and natural capital. But this too is a weak version of sustainability because it is based on the assumption that man-made capital is a substitute for natural capital, if not complete, a near perfect substitutes. But this approach ignores the vital fact that man-made capital and natural capital are complacently but not substitutable. If man-made capital is a substitute for natural capital, the reverse also must be true. But it is not. Substitution for natural capital is never complete and not even near perfect. For example, life supports services so vital for biological survival cannot be replicated or replaced. Therefore, a really strong versions of sustainability requires distinct maintenance of natural capital.

A strong version of sustainability demand that future generations are left with a legacy of the natural endowment at least equal to that of the present generation. This version emanates from the ideal that present generation has not inherited nature from its parents but borrowed it from their children. According to this view, degradation of the environment by preceding generations amounts to nothing less than robbing the future generations of their inalienable right to environmental legacy. The underlying presumption of this version is that though there is no necessary obligation for the present generation to increase the potential level of consumption of environmental resources, there is a minimum obligation of leaving behind equal opportunities to future generations.

The Brundtland Commission described sustainable development as one that does not compromise the 'ability' of future generations. This clearly enjoins a strong commitment to quality on the part of the present generation towards future people. The use of the word 'ability' indicates such an obligation as derived from situational assessment. Hence intergenerational equity involves a rang of obligations -from avoiding a catastrophy to providing an opportunity to enjoy an equal measure of environmental resources. Though theoretically, the minimal and maximal versions of sustainability might demand different courses of action, in a situation of crisis the theoretical difference between the two vanishes.

Intragenerational Equity

Equity, the inherent key element in the concept of sustainable development was sometimes interpreted as extending only to relationships between people of different generations. But such an interpretation is baseless and misleading. When sustainability implies concern for social equity between present and future generations, the concern must logically extend to equity within each generations. Since the present is the foundation for the future, sustainability looses relevance if the present is not taken care of. The Brundtland Commission's reference in its definition to the 'needs of the present' dispels any doubt about the inclusion of commitment for intragenerational equity. In fact, the Third World which was so apprehensive about earlier international efforts for protecting the environment has readily endorsed the concept of sustainable development only Because it includes intragenerational equity within its ambit as a key component.

The imperative of intragenerational equity contained in the definition of sustainable development imparts a moral obligation on the part of rich nations to assist the less obligation ones. But there are some who assert that assistance from the rich nation leads to an unhealthy dependence on the part of the poor nations and hence such moral obligations should be discouraged. For example, Garrett Hardin who advocated lifeboat ethics in the context of food assistance to the developing countries suffering from over population, contended that assistance to the poor is counterproductive because it allows excess people to survive. This, according to him, compounds the problem of survival for the rich as well as the poor. Such and similar arguments have the effect of eroding morality in general and policy in particular which is aimed at moving the world towards a better state of affairs. It should be realised that lifeboat ethics has no application in a factual context where the underdevelopment of many poor states is the result of the past and present exploitative practices of the rich ones. Moreover, ecological unity is such that intragenerational equity is not an option but an obligation.

The concept sustainable development stresses. The interdependence between economic growth and environmental quality. It also goes further in asserting that the future is uncertain unless we can deal with issues of equity and inequality throughout the whole world. Inequities in the distribution of wealth between the developed and developing countries have led to uneven development and uneven development is not compatible with sustainable development. So, sustainability cannot be secured unless developmental policies pay attention to considerations of equity such as access to resources and the distribution of costs and benefits involved in the protection of environment. Sustainability refers to the maintenance of environmental capacities throughout the world. It cannot allow one environment say in the North to be preserved over time by the simple expedient of exploiting resources in or exporting pollution to another area, such as the South.

The emphasis of the WCED's definition of sustainable development on 'needs' rather than 'wants' indicates the direction in which the notion of intragenerational equity should proceed. In emphasizing the 'needs', the Brundtland Report has offered a fundamental challenge to the materialistic and consumerist values of the rich. In this context, it is worth recalling Keynes's comment on the distinction between absolute and relative needs and the self cancelling nature of the later. Now it is true that the needs of human beings may seem insatiable. But they fall into two classes those which are absolute the sense that we feel them whatever the situation of our fellow human beings may be, and those which are relative in the sense that if their satisfaction lifts us above, makes us feel superior to our fellows. Needs of the second class, are those which satisfy the desire for superiority, may indeed be insatiable for the higher the general level," the higher still are they. But this is not so true of absolute needs--point may soon be reached, much sooner perhaps than we are all of us aware of, when those needs are satisfied in the sense that we prefer to devote our further energies to non economics purposes. The Brundtland Commission's Report in many places provided the essential linkages between intragenerational equity and sustainable development. It is now beyond any doubt that establishing equitable relationship between those needing 'needs' and those wanting 'wants', would be a most crucial factor in achieving sustainable development.

Equity applies not only to relationship between the First and Third Worlds, but also within countries between people. The absence of distributive justice due to greed and mal-distribution of power within and amongst nations led to the current sad state of affairs. So far, it is assumed that growth will automatically trickle down and thereby will lead to equity and justice within and among countries regardless of the political economic system, but the evidence is to the contrary and the 'trickle-down' never happened. The growth goes down a little way and then dries up. Sustainable development has no meaning unless it is understood as a commitment to socially equitable development of the present generations.

Sustainability in the context of intragenerational equity raises the question of the level and scale at which the goal of sustainable development to be aimed at and achieved. Is it at global level, regional or local level? The answer can only be situational. In some situations it is the global standard which makes scene while in other situations it might be national or local standards that meet the contingency. But one thing that is certain is intragenerational equity is not just equity amongst , nations alone, it is equity irrespective of nations. It is equity amongst people.

II. ENVIRONMENTAL MANAGEMENT

The irrepressible human curiosity and the unquenchable thirst for knowledge are the fundamental basis for scientific development. A major part of innovations in scientific and technological development has been directed towards generation or elevation of human comforts, thereby increasing the standard of living in the society. This led to increase in industrialisation. Some of the important improvements to our standard of living that can be attributed to the application of science and technology include:

(a) Production of more and better quality food.
(b) Elimination of many infectious diseases
(c) Invention of new faster communication systems
(d) Creation of reliable and faster transportation
(e) Supply of safe water
(f) Invention of machines to replace human and animal power.
(g) Minimising water-borne diseases through improved water technology.
(h) Mitigation of bad effects due to natural disasters e.g., droughts floods, volcanic eruptions, etc.

Consequent to these improvements, disturbing side effects such as environmental pollution, deforestations, urbanisation, loss of arable land, evolution of new organisms resistant to control, etc., have emerged. These effects are considered as potential threats to environment and to humans.

In agrarian society, people lived essentially in harmony with nature raising food, gathering firewood and making clothing and tools from the land. The wastes from animals and humans were returned to the soil as fertilizer. Hence there were no appreciable problems of air, water or land pollution. For the small settlements which grew up, the supply of water, food and other essential goods and the disposal of wastes had to be kept in balance with the changing community, but no serious environmental problems were created.

The cities of ancient times, especially those of the Roman Empire, had built systems for water supply and waste disposal. However, the municipal technologies of ancient cities seem to have been forgotten or ignored or neglected for many centuries by those who built cities throughout Europe. This resulted in the outbreaks of water-borne diseases like cholera, typhoid and dysentery. Until the middle of the 19th century, it was not realised that improper waste disposal polluted water supplies with the disease-carrying organisms. The industrial revolution in 19th century particularly in Britain. Europe and North America aggravated the environmental problems due to increased industrialisation and urbanisation. These 2 factors caused such levels of water pollution and air pollution which the cities of that time could not handle effectively. However, rapid advances in technology for the treatment of water and the partial treatment of waste-water took place in the developed countries, over the next few decades. This led to a considerable decrease in the incidence of water-borne diseases.

After the Second World War, the industrialized countries had an economic boom due to burgeoning population, advanced technology, and a rapid increase in energy consumption. These activities increased considerably from 1950 onwards thereby increasing the variety and quality of wastes discharged into the environment. New chemicals, including pesticides and insecticides, used without adequate data regarding their environmental and health effects caused and continued to cause, enormous environmental problems. Alarmingly, these problems are aggravating further since the types and amounts of such pollutants discharged to the environment are increasing inexorably as against the limited capacity of our water, air and land systems to assimilate these waste materials.

The financial crunch since 1970 forced changes in the priorities of many countries. Issues like unemployment inflation, energy, effects of globalization, resource crunch, high technology, war threats, social and political compulsions became major concerns. Thus it appears that concern about the public health and safety aspects of hazardous wastes will continue to increase for a long time. It is in this backdrop that development and implementation of effective environmental management strategies assume paramount importance.

The United Nations Conference on Human Environment was held during June 5-11, 1972 at the Swedish capital, Stockholm. This Stockholm conference was the first major international event which created global awareness about the environment by placing environmental concerns on the agenda of major international topics. The Stockholm declaration containing 26 principles reflects the range of environmental concerns of that periods. The set of 109 recommendations of the Stockholm conference, widely considered to be the "magna carta" of environment, laid down the principles and action plan to control and regulate human environment. To commemorate this great event, June 5 of every year is rightly observed as the "World Environment Day". The historic Stockholm declaration called for the world community to bear a solemn responsibility to protect and improve the environment for present and "future generation" and gave the concept of eco-development".

The United Nations Environment Programme (UNEP) convened a meeting in 1982 to commemorate the 10th anniversary of the Stockholm conference, to assess the State of environment and changes in the perception f environment since 1972 conference. In 1983, the U.N. General Assembly appointed the "World Commission on Environment and Development". This commission, having 21 countries as its members, was entrusted with the task of formulation of a "global agenda for change" and to prepare an environmental perspective upto the year 2000. After deep study, the committee submitted its report entitled "our common future" in 1987. The committee recognized that the "sustainable development" should become a central guiding principle of the

United Nations. Sustainable development implies "meeting the needs of the present without compromising the ability of the future generations to meet their own needs".

The United Nations Conference on Environment and Development (UNCED), known as the "Earth Summit" or "Eco-92", held during June 3-12, 1992 at Rio de Janeiro, Brazil, is considered as another historic event involving virtually all member states of the U.N. About 170 countries, 500 non-governmental organizations and about 2000 Journalists all over the globe participated. This conference identified the practical environmental and developmental challenges and opportunities and their inter-linkages upto the end of this century and even beyond. This conference also paved the way for creation of new partnership among the various countries. The *"earth charter"* prepared at the conference enunciates the principles, setting out rights and obligations of all nations in relation to environment and guidelines to nations in their quest for ecologically sustainable development. *"Agenda-21"* evolved at this conference provides a resume of global environment and a blue-print of action plan for solving environmental problems and the tools and resources desirable for implementing the action plan. Besides the "earth charter" and the "Agenda-21", the earth summit also came up with *"Convention of climate change"* and the "Convention on biodiversity".

The blame for deterioration of environmental quality cannot be laid only on industrial activities. Wastes from residential and commercial sources also cause environmental stress. Further, both affluence and poverty also cause pollution in their own ways. Natural calamities, illiteracy, and lack of environmental ethics are also partially responsible. Therefore, skilful and equitable environmental management often requires balancing of a number of factors some of which, at times, may be conflicting with each other.

Objectives of Environmental Management

(i) Regulating the exploitation of natural resources.

(ii) Protecting environmental degradation and maintaining environmental quality.

(iii) Balancing the ecosystem.

(iv) Preserving the biological diversity.

(v) Regulation of exploitation of natural resources.
(vi) Adopting engineered technology without creating adverse effects on environment.
(vii) Formulation of suitable environmental laws and regulations and effective implementation of the same.

Components of Environmental Management

The major components of effective environmental management are :

(i) Control of atmospheric pollution and environmental degradation.
(ii) Adopting technologies which ensure sustainable development.
(iii) Conducting environmental impact assessment to review the existing technologies and making it mandatory for clearing major projects of environmental concern.
(iv) Instilling environmental perception among people by conducting awareness programmes.
(v) Environmental education and training at schools, colleges and universities. The importance of environmental education was highlighted at the Environmental Education Conference (EEC) held at Belgrade in 1975 (called Belgrade Charter) by UNESCO and at Tblisi, U.S.S.R. in 1977 by the United Nations Education Programme (UNEP).
(vi) Controlling overpopulation.
(vii) Controlling over-consumption and craze by inculcating sublime human values such as service to society, non-material enrichment and spiritual solace.

ENVIRONMENTAL IMPACT ASSESSMENT (EIA)

"Environmental Impact Assessment (EIA) is an activity designed to identify and predict the impact on the biogeophysical environment and on human health and well-being of Legislative proposals, policies, programmes, projects, and operational procedures, and to interpret and communicate information about the impacts."

EIA is widely accepted as a tool in environmental management. It has been adopted in many countries with different degrees of enthusiasm and evolved to varying levels of sophistication.

Environmental Impact Statement (EIS)

"EIS is a public document written in a format specified by authorized national, state, and/or Local agencies".

"*Environmental Inventory* is a description of the environment as it exists in an area where a particular proposed action is being considered".

Historical Background

Whenever a new development project is planned which is likely to affect environmental quality, it is useful to carry out Environmental Impact Assessment (EIA). In many jurisdictions, EIA is mandatory before according permission to proceed with development projects such as power plants, dams, smelters, petrochemical industries, paper industries, iron and steel mills, mining, oil exploration, flood-control systems, etc.

The First comprehensive environmental legislation in the United States came into force on 1st January, 1970, in the form of National Environmental Policy Act (NEPA). The Act contained the following 3 main sections:

(1) Declaration of National Policy defining the environmental goals to be pursued by the Federal Government.
(2) Specification for the preparation of Environmental Impact Statement (EIS) on actions which significantly affect the environmental quality.
(3) Institutionalization of the Environmental Impact Assessment (EIA) in the Executive office of the president through the establishment of the council on Environmental Quality.

In Canada, the environmental assessment and review process started in 1973 for planning, decision making and implementation of new projects, programmes and activities with potentially significant environment effects.

Today, the EIA process has been accepted in many industrialized countries including United States, Canada, Japan, Australia, Netherlands, Columbia, India, Thailand, Philippines and even in some African countries like Rwanda, Botswana and Sudan. In India, the Central Ministry of Environment and Forests issued a Notification on 27th January, 1994 making EIA Statutory for 29 specified activities falling under various sectors such as

Industries, Mining, Irrigation, Power, Transport, Tourism, etc. This Notification was amended on 4th May, 1994, and the amended version includes a self-explanatory note detailing the procedure for obtaining environmental clearance, technical information, documents required to be submitted at the time of applying for environmental clearance from the Ministry of Environment and Forests (MOEF), so as to enable the submission of complete application for environmental appraisal.

The EIA process, when it started in the early 1970s, laid emphasis on evaluating the adverse/beneficial effects of development projects/activities on Physical factors, particularly those for which codes and standards are available *(e.g.,* air quality, water quality, solid waste disposal), and to incorporate suitable remedial measures at the project formulation stage. After a few years, EIA began to include biological and ecological factors, although they were difficult to quantify. Later on, the scope of EIA was further broadened to include socio-economic factors (e.g., creation of employment opportunities, recreational factors, cultural impacts, etc.). In some cases, EIA process was used to evaluate class actions such as banning a particular pesticide or regulating the lead content of gasoline.

The EIA system has been welcomed, in principle, by many people. However, in practice, some sections of people feel that it is often too technical and it does not deal with the environment in a holistic way.

Elements of the EIA Process

The first step in the Environment Impact Assessment (EIA) process is to determine whether the project under consideration falls within the Jurisdiction of the relevant Acts/Regulations, and if so, whether it is likely to create a significant environmental disruption. If so, an EIA is undertaken and the Environmental Impact Statement (EIS) is prepared. In some countries, the EIS is open to public scrutiny and reviewed at public hearings. Eventually, a political decision is taken so as to whether the development project is (a) accepted or (b) accepted with amendments or (c) an alternative proposal is accepted or (d) rejected.

Participants in EIA process: The following persons/groups/agencies usually are involved in EIA process.

(1) Proponent : Government or Private Agency which initiates the project.
(2) Decision Maker : Designated individual or Group or Body.
(3) Assessor : Individual or Agency responsible for the preparation of EIS.
(4) Reviewer: Individual/Agency/Board entrusted with the responsibility for reviewing the EIS and assuring compliance with the relevant guidelines/regulations.
(5) Other Government Agencies having special interest in the project.
(6) Expert advisers.
(7) Media and Public at large.
(8) Special interest groups: Environmental Organisations, Professional Societies, labour Union, Local Associations.

Contents of EIS: The EIS should contain the following information/ data:

(1) Description of proposed action and alternatives including that of no action:
It should include details of the construction phase, operation phase and the shut-down phase wherever applicable. Selection of alternatives to the proposed action e.g., different ways of building and operating the project, alternative sites, etc.

(2) Estimation of the nature and magnitude of the likely environmental effects of the various alternatives proposed: This is mainly done under the following 3 broad categories:
   (a) Physical factors *(e.g.,* possibility of earthquakes, possible effects on surface and groundwater quality, soil and air quality, etc.).
   (b) Biological factors (e.g., effects on vegetation, wild life, sport and commercial fish species, endangered species, etc.).
   (c) Socio-economic factors *(e.g.,* economic, demographic, social values and attitudes).

(3) Identification of the relevant human concerns.
(4) Criteria to be used in measuring the significance of environmental changes including the relative weightages to be assigned in comparing different types of changes.

(5) Estimating the significance of the predicted environmental changes and thereby the impacts of the proposed action.
(6) Recommendations for acceptance of the Project/Remedial Action/Acceptance of one or more alternatives/rejection of the project.
(7) Recommendations regarding monitoring procedures to be followed during and after implementation of the project.

Determination as to which environmental changes (including even those which are not covered by laws/regulations) are relevant is critical to the validity and credibility of the EIA process. It is also important to limit the scope of the assessment in time, space, and number of factors in order to enable the usefulness of the EIS document.

When once the factors to be included in EIS have been determined, their future magnitudes must be predicted, though qualitatively. A weighting system may then be devised to enable comparison of different impacts.

One of the single approaches to provide a visual assessment of the effects and magnitudes of the selected factors in EIA is the Leopold matrix (Fig. 6.1). The steps for using the Leopold matrix are given below:

(i) Identify all such actions that are part of the proposed project an locate them across the top of the matrix *(e.g., a, b, c,* vide Fig. 6.1).
(ii) List the relevant environmental characteristics or conditions down the side of the matrix, *(m, n, ...)*.
(iii) Under each of the proposed actions (namely, *a, b, ...*), place slash at the intersection with each item on the side of the matrix if an impact is possible.
(iv) After completing the matrix, write, in the upper left-hand corner of each box with a slash, a number in the scale of 1 to 10, which indicates the *magnitude* of the possible impact. In the scale of 1 to 10 chosen above, number 10 represents the *greatest magnitude* of impact whereas number 1 represents the least. Before each number so placed, put + if the impact would be beneficial.
(v) In the lower-right-hand corner of each box with a slash, write a number, in the scale 1 to 10, which indicates the *importance* of the possible impact *(e.g..* Regional vs. Local). Here also,

number 10 represents the *greatest importance* of the impact, while number 1, represents the least importance.

(vi) The text that accompanies the matrix should provide discussion of the significant impacts, of those columns and rows with large numbers of boxes marked, and of those individual boxes which have larger numbers.

	a	b	c	d	e
m	4/2			7/3	8/5
n	6/1	8/8		+1/+1	9/6
.	.	.	.	.	.

Fig. 6.1. Leopold matrix (Leopold *et al.* 1971)

Design of EIA

There are several approaches for EIA documented in literature which should be consulted.

Further, it will be useful to study Environmental Impact Statements that have been prepared for similar developments/ projects. Some of the important aspects that should be considered for the design of an EIA process are listed below:

(A) Project Design and Construction

(i) Type of project under consideration.
(ii) Physical dimensions of the area being considered.
(iii) Whether there is an irretrievable commitment of land?
(iv) Whether there will be serious environmental disruptions during construction?
(v) Whether the resources will be used optimally?
(vi) Whether the project is a critical phase of a larger development?
(vii) What are the long-term plans of the proponent?

(B) Project Operation

(i) How will the hazardous waste products be handled?
(ii) What provisions have been made for training the employees for environmental protection?

(iii) What are the contingency plans developed to cope up with the possible accidents?
(iv) What plans have been made for environmental monitoring?
(v) What provisions have been made to check the safety equipment regularly?

(C) Site Characteristics

(i) Whether the terrain is creating problems in predicting groundwater characteristics, air pollution, etc.?
(ii) Whether the site is susceptible to floods, earthquakes and other natural disasters?
(iii) How many people are likely to be displaced because of the project?
(iv) Whether the local environment is conducive for the success of the project?
(v) Whether the project will interfere with the movements of fish population and important migratory animals?
(vi) What are the main attributes (*e.g.*, protein content, calorie content, weed or pest status, domesticity, carnivorousness, rarity of species, etc.) of the local fauna and flora?
(vii) Whether any historic sites or traditional thoroughfares are likely to be endangered because of the project?

(D) Possible Environmental Impacts

(i) What are the possible short-term and long-term environmental impacts for this type of projects during construction and after construction?
(ii) Who would be effected because of these impacts?

(E) Socio-Economic Factors

(i) Who are the expected Gainers and losers by the projects?
(ii) Where are the expected trade-offs?
(iii) Will the project interfere (blend, increase or reduce) the existing inequalities between occupational, ethnic, sex, and age groups?
(iv) Will it effect the patterns of local/regional/national culture?

(F) Socio-Political Factors

(i) What are the relevant political factors that have to be considered?

(ii) What will be the difficulties in implementation of construction and operation phase of the project?

(iii) What are the relevant governmental regulations and procedures?

(G) Availability of Information and Resources

(i) Whether local and outside experts are available to consult regarding specific impacts of the project?

(ii) Whether the relevant guidelines, technical information, and other publications are available to identify and deal with the possible impacts of similar projects.

(iii) Whether relevant environmental standards, by-laws etc. are considered?

(iv) Whether the sources of relevant environmental data are identified and whether they are accessible?

(v) Whether the views of the specialist groups and general public regarding the project have been ascertained and considered?

(vi) Whether the competent technical manpower available to handle the project and the possible impacts?

When once the magnitude and the significance of the impacts have been determined, the EIA may be considered to be essentially complete. Then follows the political decision as to whether the project be accepted, accepted with alteration or rejected.

The EIA is a potentially useful component of good environmental management. It represents a reconciliation of environmental and socio-economic factors with respect to the proposed development. However, some people still seem to have reservations about this system on the plea that the EIA is expensive, it delays the projects and the predictions are uncertain.

Environmental Audit for Sustainable Development

Environmental Management is absolutely essential for sustainable development because it minimizes the environmental disturbances and ensures unhampered pace of industrial development and economic growth. Environmental audit is an important tool for environmental management because it enables the environmental Pollution/Control Agencies to ensure the compliance with the environmental protection laws. It also motivates the mining and processing industries to demonstrate their concern and greater overall awareness towards their social obligation for environmental protection and to adapt eco-friendly technologies.

Environment Audit is an important management tool comprising of a systematic, periodic, objective and documented evaluation and assessment as to how well the environmental management systems are organised to facilitate control of environmental practices and how well the company policies are complying with regulatory requirements. ***Environmental review, environmental surveillance*** or ***environmental assurance*** are the other synonyms used for environmental audit.

Environmental auditing comprises of the following steps:

(i) Compiling of all relevant information on environmental management
(ii) Evaluation of the information collected.
(iii) Formulation of conclusions and identification of areas that warrant improvement.
(iv) Active follow-up of the points raised and recommendations made to achieve the objectives of environmental management.

Environmental auditing may be practised in different organisations having different audit objectives. However the following interrelated characteristics are identified for an effective audit programme:

(i) Scope (ii) Organisation (iii) Objectives (iv) Resources (v) Approach and (vi) Coverage.

Environmental audit may have flexible methodology depending on the type of industry and its situation. Environmental audit should be preceded by well-defined pre-audit activities, on-site visits and post-visit activities, followed by preparation of report and suitable follow-up action plan.

Environmental audit is regarded as a voluntary responsibility of an industry or organisation. The following benefits will be accrued for an industry/organisation that practises environmental audit:

(1) It helps in assessing whether the existing environmental practices being followed are satisfactory and whether the environmental protection regulations are complied with.

(2) It provides an opportunity for comprehensive review of environmental policies, management systems, organisations and practices and to assess whether introduction of new innovative practices are necessary to comply with the stringent regulations from time to time.
(3) It protects against possible penalties, litigations or regulatory risks.
(4) It contributes its modest share towards sustainable development and gives due credit for environmental management to the Management.
(5) It provides an up-to-date environmental data base which may be useful in emergencies and also while making decision on plant modifications.

In India, environmental audit system has been introduced for the assessment of development projects on a regular basis. The system envisages submission of an Environmental Audit Report in a prescribed form for the financial year ending 31st March on or before 15th May starting from 1993 to the State Pollution Board by every person carrying on an industry, operation or process requiring approval under the water (Prevention and Control of Pollution) Act, 1974 or the Air (Prevention and Control of Pollution) Act, 1981 or both or an authorization under the Hazardous waste (Management and Handling) Rules, 1989. The report should clearly spell out the steps taken or proposed to be taken for adoption of clean technologies, waste minimization, recycling and reuse, pollution control investment in environmental protection and resource conservation measures taken.

Environmental audit is an effective tool for environmental management and sustainable development. The concept of sustainable development is based on

(i) Symbiotic relationship between consumer human race and producer natural systems, and
(ii) Compatibility between ecology and economy.

The principle of sustainable development is basically related to the carrying capacity of the ecosystem. The carrying capacity may be defined as the rate of maximum resource consumption and waste discharge into the environment that can be sustained indefinitely in a defined region with progressively impairing bioproductivity and ecological integrity. Thus, "carrying capacity" provides the physical limits to economic development.

United Nations World Commission on Environment and Development defined sustainable development as a new path of progress in which the needs and aspirations of the present generation can be met without compromising the ability of the future generations to meet their own needs and aspirations. Sustainable development is a process in which the exploitation of natural resources, the pattern of investments, and the institutional changes are all made keeping in view of the present as well as future needs of the human race on our planet.

In order to minimize the cost of environmental protection while maintaining the cost of natural resource exploitation within acceptable limits, it is essential to make structural economic changes to raise the levels of ecological and economic efficiency. This involves the following steps:

(1) Resorting to eco-friendly technologies which conserve nature resources, generate least pollution, promote sustainable industrial growth and provide direct economic benefits to society.
(2) Pursuing environmentally compatible areas of economy.

Some examples of such structural changes in different facets of human activity are as follows:

(1) Energy Sector

(a) Rational use of energy
(b) Conservation of energy and (c) harnessing renewable sources of energy.

(2) Agriculture Sector

(a) Development of land-use plans that are compatible with the eco system.
(b) Eco-cultivation.
(c) Harnessing modern developments in biotechnology and genetic engineering in agricultural practices.
(d) Use of organic manures, and
(e) Use of biopesticides.

(3) Transport Sector

(a) Providing efficient public transport system.
(b) Replacing diesel and petrol with LPG.
(c) Reduction in the specific energy consumption of automobiles.

(d) Convincing people to use private vehicles selectively for long distances and for emergencies.
(e) Encouraging sharing of vehicles among people going to same of nearby destinations.
(f) Proper maintenance of vehicles for high efficiency and less pollution.

(4) Manufacturing Industries

(a) Utilizing highly efficient and eco-friendly technologies with built in waste treatment systems.
(b) Recycling and Reuse of raw materials.

(5) Construction Sector

(a) Use of environmentally compatible building materials
(b) Efficient use of land, water and energy.
(c) Use of renewable energy sources.
(d) Replacing energy-intensive design with labour intensive designs.

Apart from the above, other measures such as imparting environments education and environmental consciousness among masses and inculcating the values of non-material enrichment and spiritual solace among people, go a long way in our endeavours to achieve sustainable development.

Eco-labelling of Environment Friendly Products

The scheme of eco-labelling of environment friendly products provide accreditation and labelling for consumer products which meet certain environmental criteria along with quality requirements of the Indian standards for that product. This scheme was introduced by the Ministry of Environment and Forests by a resolution dated 20th February, 1991, which was publishes in the Gazette of India (Extraordinary). According to this, an "environment friendly product" is defined as any product which is made, used or disposed off in a way that significantly reduced the harm it would otherwise cause to the environment. Such products which meet the stipulated environmental and quality requirements are labelled with "Eco-mark" of the notified design. The major objectives of the scheme are as follows:

(a) To provide incentive for manufacturers and importers to minimise adverse environmental impact of products.
(b) To reward genuine initiatives by manufacturers to reduce adverse environmental impact of their products.

(c) To encourage people to purchase products having less harmful environmental impacts, thereby contributing to their environmental responsibility.

(d) To contribute towards improvement of quality of the environment and to encourage sustainable resource management.

The Resolution regarding Eco-labelling also provides for the following mechanism for implementation of the scheme:

(i) A Steering Committee of the Ministry of Environment and Forests to determine the category of products to be covered under this scheme and priority wise criteria to be adopted.

(ii) A Technical Committee constituted by the Central Pollution Control Board (CPCB) to identify the specific product to be selected and criteria to be adopted. including "inter se" priority between the criteria.

(iii) The Bureau of Indian Standards (BIS) would assess and certify the products and draw-up a contract with the manufacturers, allowing the use of the label, on payment of a fee.

The scheme also includes launching of countrywide mass awareness campaign to encourage consumers to purchase products having less harmful environmental impacts. The scheme also envisages assistance to consumer organisations for comparative testing of products and dissemination of information to the public. Obviously, the major purpose is to direct the market forces to bring about a change in the manufacturers to produce products that are environmentally more benign.

Sixteen consumer product categories have been identified for this purpose in the first phase and the criteria for awarding Eco-mark to products like toilet soaps and detergents have already been notified.

Judicial Response for Environmental Protection

The judiciary in our country has also been exhibiting exemplary concern and appreciation towards environmental protection and ecological conservation. The Supreme Court of India, in its historical Judgement on 12th March, 1985, in a conflict between environmental conservation and industrial development, observed as follows:

"This is the first case of its kind in the country involving/issues relating to environment and ecological balance and the questions arising for consideration are of grave movement and significance not only to the people residing in the Mussorie Hill range forming part of the Himalayas but also in their implications to the Welfare of the generality of people living in the country. It brings into sharp focus the conflict between development and conservation and serves to emphasize the need for reconciling the two in the larger interest of the country.

The consequence of this order made by us would be that most of lime stone quarries would be thrown out of business. This would undoubtedly cause hardship to them, but it is a price that has to be paid for protecting and safeguarding ecological balance and for protecting their cattle, homes and agricultural land.

We are conscious that as a result of this order made by us, the workmen employed in the lime stone quarries will be thrown out of employment. We would, therefore, direct that immediate steps shall be taken for reclamation of the areas forming part of such lime stone quarries with the help of the already available Eco-task Force of the Department of Environment, Government of India and the workmen who are thrown out of employment in consequence of this order shall, as far as practicable and in the shortest possible time, be provided employment in the afforestation and soil conservation programmes to be taken up in this area".

In another case, the Rural Litigation and Entitlement Kendra, Dehradun brought to the notice of the Supreme Court that in Doon Valley, the mine owners and lessees were carrying on indiscriminate quarrying of limestone. The mining operations led to deforestation resulting in stripping the landscape base of its Verdant cover. The waste produced due to digging was carried down by rain-water to the villages and the agricultural lands located at lower levels. Some of the naturally formed streams were blocked. Blasting of the lime stone in the mines loosened the rocky structures by shaking the oil, thereby disturbing the entire ecology of the area. It was also apprehended that if the mining continued further, it may lead to dearth of water in the entire belt. The Supreme Court treated the letter received from Rural Litigation and Entitlement Kendra, Dehradun as a Writ Petition. After hearing the case, the Supreme Court ordered that all fresh quarrying be stopped. It also directed that some of the quarries be closed down. The District Magistrate and the Superintendent of Police, Dehradun were directed to enforce the order.

In a landmark Judgement issued on July, 1987 (AIR 1987, AP 171), the Andhra Pradesh High Court observed as follows:

"The protection of Environment is not only the duty of the citizen but it is also the obligation of the State and all other State organs including courts. In this context, environment law has succeeded in unshackling man's right to life and personal liberty from the clutches of common law theory of individual ownership. The enjoyment of life and its attainment and fulfilment guaranteed by Article 21 of the Constitution embraces the protection and preservation of nature's gift without which life cannot be enjoyed. There can be no reason why practice of violent extinguishment of life alone should be regarded as violative of Article 21 of the Constitution. The slow poisoning by the polluted atmosphere caused by environmental pollution and spoilation should also be regarded as amounting to violation of Article 21 of the Constitution".

India's activist Judiciary has taken up the task of environmental protection. The Judiciary's concern for protection of the environment and its efforts to ensure prompt action by the Authorities are the remarkable features of Judicial activism in our country. In all such endeavours, the approach of the Judiciary is not that of agitating environmentalists, but that of an objective arbiter. This is how, the Judiciary is keeping a balance between environmental protection on one hand and development on the other.

ENVIRONMENTAL LEGISLATIONS IN INDIA

Our country also demonstrated its concern for pollution control and environmental protection by enacting several legislations and constituting Statutory Bodies dedicated for this cause. A list of Environment related legislations is given in Table 6.1.

Table 6.1. Select list of Environment Related Legislations in India.

A. Central Enactments

1. Air Pollution

(a) The Indian Boilers' Act, 1923
(b) The Factories Act, 1948
(c) The Mines and Minerals (Regulation and Development) Act, 1947
(d) The Industries (Development and Regulation) Act, 1951
(e) The Air (Prevention and Control of Pollution) Act, 1981, amended in 1987.

2. Water Pollution

(a) The River Boards Act, 1956
(b) The Merchant Shipping (Amendment) Act, 1970
(c) The Water (Prevention and Control of Pollution) Act, 1974 amended in 1988
(d) The Water (Prevention and Control of Pollution) Cess Act, 1977 amended in 1991.

3. Pesticides

(a) The Factories Act, 1948
(b) The Insecticides Act, 1968
(c) The Poison Act, 1991

4. Radiation

(a) The Atomic Energy Act, 1962
(b) Radiation Protection Rules, Act, 1972

5. Forests, Fisheries, and Others

(a) The Indian Fisheries Act, 1897
(b) The Indian Forest Act, 1927
(c) The Prevention of Food Adulteration Act, 1954
(d) The Ancient Monuments and Archaeological Sites and Remains Act, 1958
(e) The Wild Life (Protection) Act, 1972, amended in 1983, 1986 and 1991
(f) The Urban Land (Ceiling & Regulation) Act, 1976
(g) The Forest Conservation Act, 1980, amended in 1988.

6. Comprehensive Environment Protection

The Environment (Protection) Act, 1986

B. State Enactments

1. Smoke Control

(a) The Bengal Smoke Nuisance Act, 1905
(b) The Bombay Smoke Nuisance Act, 1912
(c) The Gujarat Smoke Nuisance Act, 1963

2. Water Pollution

(a) Orissa River Pollution Prevention Act, 1953
(b) Maharashtra Prevention of Water Pollution Act, 1969

3. Land Use

(a) The Bihar Waste Land (Reclamation, Cultivation and Improvement) Act, 1946
(b) The Andhra Pradesh Improvement Scheme Act, 1949
(c) The Acquisition of Land for Flood Control and Prevention of Erosion Act, 1955
(d) The Delhi Restriction of uses of Land Act, 1964

4. Pest Control

(a) The Mysore Destructive Insects and Pests Act, 1917
(b) The Andhra Pradesh Agricultural Pests and Disease Act, 1919
(c) The U.P. Agricultural Disease and Pests Act, 1954
(d) The Assam Agricultural Pests and Disease Act, 1954
(e) The Kerala Agricultural Pests and Disease Act, 1958

SALIENT FEATURES OF SOME IMPORTANT LAWS

The Wild Life Protection Act, 1972 (Amended in 1983,1986 and 1991)

This Act was enacted for providing protection to wild animals and birds. The Act also provides for the constitution of a Wild Life Advisory Board, appointment of Chief Wild Life Warden, Wild Life Wardens and other employees by the State Governments for the protection of Wild Life. Regulation of hunting of wild animals and birds, laying down the procedures for declaring areas as sanctuaries, national parks and biosphere reserves, and regulation of trade in wild animals were also provided by the Act, List of endangered species, which is revised from time to time, is also included in the schedule of the Act.

As per the provisions of this Act, no one is permitted to hunt any wild animal, except Vermin, without a licence from the chief Wild Life Warden. *A* record of Wild Life animals hunted or captured has to be maintained. A special permit may be granted to hunt a wild life animal for education, scientific research, scientific management and collection of specimens for Zoological gardens, museums, etc. The Act provides for the establishment of sanctuaries national parks, game reserves, and closed areas.

All wild life animals are the property of the Government. Trade or Commerce in wild animals and animal articles and trophies is strictly regulated. No person can cook or serve meat of wild animals in any eating house without a licence. Penalties for violating the provisions of the Act have also been laid down in the Act.

The Forest Conservation Act, 1980 (Amended in 1988)

As per this Act, no forest land or any portion thereof may be used for any non-forest purposes without the prior permission of the central Government. The Act has been amended in 1988 for incorporating more stringent penal provisions against violators of the Act. The scope of the definition of "non-forest purposes" was extended to include cultivation of tea, coffee, rubber, palms, oil-bearings plants, horticultural crops and medicinal plants. No State Government or other authority may issue order directing that any forest land or any portion thereof may be assigned by way of lease or otherwise to any private person or to any authority, corporation, agency or any other organisation not owned, managed or controlled by Government without prior approval of Central Government.

The Water (Prevention and Control of Pollution) Act, 1974 (Amended in 1988)

This Act provides for the prevention and control of water pollution and for maintaining or resorting of wholesomeness of water. The Act stipulates establishment of the Central and State Boards for this purpose, and also stipulates how these Boards are to be constituted.

The Act defines terms like pollution, sewage effluent, trade effluent stream and boards. The Act also assigns the functions to be carried out by the Central and State Boards.

The Water Boards have power to obtain information, to take samples of effluents from any industry/establishment and to make survey of any area and gauge and keep record of the flow or volume and other characteristics of any stream or well.

A person empowered by the Board has the right to enter, and inspect any place and examine any plant, record register, document or any other material object, or for conducting a search of any place where he has reason to believe that no offence of water pollution is committed. The Board has wide powers to prohibit the use of any stream or well for discharging any pollutant in it. The Board has powers to restructure the outlets for dumping pollutants.

The Act prohibits disposal of any poisonous, noxious or polluting matter or any matter causing obstruction to the proper flow of water in a stream. However, dumping of any material into a stream for the purpose of reclamation of land is not considered an offence.

The Act provides for severe and deterrent punishments for violation of the Act which includes fine and imprisonment.

The Water (Prevention and Control of Pollution) Cess Act, 1977 (Amended in 1991)

This Act empowers the Central Water Board to collect cess on water consumed by persons carrying on certain scheduled industries and by local Authorities responsible for supplying water. The cess and the consent fees form the major sources of revenue to run the Central and State Water Boards. The Act has been amended in 1991 with a view to augment the resources of the Boards by removing the lacunae in the Act and to provide rebate to the Industries for complying with the consumption and effluent quality standard.

Air (Prevention and Control of Pollution) Act, 1981 (Amended in 1987)

This Act was passed under Article 253 of the Constitution of India and in pursuance of decisions of Stockholm Conference. The objective of this Act is to provide for the prevention, control and abatement of air pollution in order to preserve the quality of air.

The Act defines relevant terms such as air pollution, air pollutant, auto mobile, industrial plant, etc. Air pollution is defined as the "presence of any liquid or gaseous substances in the atmosphere in such concentration as may be or tend to be injurious to human beings or other living creatures or plants or property or environment".

All sources of pollution such as automobiles, diesel vehicles, industries, transport, railways and domestic fuels.

The Central and State Water Boards have been entrusted with the task of controlling and preventing air pollution and accordingly they have been redesignated as Central Pollution Control Board and State Pollution Centre Board respectively. The functions of these Boards were clearly defined.

The Act provides the declaration of certain heavily polluted areas a "Air Pollution Control Area" and no industrial plant shall be operated in these areas without prior consent of the State Pollution Control Board.

The State Boards have to pay down and enforce standards for prevention and control of air pollution. The State Government in consultation with the respective Board may give instructions to the concerned Authority in charge of Registrations under the Motor Vehicles Act 1939, to ensure emission standards from automobiles. Failure to comply with the conditions prescribe for this purpose is punishable with fine and imprisonment.

The State Boards have powers to sue a polluter in a court of law to prevent him from polluting the air, and the expenses incurred by the Board for doing so will be recovered from the polluter. Further, the Boards have powers to authorize any person to enter and inspect the premises of the polluter and to collect samples of emissions from Chimneys, flues, ducts or an other outlets for analysis of the pollutants.

The Act has been amended comprehensively in 1987 to render it more effective and to include "noise" also under the definition of air pollutants.

The Environment (Protection) Act, 1986

The Environment (Protection) Act, 1986 was enacted as per the spirit of the Stockholm Conference held in June, 1972 to take appropriate steps for the protection and improvement to prevent hazards to human beings, living creatures and property. This is a landmark legislation to provide a single focus in the country for the protection of environment and to plug the loopholes in the earlier laws.

This Act ensures enforcement of several Acts/ Regulations concerning pollution control and environmental protection/safety.

This Act confers powers to the Central Government to take all such measures as it deems necessary or expedient for the purpose of protecting and improving the quality of the environment and preventing, controlling and abating environmental pollution. It empowers the Central Government to issue directions (a) for the closure prohibition or regulation of any industry, operation and process and (b) for the stoppage or regulation of the supply of water, power or any other service even without obtaining court orders.

The Act also empowers the Central Government to make rules for the first time or the (i) Standard of quality of air, water and soil for various areas and for various purposes (ii) Maximum permissible limits of concentration for various environmental pollutants (including noise) for different areas (iii) procedures and safeguards for handling of hazardous substances (iv) prohibition and restrictions on the location of industries and carrying out processes and operations in different areas (v) Procedures and safeguards for prevention of accidents which may cause environmental pollution and (vi) providing for remedial measures in case of accidents.

It is mandatory for persons carrying on any industry, operation etc. not to allow emission or discharge of environmental pollutants over and above the limits stipulated by the relevant standards. The Act provides for stringent penalties for defaulters. Any person can make a complaint of violation of provisions of the Act to the Central Government or Authority or Officer authorized for this purpose.

ENVIRONMENTAL MANAGEMENT SYSTEM (EMS) STANDARD—ISO 14000 SERIES

The preparatory work on the first international standards on environmental management systems was undertaken by the International Organisations for Standardisation (ISO). These norms come under the ISO 14000 series. The major objectives of these series is "to promote more effective and efficient environmental management in organisations and to provide useful and usable tools – ones that are cost-effective, system-based, flexible and reflect the best organisations and the best organisational practices available for gathering, interpreting and communicating environmentally relevant information". The intended end-result is the improvement of environmental performance.

ISO 14000 series of standards represents an opportunity for industrial organisations and enterprises in developing countries for technology transfer.

Further, it offers a source of guidance for introducing and adopting environmental management systems based on the best universal practices, in the same way that the ISO 9000 series onquality management systems, which is now widely applied, represents a tool for technology transfer of the best available quality management practices.

The basic standard in this series is ISO 14001 Environmental Management Systems – specification with guidelines for use. It

The subjects covered under the various ISO numbers are given below:

ISO Number Range	*Subject*
ISO 14000 - 1409	– Environmental Management Systems
ISO 14010 - 14015	– Environmental Auditing
ISO 14020 - 14029	– Environmental Labelling
ISO 14030 - 14039	– Environmental Performance Evaluation
ISO 14040 - 14049	– Life Cycle Assessment
ISO 14050 - 14059	– Terms and Definitions
ISO 14060	– Environmental Aspects in Product Standards

contains the requirements of a sound environmental management system, in the same way that ISO 9001 contains the requirements of a sound quality management system.

The ISO 14001 standard is applicable to any organisation (*e.g.*, a public or private enterprise, company, firm, institution or operational unit within an organisation) aiming at

(a) Implementing, maintaining and improving an environmental management system.
(b) Ensuring its conformance with its stated environmental policy.
(c) Demonstrating such conformance to others, either through an independent, third party certification or registrations of its

environmental management system or a self-declaration or conformance with the standard.

ISO 9000 series on quality management system is focused mainly on ensuring customer satisfaction and specifies 20 requirements in ISO 9001 to achieve this objective. Similarly, ISO 14000 series on environmental management systems addresses the needs of a wide cross-section of people in the society to protect the environment and for this purpose the following key requirements were specified under ISO 14000:

(1) Environment Policy
(2) Planning
  (2.1) Environmental aspects
  (2.2) Legal and other requirements
  (2.3) Objectives and targets
  (2.4) Environmental management programmes
(3) Implementation and Operation
  (3.1) Structure and Responsibility
  (3.2) Training, Awareness and Competence
  (3.3) Communication
  (3.4) Environmental Management System Documentation
  (3.5) Document Control
  (3.6) Operational Control
  (3.7) Emergency preparedness and response
(4) Checking and corrective action
  (4.1) Monitoring and Measurement
  (4.2) Non-conformance and corrective and preventive action
  (4.3) Records
  (4.4) Environment Management System Audit
(5) Management Review

ISO 14001 explains each of the above requirements giving details of what should be done but not on how it should be done. For instance, under 2.3 (objectives and targets), it states as follows:

"The organisation shall establish and maintain documented environmental objectives and targets, at each relevant function and level within the organisation.

The objectives and targets shall be consistent with the environmental policy, including the commitment to prevention of pollution.

When establishing and reviewing its objectives, an organisation shall consider the legal and other requirements, its significant environmental aspects, its technological options and its financial, operational and business requirements, and the views of interested parties".

Adoption of ISO 14000 series of Environmental Management Standards by any organisation/company is voluntary. However, exporters from developing countries are already subjected to pressure from buyers in industrialized countries who are insisting on ISO 14000 certification for the products being exported. However, ISO 14000 should not be considered only as an export requirement but as a practical demonstration of environmental concern of an organisation. Thus, the overall aim of the ISO 14000 environmental management standards is to support environmental protection in balance with socio-economic needs.

III. ACID RAIN

Rain has traditionally been regarded as the harbinger of growth and productivity and prosperity. Mankind has always acknowledged "the useful trouble of the rain". The 110,000 cubic kilometres of rain that fall each year have kept much of the earth's life support system alive. Rain is the most important source of fresh water, available to all, free of cost. This source of joy, provide a new lease of life by greening the landscape (Varshney, 1983).

However, the rain has taken on a new and threatening complexity. Describing the alarming phenomenon in its State of Environment Report, the United Nations Environment Programme (UNEP) notes the way in which the rain reacts with the sulphur and nitrogen oxides that pollute the air to produce "acid-rain". Now-a-days one hears much about the so-called "acid-rain". A few years, ago, it became a jargon term among scientists to describe certain changes in atmospheric chemistry and caught the fancy of the layman and the press alike.

The term acid-rain is not new. It was first coined in Menchester, England over a century ago. What is new is the realization that it is perhaps the most pernicious Global problem. Subtle bulethal, acid rain is rising up as the most crucial ecological issue, particularly in European countries. It is a modern, post industrial form of ruination. Although acid rain has become in recent years a much used and easily understandable catchword, it does not describe adequately some of the most important properties of this phenomenal.

In the first place, the phrase "acid-rain" is not sufficiently precise. The neutrality point with pH=7.00 is not very useful as a reference point for acid rain. The reason is that neutral rain is usually on acid side (pH values than 7.00). It is well known that rain water in equilibrium with the atmospheric carbon dioxide will have a pH of 5.60 and this value has often been used to define acid rain. This, however, does not result in a reliable definition of acid rain, since the atmosphere everywhere contains chemical compounds of natural origin (e.g., seasalts, dust particles, volcanic ash and gases etc.) which appreciably influence the pH of precipitation.

Secondly, acid rain is not the only important component responsible for the acidification of soil and water. In the northern colder climates, for example, much of the annual precipitation occurs as snow and not as rainfall. Liquid of frozen water in precipitation does carry acidifying chemical compounds, but one must also take into account the process of "dry deposition". In fact, dry deposition are the main sources for acid formation and dominates over wet deposition close to the sources of emissions.

In the third place, acidity (i.e., hydrogen ion concentration) in the precipitation mayor may not be the parameter that really matters. It all depends upon the effects being considered. For some direct effects e.g., foliar damage to vegetation, corrosion or acid inputs directly on water surfaces. But when one looks at short term variations in the runoff acidity, the more relevant parameter may be the content of mobile anions, rather than hydrogen ion concentration, in the atmospheric deposition.

Moreover, in addition to its acidity; sulphur and nitrogen contents, acid deposition often contains higher concentration of heavy metals, other trace metals and organic pollutants. As an example, Table 6.2 shows the data on the annual volume - weighted mean concentrations of major ions in the precipitation in southern Sweden, Birkenes (Norway) and New Hampshire (USA).

Table 6.2 : Annual volume—weighted mean concentrations (ueq/l) of major ions in precipitation (Seip and Tollan, 1985)

Ion	*Southern Sweden*	*Birkenes (Norway)*	*New Hampshire (USA)*
H^{+}	52	69	70
SO^{2-}	70	71	57
NO^{-4}	31	41	24
NH^{+3}	31	42	11
Na+4	15	53	5
Ca^{2+}	14	9	7
Mg^{2+}	8	12	3
K^{+}	3	3	2
Cl^{-}	18	63	12

Acid rain, as the term implies, is the deposition on earth of the dilute solutions of acids (namely sulphuric and nitric) with rainfall. Sulphur dioxide mainly comes from coal and oil power generating industries, industrial boilers and smelters. Oxides of nitrogen come from automobile exhaust, high temperature combustion engines and chemical fertilizer factories. Sulphur dioxide undergoes a variety of photochemical and catalytic reactions in the atmosphere. During day time and under conditions of low humidity in the presence of hydrocarbons and nitrogen oxides, the predominant reaction is oxidation of sulphur to sulphuric acid, Soot particles are strongly involved in catalyzing the oxidation of sulphur dioxide. Metallic ions such as Mn(II), Fe(II), Ni(II) and Cu(II) are also known to catalyze the oxidation reaction. At night, under foggy conditions, rainy or high humidity conditions, oxidation of sulphur dioxide to sulphate ion is accelerated by ammonia ions (Robinson and Robbins, 1970), Subsequently, these compounds are deposited on the earth as aerosols and particulates (dry deposition) or they are carried to the earth by the rainfall, snow or dew (wet deposition),

Its destructiveness is most obviously felt in lakes and rivers, killing fishes and destroying phosphate resources on which phytoplankton and other aquatic plants depend for nutrients. Soils are likewise affected as acidification due to acid deposition and certain biological processes within the soil erodes fertility. It is suspected to spirit away mineral nutrients from the poor soil upon which forests depend. It has been seen to damage various kinds of vegetation including crop-lands and forests. It assaults historic buildings, bridges, dams, industrial equipments, water supply network, power and telecommunication cables and cultural treasures by accelerating erosion. It is known to contaminate public drinking water (Melkania and Melkania, 1987). In addition it is known to pose a substantial threat to health by causing skin burns and making holes in cotton clothing.

This account deals with the current understanding of the phenomenon acid rain, its adverse effects on the ecosystem and strategies for dealing with the problem.

DEFINITION OF ACID RAIN

Acid rain is an umbrella term, which is used to cover a number of different pollution processes. It is used to describe all precipitation -rain, snow, fog, dew and dry acidic deposition - which are more acidic than normal. It refers to any precipitation which has a pH valueless than 5.6, the pH of uncontaminated rain water. The pH of acid rain can be compared with that of common acid and alkaline substances (Fig. 6.2). Because the pH scale is logarithmic, there is tenfold difference between numbers. Thus, water at pH 4 is ten times more acidic than at pH 5, and 100 times more acidic than at pH 6. Acid precipitation generally ranks between about 5.6 and 3.5, and in some cases even lower. Because acid rain includes other forms of precipitation as snow, sleet, hail, dew, frost and dry fall, it may be appropriate to refer the phenomenon as acid deposition and acid precipitation or "atmospheric deposition" collectively. The real acid deposition, often turn up days later and hundreds of kilometres from the source of emission.

EMISSION OF SULPHUR DIOXIDE AND NITROGEN OXIDES

Sulphur dioxide is released when fossil fuels are burned, or sulphur ores are smelted to extract metal. The sulphur content of oil and coal varies, depending largely on where it comes from. Crude oil can have anything from 0.1-3.0 per cent sulphur. Coal can have sulphur content from 0.45-5 per cent; sulphur content of coal

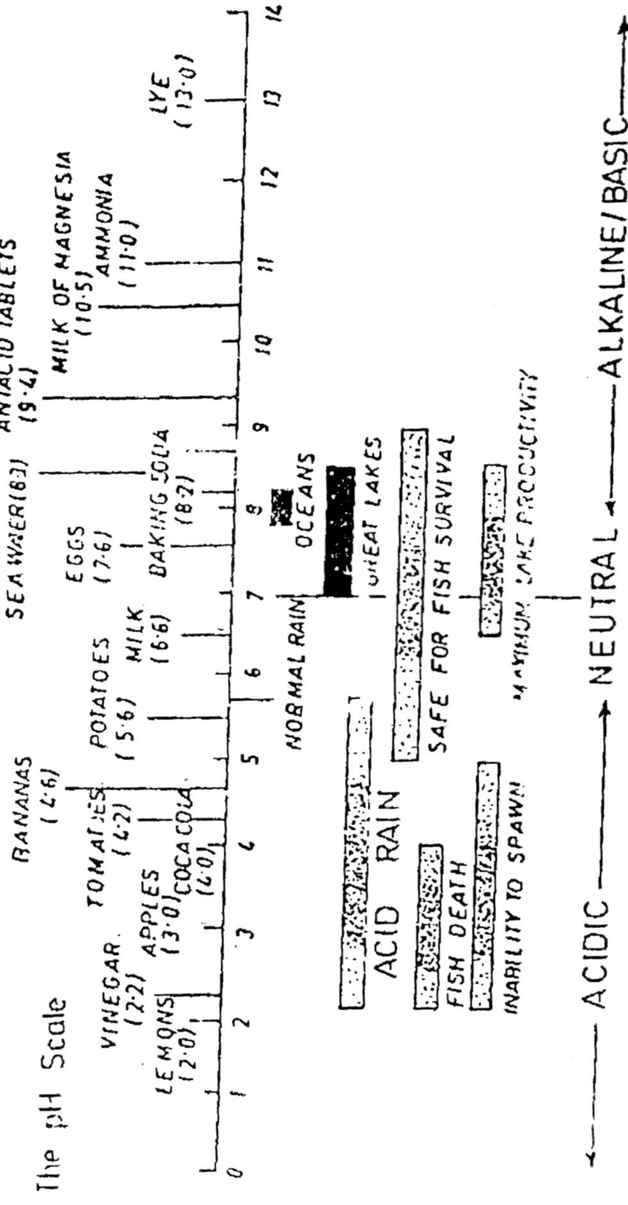

Fig. 6.2 : pH value of common acid and alkaline substances.

varies from one grade to another, and even from one seam to another in the same mine, reflecting local variations in ecological conditions millions of years ago. Hard coals (like anthracite) are normally low in sulphur, while soft coals (like bituminous, sub-bituminous and lignite/brown) are high sulphur coals.

More than half of the global acid-rain, snow and fog may be due to natural sources such as sulphur dioxide emission from volcanoes, and nitrogen oxide emission from chemical and bacterial nitrification in soil. However, about 120 million tonnes of sulphur dioxide are emitted every year from coal or oil-fired power stations, industries that use fossil fuels, and metal smelting. The geographical concentration of non-natural sulphur dioxide varies from one region to another (Table 6.3). In Europe, about 90 percent of the atmospheric sulphur dioxide is man-made (Overrein, 1983). In the United States total sulphur dioxide emission measured between 1960 and 1970, pollution controlled to decrease in urban sulphur dioxide concentration (Altschuller, 1980). In the United Kingdom and the United States, coal fired power stations account for the two-third of man-made sulphur dioxide emission. In West Germany, they account for half. In Canada, smelters account for 63 per cent of emissions. The largest single source of man-made sulphur, dioxide in the world is the Copper and Nickel smelting complex at Sudbury, Ontario, which annually produced 632000 tonnes of sulphur dioxide in the early 1980's.

Table 6.3 : Annual emissions of sulphur dioxide (in thousand tonnes) and deposition (%).

	1980	*1986*	*Sulphur deposition*		
			Foreign	*Domestic*	*Undecided*
1	*2*	*3*	*4*	*5*	*6*
USSR	25,000	24,000	32	53	15
USA	23,200	20,800	-	-	-
China	-	18,000	-	90	10
Poland	4,100	4,300	52	42	6
East Germany	4,000	4,100	32	65	3
Canada	4,650	3,72	50	50	-
UK	4,670	3,540	12	79	9

1	2	3	4	5	6
Spain	3,250	3,250	18	63	19
Italy	3,800	3,150	22	70	8
India	32,00	-	-	100	-
Czechoslovakia	3,100	3,050	56	37	7
West Germany	3,200	2,400	45	48	7
France	3,558	1,845	32	54	14
Yugoslavia	1,175	1,800	41	51	8
Hungary	1,633	1.420	52	42	4
Bulgaria	1,034	1,140	-	-	-
South Africa	1,000	-	-	100	-
Greece	800	720	-	-	-
Belgium	799	467	-	-	-
Finland	584	370	55	26	19
Denmark	438	326	-	-	-
Netherland	487	315	-	-	
Portugal	266	305	-	-	-
Turkey		276	-	-	-
Sweden	483	272	58	18	24
Austria	354	170	-	-	-
Ireland	219	138	-	-	-
Romania	200 -	-	-	-	
Norway	141	100	63	8	29
Switzerland	126	63	-	-	-
Luxembourg	23	13	-	-	-
World Total Approx 120 million tonnes in 1980.					

Sources:

1. EMEP and UNECE.
2. From United Nations, National Strategies and Policies for Air Pollution Abatement (New York: United Nations, 1987).
3. Ross, Lester, Environmental Policy in China (Bloomington: Indiana University Press 1988).
4. Centre for Science and Environment, India, 1985.

Burning fossil fuel is also the main source of man-made nitrogen oxides, the second key element in acid-rain. Combustion oxidizes both the nitrogen present in the fuel and some of the nitrogen which is naturally present in the air. The quantity and composition of the nitrogen oxides depends on the method and temperature of combustion. Burning will normally oxidize 5-40 per cent of the nitrogen in coal, 40-50 per cent of that in heavy oil, and nearly 100 per cent of that in the light, and gas. The higher the temperature, the more nitrogen oxides are formed. The major source of nitrogen oxides in industrialize, countries are the power stations which burn fossil fuels. Automobiles also makes a big countries in nitrogen oxide emissions.

About half the nitrogen oxide emission in the earth's atmosphere comes from anthropogenic sources. In Europe and North America, man-made nitrogen oxide outweigh's natural nitrogen oxide by 5-10 times (Galloway and Dillon, 1982). The proportion of nitrogen oxide in the atmosphere in Europe have grown by 40-50 per cent during the 1970's, but rates vary (Grennefelt, 1982). In some countries, such as West Germany; the United Kingdom and France, nitrogen oxide emissions are falling from 1980 to 1986 (Table 6.4).

TRANSNATIONAL MENACE OF ACID RAIN

Twenty or thirty years ago, most sources of air pollution were close to the ground and only transported to short distances. The result was heavy pollution in urban and industrial pockets, while rural areas generally were not affected. A cheap and quick solution to the local air pollution problems employed was to raise the height of the chimneys and stacks to diffuse the pollutants. The "spread it farther and wider" method improved the air quality locally, but transported this ecological problem into a regional or perhaps a global one. Now pollution is carried to tremendous distances in the atmosphere. Moreover, during the longer journey, significant chemical and physical changes occur to produce more hazardous products. Acid rain is a result of the combination of long-distance atmospheric transport and transformation.

Like the wind that disperses it, acid rain has the potential of being widely distributed, Sulphur and nitrogen compounds omitted by burning fossil fuels can be blown far from their source to a thousands of kilometers by wind, causing acid rain in countries far from their point of origin. The strange aspect of the acid rain is that the victimiser is one locality but the victim is another. The victims of acid rain are perhaps Sweden and Canada. While Canada gets

Table 6.4: Annual nitrogen oxides emissions (in thousand tonnes).

	1980	*1986*
United States	20300	19400
West Germany	3100	2900
USSR	2790	2930
UK	1916	1690
France	1867	1693
Canada	1725	1785
Italy	1550	1537
Czechoslovakia	1204	1100
Spain	-	1122
Poland	-	840
Netherlands	535	522
Belgium	442	385
Sweden	328	305
Hungary	-	300
Finland	280	250
Denmark	251	238
Austria	216	216
Norway	-	215
Switzerland	196	187
Portugal	166	192
Bulgaria (est)	-	150
Greece	127	150
Ireland	67	68
Luxembourg	23	22

Source: United Nations, National Strategies and Policies for Air Pollution Abatement (New York; United Nations, 1987).

acid rain from a Petrochemical industry located in Northern America, heavy winds take acid rain from factories located in Bretain and France to Sweden. Equally grim is the acid rain in Norway, Denmark and West Germany. About 90 per cent of the acid rain of Norway and 75 per cent of the Sweden, occur due to the drifted acid rain pollutants. Although the exchange of pollutants is a two-way process, and each country should receive an amount of pollutant that it sends out, but actually, some countries receive pollutants more than they send out (Table 6.5). For example, Canada receives more than twice a much sulphur than it sends out. Acid rain is therefore, becoming a major political issue between nations. It is an international nightmare; in fact a more effective form of undeclared chemical warfare that has not yet been discovered.

The tropical region contains a large population and where energy usage, industrial capacity; and agricultural practices are is

Table 6.5: Emission and deposition of sulphur in Europe (x 1000 metric tonnes). -

Country	*Emission*	*Deposition*
Albania	50	67
Austria	215	,341
Belgium	404	161
Bulgaria	500	346
Czechoslovakia	1500	1301
Denmark	228	109
Finland	270	293
France	1800	1272
German Democratic Republic	2000	778
Germany (F.G.R.)	1815	1158
Greece	352	253
Hungary	750	467
Iceland	6	74
Ireland	87	65
Italy	2200	1132
Luxembourg	24	11
Netherlands	240	173
Norway	75	255
Poland	2150	1330
Portugal	84	73
Romania	100	797
Spain	1000	583
Sweden	275	472
Switzerland	58	171
United Kingdom	2560	847
Yugoslavia	1475	1093

currently most rapid. It is now very clear that activities such as generation of electricity from fossil fuels, use of liquid fuels for transportation, and smelting/refinement of metals are major sources of the atmospheric oxides of nitrogen and sulphur, which are involved in atmospheric acidification. Acid rain has been well documented in southwestern China (Table 6.6).

Table 6.6: Precipitation pH data (monthly mean) from four locations in Guizhou Province, China in 1984 (Zhao and Xiong, 1988).

Month	*Guiyang (urban)*	*Luizhang (rural)*	*Keyang (rural)*	*Shisun (rural)*
Jan.	3.9	4.3	4.3	5.9
Feb.	4.0	4.4	4.1	4.3
March	3.8	4.4	4.2	3.9
April	4.1	4.2	4.1	4.7
May	4.0	4.5	4.6	4.3
Jun	4.5	4.9	5.4	5.1
July	4.5	4.9	5.2	4.7
Aug.	4.1	4.6	4.5	4.5
Sep.	3.7	4.8	4.5	4.4
Oct.	3.8	4.3	4.6	4.5
Nov.	3.7	5.4	4.5	7
Dec.	3.4	5.4	4.3	4.7

In general tropics are characterised by agricultural activities and practices such as rice farming and biomass burning, which are the sources of CH_4, NMHC, CO, NO, and N_2O to the atmosphere. In China, large quantities of fuel are also burned in numerous widely disturbed, low level sources giving consequently large area source strengths for sulphur dioxide, which cause acid precipitation.

In the desert areas of central Asia, dust particles become airborne in the dust storms during the summer months. This dust is transported by prevailing westerlies even upto north-west Pacific region. These dust particles become major contributors to the alkaline rainfall that has been observed in parts of India (Jodhpur) and perhaps Thailand (Khemani et al, 1989 a, b; Verma, 1989).

There have been many rainwater studies from the Indian subcontinent, consistently showing rainfall to have relatively alkaline

Table 6.7. Monthly volume-wighted men pH data from 10 stations in India and one in Thailand (Source: World Meteorological Organisation).

BAPMoN site	*mean pH*	*Length of record .*
Allahabad	6.82	10/06/77-27/12/85
jodhpur	7.25	01/05/74-08/10/85
Kodaikanal	602	04/02/77-01/12/85
Minicoy	6.50	01/09/77-03/12/85
Mohanbari	6.11	01/10/74-19/12/85"
Nagpur	6.12	05/06/77 -16/12/85
Pune (weekly)	6.60	12/02/84-10/11/85
Port Blair	6.12	27/02/75 -01/11/85
Sri nagar	7.06	05/01/77-08/12/85
Visakhapatnant	6.42	03/05/77.19/11/85
Kosichang	6.54	15/11/83-31/07/85;-

pH value of 6 or above (Table 6.7), while at the other extreme Australia yield acidic rain, with pH value at the start of the wet season near or below 4 (Galloway et al, 1982). However, in these cases sulphate and nitrate concentrations are each only a few micro-equivalent per litre, with simple organic acids contributing most of the acidity. The sources of these organic acids remain uncertain (Sanhueza et al, 1989).

There have also been a number of studies in Southeast Asia looking at the cycling of nitrogen through fertilised and unfertilised rice paddies. These have demonstrated clearly the emission of reactive gases from these systems, including large losses of ammonia in Chinese experiments where ammonium bicarbonate is used as fertilizer (Ayers, 1990).

CHEMISTRY OF ACID RAIN

Most rainfall is slightly acidic because carbon dioxide in the atmosphere reacts with rain water to form mild carbonic acid. As carbon dioxide is present in the clean (unpolluted) air in traces (10,000 cubic metres of air contains only about 3 cubic metres of carbon dioxide) the acidity of natural rain is indeed extremely mild. For all practical purposes, it may be regarded as neutral.

Natural processes such as volcanic eruptions, forest fires and bacterial decomposition of organic matter can produce acidic sulphur and nitrogen compounds that produce acid rain. Annual estimates vary between 78–284 million tonnes of sulphur oxides and between 20–90 million tonnes of nitrogen oxides from natural sources. Man-made sources -electrical generating plants, industrial boilers and smelting plants–on the other hand release 100 million tonnes of sulphur dioxide as well as acidic soot, with the burning of coal accounting for 60 per cent of the emissions, the burning of petroleum products 30 per cent, and various industrial processes 10 per cent. As for nitrogen oxides, fossil fuel combustion yields about 20 million tonnes 01 nitrogen a year.

Precisely, how acid rain is formed in the atmosphere is still a mystery. When these pollutants are vented into the atmosphere by tall smoke stakes, molecules of sulphur dioxide and nitrogen oxides are caught up in the prevailing winds where they interact, in the presence of sunlight, with vapours to form sulphuric acid and nitric acid mist. Possible chemical transformation reactions of the atmospheric sulphur dioxide, nitrogen oxides and ozone are presented in Fig. 6.3.

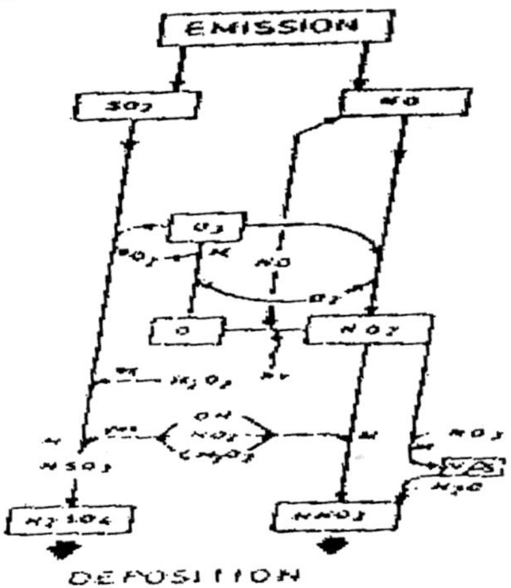

Fig. 6.3 : Possible chemical transformation reactions of in acid rain formation.

The sulphuric acid and nitric acid thus formed remain in vapour state under the prevalent high temperature conditions and begins to condense slowly as the temperature falls. The resulting condensate takes the form of aerosol droplets which owing to the presence of unburnt carbon particles, will be black, acidic and carbonaceous in nature. This matter is called "acid - smut" (Agarwal, 1980).

It is believed that oxidation of sulphur dioxide is much slower than that of nitrogen oxides, so that sulphur dioxide may remain airborne for three to four days, compared with nitrogen oxides which may persist for only half a day. The significant consequences of this phenomenon is that acid rain derived from sulphur dioxide is transported much further than nitrogen oxides.

Acid rain is a global environmental problem. It has been studied in details in the following regions of the world–Southern Norway, Southern Sweden, Southern Scotland, Ontario (Canada), Nova Scotia (Canada), Western Adirondaks (USA) and Maine (USA). In some areas acidification seems to have occurred more recently, such regions are–Thuringer Wald (COR), Bohmer Wald (Czechoslovakia), Erzebirge (COR), Jutland (Denmark), Belgium, Netherlands and Finland. Such information is not available for other parts of the world due to lack of the long-term records (Dickson, 1981).

In European countries the cause of acid precipitation is largely sulphur oxides followed by nitrogen oxides. Sequeira (1987) recorded trace to slight weighted average free-acid levels (pH 5.8-4.7) south of about 47°N, except in Northern Italy. Moderate or rather high free acid contents (pH 4.5-4.2) was recorded north of this latitude.

The cause of acid precipitation is not as apparent in the United States as in Europe. The discrepancy between sulphate concentration and acidity of precipitation may be due to changes in emission chemistry; before 1950 much of the sulphate may have been in unionised particles or neutralised by bases from smoke particles (Likens and Bormann, 1974). This statement is consistent with the shift from coal to other fuels that began in the early 1900's (Singer, 1970), and with the increased use of particle precipitators in smoke stacks, since about 1950 {Patrick et al, 1981). The increased acidity of precipitation since 1950 may be as a result of increased contribution from nitric acid (Haines, 1981).

HISTORICAL EVIDENCE

The first incidence of acid rain seems to have coincided with the onset of the industrial revolution in the mid-19th century. British Chemist Robert Angus Smith first coined the term "acid rain" in 1872, in a 600 page treatise entitled *"Air and Rain: The beginning of Chemical; Climatology"* which was published in 1872, examining links between the sooty skies over Manchester, and the acidity he discovered in local precipitation. His pioneering work was forgotten until the post-war surge of industrialization which brought about an increasing use of fossil fuels and greater public awareness of the dangers of secondary air pollution. Later Gorham (1955) described this phenomenon in England and as a regional phenomenon in Scandinavia in the late 1960's. While studying the Hubbard Berok Environmental Forests in New Hampshire in 1963. Gene Likens of the Cornell University noticed remarkably acidic rainfall from remote areas. But it was not until 1967 that Swedish soil scientists Svante Oden observed acidic rainfall in certain area's anti etched acidic rain into the consciousness of the scientific community by dramatically labelling it as man's "Chemical War" on nature.

Increasing acidity of precipitation in Europe, as documented by the International Meteorological Institute, Stockholm was reported in 1967 (Oden, 1967). The weighted mean annual pH of precipitation in Southern Norway, for example, declined from 5.0 to 5.5 in theatre 1950's to 4.2 to 4.4 in the mid 1970's. The air quality in Europe about 30 years ago had deteriorated over heavily industrialized centres to the point where women's stockings were known to disintegrate during rain spells because the level of acidity was high enough to attack man-made fibres.

Sweden system of atmospheric monitoring stations set up in 1952, was further extended to most of the western Europe including the United Kingdom. These stations began accumulating evidence of increasing sulphate and inorganic nitrogen compounds in rain water and lowering pH values of rain. Oden showed that a central area in Europe with precipitation more acidic, than surrounding areas, had been expanding from year to year. He clearly demonstrated that the long range transport of sulphur dioxide and nitrogen oxides from United Kingdom and Central and Eastern Europe to Scandinavia.

Based on Oden's report, the organisation for Economic Cooperation and Development (OECD) launched the Long Range Transport of Air Pollutants (LRTAP) programme. Later on the United Nations Economic Commission for Europe (ECE) in Geneva was set up to include all the OECD member and non-member countries for LRTAP. ECE also set up an European Monitoring and Evaluation Programme (EMEP) for short sulphur deposition data. The Trans-European Monitoring Programme developed the concept of sulphur budgets. All countries emit sulphur dioxide from power stations, industries and domestic sources but some export more than they receive because of the prevailing wind patterns.

EMEP data in 1983 for United Kingdom had an average monthly sulphur deposition of 84700 MT, only 20 percent of which came from outside sources. In Norway', the direction of prevailing winds resulted in 92 percent of its 255000 MT average monthly sulphur deposition that year coming from mainly United Kingdom and Central Europe. Data for these countries are presented in Fig. 6.4.

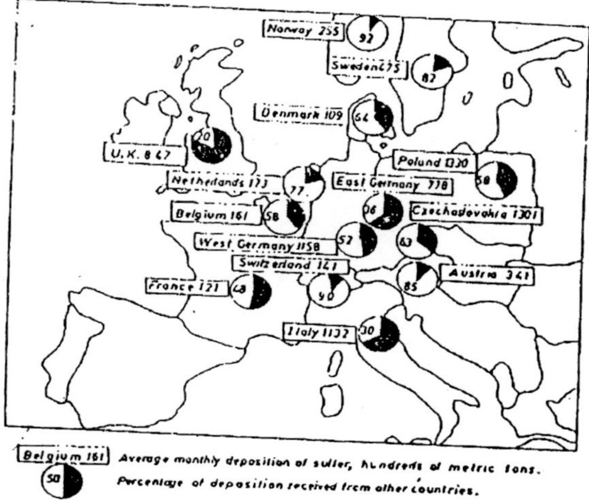

Fig. 6.4 : Most European countries import more sulphur pollutants than they produce.

In the United States, about two-third of the land area receives acid precipitation. The pH of the average annual rainfall in most states along the Mississippi river is pleasantly less than 4.5. In Adirondack state park in New York state, the pH of the average annual rainfall is less than 4.3.

CONSEQUENCES OF ACID RAIN

There have been widespread debates on the extent to which acid deposition helps or harms the terrestrial and aquatic systems. Acid rain contains nutrients, such as nitrogen, potash and trace elements, often in a form readily available or plant use. The sulphur contained in the deposition is beneficial to those soils which are sulphur deficient. But too much of nutrients, however, are as bad as not enough. Elevated nitrogen content may predispose trees to winter damage. Nitrogen supplied to high elevational trees through cloud water may result in increased susceptibility to damage from early frost or desiccation.

The effects of acid precipitation are determined by physical factors, such as stream flow, evaporation, stream size, soil and bed-rock type and depth.

The damage to nature caused by acid rain is "already unacceptably high", but the future effect is "potential enormous". It has been shown "beyond any reasonable doubt that forests and fresh waters are seriously and detrimentally affected". As industrialization proceeds to developing countries acid rain is spreading to tropical forests. This means increasing awareness of the threats to the stability of the environment and mobilising effective moral and financial support worldwide for the safeguard of the living world.

EFFECTS ON THE AQUATIC ENVIRONMENTS

In aquatic ecosystems, as the water becomes increasingly acidic with successive rain falls the lake or river becomes oddly clear and bluish. The subtle process of acidification often remains undetected until damage has occurred. All atrophic levels contain organisms that are sensitive to acidification. As a result, the number of organisms may not be declined because acid tolerant species are able to use available resources. The limiting factor for production may be a primary nutrient, such as phosphorus, rather than toxic effects of pH or metal leached by disintegration and release from soil. Acidification of lakes and streams to pH 5 or less appears to reduce decomposition because of the elimination of some species of bacteria, which may be replaced by other species of bacteria or fungi (Gorham, 1976).

Primary producers

Acidification affects communities of planktonic and benthic algae in lakes and streams. The number of species of phytoplankton in acidic lakes decreases as pH declines (Aimer et al, 1974, 1978). Diatoms collected from a series of acidophillic species increased between 1949 and the mid-1970's (Leivested, 1976). In Dutch moorland pools, diatom species diversity declined from 1970 to 1978 in acidified pools.

Aquatic macrophytes also show changes related to acidification. In Swedish lakes, macrophytic communities once dominated by *Lobella* species were later dominated by *Sphagnum* species (Grahm et al, 1974; Grahm, 1978) or *Juncua bulbosus* (Nilssen, 1980). A similar dominance of *Sphagnum* was evident in an acidic lake in New York (Hendrey and Vertucci, 1980).

Invertebrates

In Ontario and Sweden, zooplankton biomass was lower in acidic than in similar non acidic lakes (Roff and Kwiatkowski, 1977). The number of zooplankton species present in a lake decreased as pH decreased in Sweden (Aimer et al, 1974, 1978), Norway (Hendrey and Wright, 1976; Raddum et al, 1900) and Ontario (Sprules, 1975 a, 1975 b).

Mollusks are highly sensitive to acidification, as would be expected due to the high calcium carbonate requirements of this group for shell formation. Snails were found in Norwegian lakes with pH at or below 5.2, and were rare or reduced at pH 5.2 and 6.6 (Okland, 1969; Raddum, 1980). The snail *Ancylus* species was not found below pH 5.7 in the river Duddon, England (Sutcliffe and Carrick, 1973). Mollusks are not found in Ontario lakes with pH at or below 5 (Scheider et al, 1975; Roff and Kwiatkowski, 1977).

The crustaceans *Gammarus lacustris* and *Lepidurus arcticus* are widespread in Norway and are important fish food organisms where they occur. They are not found in lakes with pH less than 6.0 (Leivested et al., 1976 Okland, 1980). In Sweden the Crayfish I *(Astacus astacus)* is common in lakes with pH less than 6.0. Some groups of aquatic insects are reduced at low pH while others flourish. Many species of Ephemeruptera and Plecoptera disappear as pH declines. Raddum et al (1980) found that Coleoptera, Hemiptera and Megaloptera were more abundant in lakes with pH below 4.8.

Amphibians

Glass and Loucks (1980) observed the decline of the frog *(Rana temporaria)* and the toad *(Bufo bufo)* from a Swedish lake where the pH had declined to 4.0 -4.5 and from which all fish had disappeared. Exposure to acid decreased sodium influx in isolated frog skin, and thereby reduced active sodium transport. However, there was no significant change in osmotic permeability of intact frog *(Rana pipens)* exposed to low pH.

Ichthyofauna

Effects of acid precipitation on fish include mortality, reproductive failure, reduced growth rate, skeletal deformities and increased uptake of heavy metals. The earliest recorded impact was on Atlantic Salmon in few southern Norway rivers Jensen and Snekvik, 1972). Death of fish at low pH has been attributed to the failure of ion regulation or asphyxiation, or to elevated metal concentration associated with low pH. Exposure to low pH water causes edema between outer lamellar cells and remaining tissue, erosion of lamellar and swelling of filaments (McKenna Duerr, 1976). Primary mode of acid toxicity in fish is gill damage, which impairs respiratory, excretory all diver functions. Liver impairment reduces tolerance of fish to other toxicants. Dyne (1981) believed that the death of fish embryos is the result of corrosion of epidermal cells by acid, which interferes with respiration and osmoregulation.

The low pH interferes with respiration through several mechanisms. Elevated hydrogen ion concentration may cause excessive secretion of mucus from the gills thereby reducing the rate of oxygen diffusion across the gill surface (Dively et al, 1977). At extremely low pH, an increasing flux of hydrogen ions reduces blood which in turn reduces the oxygen carrying capacity of heaemolobin (Spry et al, 1981) Packer (1979) reported reduced oxygen consumption in Brook Trout exposed to acutely lethal pH. The reduced consumption was caused by decreased oxygen transfer and reduced blood oxygen capacity.

The most likely explanation of the physiological effect of hydrogen ion on fish is that at moderately low pH (4-5) failure of ion regulation is the primary response. At very low pH <3.5), respiratory failure occurs.

Acidification of surface waters is accompanied by increases in concentration of some metals. Aluminium concentration appears to be very important in determining the effect of acidification on fish. Cronan and Scholield (1979), Baker and Schofield (1980) and Schofield and Trojnar (1980) showed that mortality of Brook Trout in New York was caused by aluminium and pH in combination, rather than by either factor singly.

EFFECTS ON TERRESTRIAL ECOSYSTEMS

On land, acid rain is absorbed by soil where it can break down natural minerals as calcium, potassium and aluminium and carry them into the substrate, drawing away or leaching a key source of nutrients for vegetation.

Soils

Soils are normally much better able to resist acidification than lakes and rivers. Nevertheless, acidification may cause nutrients like potassium, magnesium, calcium and other micro nutrients to leak more rapidly out of the soil. Thus, decreasing soil fertility. Aluminium concentration would rise just as they in water, damaging plants and reducing the availability of phosphorus to them.

Acid rain does seem to have a distinct effect on soil microbiology; chemistry and fauna but the effects on growth of plants including trees are far less clear. Indeed, deposition of nitrogen may even have a fertility effect and increase productivity significantly atleast in the short run.

The principal cause of the acidification of farmland are acid rains, acidifying fertilisers, an insufficiency of plant matter that can be broken down by natural processes and the leaching out of plant nutrients. Nearly 15-50 percent acidification of arable land in Europe may be attributed to modern nitrogen fertilisers, while : 10-20 percent acidification is due to acid rain (Annonymous, 1984/ 1985).

The forest soils are particularly fragile because unlike agricultural soils they are not limed. The leaching calcium, magnesium and potassium encouraged by highly acid rain falling on soil accelerates loss of nutrients, thereby reducing soil productivity and the tree growth remarkably. Trace metals, elevated hydrogen levels and other contaminants in the forest floor may reduce seed germination and seedling growth.

Forest and Crops

Forest damage has become a major topic of public and scientific discussion in recent years. Species exhibiting most serious damage of acid rain are conifers such as fir, pine and spruce. However, deciduous trees like beech are also affected adversely. Acid rain may damage trees in many ways (Fig 6.5). One of the most damaged area in Germany is the famous 'Black Forest', which is largely covered with conifers. Red spruce is declining in Eastern North America and sugar maples of Quebec in Canada are also showing signs of decline (Vogelmann, 1982). In the yearly inventories of damage, reported by Federal Ministry for Food, Agriculture and Forestry, the portion of damage to German forest has risen from 8 percent in 1982 to 52 percent in 1985.

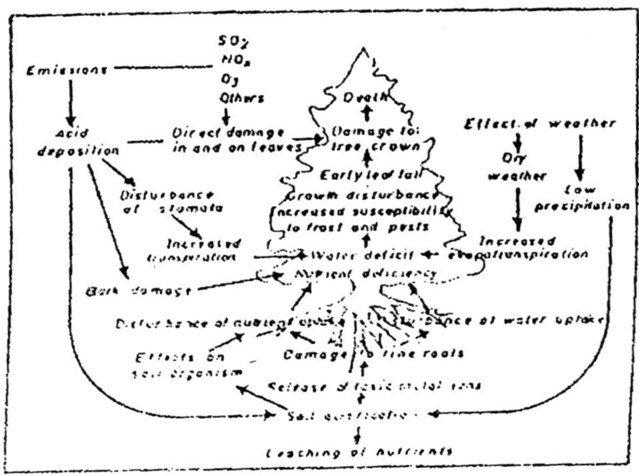

Fig. 6.5 : Possible pathways of damages in forest trees by acid rain

Forest damages may vary from minor foliar injury to large scale destruction and death of forest stands. In 'Crown die back' injury, leaves or needles at the tree top turns yellow, then brown and ultimately drop off. The tree growth may be decreased in association with annual concentration as small as 25-30 ug of in sulphur dioxide.

Acid rain can have a variety of effects on terrestrial vegetation including agricultural crops (Lee and Weber, 1979; Hileman, 1981; Maiti, 1982; Chevone et al, 1986). Through acid rain nutrients from foliage may be leached, causing lesions on leaves (chlorosis and necrosis) and erosion of external surface of young plant parts. It may also affect photosynthesis and growth of young tissues. Acid precipitation may exert a subtle effect on the reproductive potential of plants. Due to acid rain plant development power is diminished; gross deformation occurs, vitality is snapped and chances for survival are lessened. The direct effects of acid-rain on vegetation are:

1. Damage to protective surface structures;
2. Disturbance of gas exchange and metabolic and growth processes;
3. Poisoning of plant cells, resulting in necrotic lesions;
4. Alternation of leaf and root exudation processes affecting associated microbes;
5. Synergistic interaction with other environmental stresses;

In a plant canopy, acid rain must penetrate through cuticle or stomata to reach leaf cells. It has been that postulated that cuticle's are perforated with micropores (Crafts, 1961). Cuticular pores may be numerous in specialized areas such as at bases of trichomes, hydathodes, glandular hairs (Schnepf, 1965), water, absorbing scales of Bromeliaceae (Haberlandt, 1914) and stigmas (Konar and Linskens, 1966). These results may explain why injury from acidic precipitation occurs more frequently at bases of trichomes and hydathodes. Penetration of acid rain through stomata is thought to be infrequent if it occurs at all (Adam, 1948; Gustafson, 1956, 1957; Sargent and Blackman, 1962), Generally, spontaneous water infiltration of stomata will occur if the contact angle is smaller than the angle of the a Future wall. The degree of stomatal opening from 4-10 turn is little importance in penetration. However, cuticular ledges present at the entrance to the outer vestibule and between the inner vestibule and substomatal chamber resulted in very small wall angles. These small wall angles may not be adequate enough for water penetration of stomata (Schonherr and Bukovac, 1972). All evidences suggest that the main route of substances should be through the cuticle and not through stomata (Evans, 1988),

The acid deposition on the leaf surface reduce the rate of photosynthesis. In addition, the nitrate content of acid deposition combines with the potassium content of leaf's nutrient system which then is leached or chemically washed from the leaf. If this vital potassium is not supplied by the roots, the entire tree may suffer from nutrient stress, Alteration in leaf surface chemistry, may increase vulnerability of the leaf to disease agents and toxic materials.

In Sweden, acid rain in conjunction with ozone are being named as responsible for forest damage. The forests in north-west Czechoslavakia in the proximity of the Sokolow coal basin around which several mammoth solid fuel power stations were built have been effected by acid rain. By he end of 1978 forests around 41 per cent of woodlands in the Ore Mountains and Slavkov forests had been damaged 36000 hectares of land, with 1000 contiguous hectares entirely dead. The damage continues to grow each year, with the most serious situation developing in forests hit by sulphur dioxide emissions from power stations.

There is a possibility that the forests of the world may be destroyed completely if pollution persists. In contrast with natural degradation in which the ecosystem rebuilds itself, trees affected by acid rain will take centuries or even millennia to achieve the predisturbance level of productivity structure and function. Degrading forests not only result in ecological imbalance, but also posses a threat to the economics of European countries but also posses a threat to the economics of European countries (Elsworth, 1984).

Wild-life

The effects on wild-life of acid rain are not as apparent as those on fisheries. Nevertheless, several direct and indirect effects of acid rain on the productivity and survival of wildlife population have been reported (Mayer et al, 1984).

Acid rain can directly affect the eggs and tadpoles of frogs and salamanders that breed in small forest ponds. Increases in acidity of these waters may impair hatching success or survival of young animals. It has been postulated that acid rain can indirectly affect wildlife by allowing metals bound in soil and sediments to be released into the aquatic systems where toxic amounts may be invested by wildlife. Other indirect effects of acid rain on wildlife are loss or alteration of plants food and habitat resources, because abundance and diversity of plants is directly related to that of wildlife, any affect on the former will also affect the latter. For example, lichens are extremely sensitive to pollution and acid rain; the decline of a particular lichen often used as a nesting material by the northern perula *(Perula americana)* may be partly responsible for this species increased rarity.

Health hazards

The effects of acid rain on human health have not received comparable attention. However, acid rain is known to be highly corrosive in action and results in skin diseases and even severe deaths. The best example of such effects have been reported by Kitagava (1985). Some 600 cases of severe lung diseases that occurred over a period of 8 years were studied. All the victimised relatively close to a titanium dioxide pigment factory that emitted 100 -300 tonnes of sulphuric acid aerosols per month. In one area with the highest incidence of the disease, measurements have indicated that acid concentrations averaged 160 microgram per cubic meter ($\mu g/m^3$). Average concentration of acid rain and the incidence of lung disease declined with increasing distance from the plant. The incidence of lung disease dropped sharply, furthermore when the plant installed controls to remove sulphuric acid from its emissions.

Material effects

Acid rain accelerates corrosion in most construction materials in buildings, bridges, dams, industrial equipments, water supply networks, underground storage tanks, hydroelectric turbines and power and telecommunication cables. It also assaults historic buildings and cultural treasures by accelerating corrosion. For example, substantial damages to the Parthenon in Athens and Trojans Column in Rome are largely due to acid fallout.

ACID RAIN CONTROL

1. Liming

The damage to lakes and other water bodies can be eliminated by adding lime. Many chemicals such as caustic soda, soda, sodium carbonate, slacked lime and limestone are most popular for raising pH of those water. But while liming, eliminates some of the symptoms of acidification it is expansive and not practical for many lakes and running water. Most of all, it is not real cure it does not attack the causes of problem. It nevertheless is a good interim measure, and can concurrently be used to counterbalance the increasing acidification of cropland. Although liming can restore many species and improve water quality in lakes and streams, it must be repeated periodically (every 3-6 years) to remain effective.

2. Emission control

The obvious and only lasting solution to the acid rain problem is a reduction of sulphur and nitrogen oxides emissions. The use of fuel that are low in sulphur is not really practical because the world supply in these fuels is believed to be limited. Various techniques are also available for scrubbing fuel gases before they are released into the atmosphere. Methods are also available to reduce sulphur emissions from non-ferrous smelters. Oxides of nitrogen can also be reduced through reduction or better control of combustion temperature. Reduction in the emission of acidic gases is necessary and should be part of all new projects. Investment in controlling pollutants at the source is most effective and often the east expensive.

The Stockholm conference in 1982 on acidification of the environment was an important step in the right direction. In March 1983, the long range transboundary air pollution convention accepted an obligation to reduce their 1980 base emission levels by 1993 at the latest.

Alternatively there are other potential sources of energy instead of fossil fuel, there are many ways of improving energy conservation measures.

POTENTIALS OF ACID RAIN IN INDIA

It is often claimed that, acid rain is not yet a serious environmental problem in India, partly due to the low level of industrial activities and partly due to the presence of alkaline dust in the atmosphere. Acid deposition may or may not be a serious problem in India now but the potential for this environmental menace is very much there and the situation is becoming worse with the passage of time. There is a need to be on the alert are to conduct systematic investigations to detect the presence of atmospheric acidity in industrial localities of the country. No remedial action will be possible in this respect, unless we have accurate data on the extent of this problem.

The studies conductivity the Department of Meteorology have shown no evidence of any sufficient problem of acid rain in India. However, occasionally before the onset of monsoon, samples of rain water collected from highly industrialised areas of the country did indicate some acidity. On the other hand, several independent investigations have reported the occurrence of acid rain at Agra, Pune and Unnao. Thus, it appears that the monster of acid rain has started to raise its ugly had in India too and now is the time to take the necessary steps to tame it.

Acid rain problem is very much in the making in the country. Acid rain was first noticed in Bombay in 1974. Observations of acid rain in Trombay and Chembur, two adjacent suburbs of Bombay revealed rain of pH 4.5 and 4.8 respectively. Later on, acid rain was also recorded from Vashi (Bombay), Delhi, Nagpur and Pune. Although harmful effects have not been recorded so far on human beings, vegetation and crops, we must be aware of such dangers as evident in European countries.

Udyogmandal industrial area is one of the largest of its kind in south India hosting a number of chemical industries. Rain water collected from the area showed the pH range from 3.6- 6.8. The maximum frequency of occurrence was in the range 3.6-4.5 (48%). The average pH based on mean hydrogens ion concentration worked out to 4.4. The pH of rain water collected from background area (30 km) varies from 5.7 -6.4 (Table 6.8). Data available from studies carried out elsewhere in the country .have shown pH variation from 4.5-6.3 in Bombay and 5.1-6.6 for country wide sampling network covering stations (Kelkar, 1982). The sulphuric acid unit of the fertilizer factory in the industrial complex emits sulphur dioxide and sulphur trioxide and acid mist which contribute to the higher acidity of rain water in this industrial area. On the other hand, lower acidity in the background area might be due to the fact that pure rain water in equilibrium with atmospheric carbon dioxide gives a pH 5.7. The dissolved solids showed maximum frequently (34.4%) in the range 10.1-20.0 mg/l. Two fold increase in the dissolved salt content have been observed in the study area as compared to control area. Chlorides in the rain water varied from 1.1-13.2 mg/l. The maximum frequency (78.2%) was observed in the range 1-5 mg/l. Sulphite varied from 1.0-32.3 mg/l. The maximum frequency of occurrence (59.7%) was observed in the range 1.10 mg/l.

Table 6.8: Characteristics of rain water in Udyogmandal industrial area, Kerala (Pillai, 1988).

Characteristics	*Sample*	*Concentration range*	*% frequency occurrence*	*Mean± S.D*
pH	Industrial	3.6-4.5	48.1	-
		4.6-5.5	25.9	4.4
		5.6-6.8	26.0	-
	Control	5.7-6.4	-	-
Dissolved solids(mg/l)	Industrial	1.0-10.0	20.7	-
		10.1-1-20.0	34.4	20.6
		20.1-30.0	24.0	±24.4
		30.1-59.2	20.9	-
	Control	8.4-23.2	-	14.0
Chlorides (mg/l)	Industrial	1.0-5.0	78.2	-
		5.1-10.0	17.8	4.2
		10.1-14.0	4.0	±2.2
	control	2.1-2.3	-	2.2
Sulphate (mg/l)	Industrial	1.0-10.0	59.7	-
		10.1-20.0	31.4	9.9
		20.1-32.3	8.9	±6.4
	Control	3.4-4.1	-	3.8

The trace elements in the rain water showed variable concentrations (Table 6.9). Maximum value (50-1500 µg/l with a mean value of 453 ± 351 µg/l) was observed for zinc, while minimum value (5-33 with a mean ±10 µg/l), was observed, for nickel. The order of predominance of trace elements in rain water was :

Zn > Fe > Cd > Pb > Mn > Ni

Table 6.9 : Trace element concentration (μg/l) in rain water over Udyogmandal industrial area, Kerala (Pillai et aI, 1988).

Element (μg/l)	*Background area*	*Industrial area range*	*mean ±S.D.*
Zinc	40.0	50 -1500	453±351
Cadmium	10.0	10 -150	41±361
Manganese	4.4	5 -36	14±10
Copper	2.2	3 -370	40±82
Iron	8.0	21 -633	122±147
Nickel	5.0	5-33	10±8
Lead		11-84	25±23

A similar study to detect some common pollutants, including acidity, at Rourkela—an industrial town having an integrated steel plant and surrounded by numerous medium and small scal industries, indicate that Rourkela still does not have the problem of acid rain (Table 6.10). In fact, the sample, were always alkaline (pH greater than 7.00), except for some samples collected in May and September 1987. On the other hand, if we compare the pH values for the month of June 1986 and 1989, we find that the average value and the range of pH both have decreased over the past three years, indicating an increase in the concentration of acidic compounds in the atmosphere of Rourkela.

Table 6.10: pH values of rain water samples from Rourkela (Tiwari et al., 1987; Patel and Tiwari, 1989).

Year	*Month*	*pH value* Average	range
1986	April	7.42	7.05-7.76
	May	7.44	6.32-8.33
	June	7.94	7.34-8.28
	July	7.72	7.50-7.93
	August	7.65	7.38-7.91
	September	6.60	6.08-7.12
1989	June	7.2	7.04-7.33

The problem of acid rain in this country will grow with the increased industrialisation if the atmospheric emissions are not controlled substantially. Fossil fuel consumption is also increasing steadily. It has been estimated that the total mission of sulphur in the country from burning of fossil fuel increased from 1.4 million tonnes in 1966 to 3.2 million tonnes in 1979, and will Increase remarkably around 2000 AD (Varshney, 1983). Thus, within a period of only 13 years, the increase in total sulphur dioxide emission is 21 percent as compared the increase of 8.4 percent in the United States of America during the same period. A vast expansion of Thermal Power generation and rapid pace of industrialisation would certainly effect the biogeochemical cycling both qualitatively and quantitatively.

Due to quick and urgent need of power, India is increasingly turning to super Thermal Power stations served by coal mines. The concept of energy-park clusters if high capacity power stations built as initiated by the planning commission in the last decade, may pose serious and water pollution problems, if not managed properly. Ex-deputy Environment Minister Mr. Dig Vijay Singh once warned that if pollution control measures are not taken by Thermal Power stations, acid rain would hit India within 10 years.

The slow turning to yellow of the white marble of the Taj Mahal of Agra has been discussed much. The pollutants that can damage this monument are sulphur dioxide and nitrogen oxides which corrode the marble by producing corrosive acids, 85 per cent of the total air pollution in Chembur came from the Thermal Power station located at Trombay. In Agra, shutting down the two old Thermal power stations has reduced the sulphur dioxide levels in the air by 75 percent, the experience of Agra confirm what the residents of Delhi are reminded of every year in winter that most of the contribution to deteriorating air quality comes from the coal-based Thermal Power stations at Indraprastha and Badarpur.

Rampant air pollution in the Jharia-Raniganj coal belt bordering Bihar and West Bengal at Dhanbad is exposing both inhabitants and miners to hazards of respiratory disorders. The conversion of effluents into sulphuric acid, was found to be higher during the winter and lower during the summer.

So far pollution control in the public sector has consisted of merely building tall chimney stacks in order to disperse the emissions over a wide area. Dispersal is not control. In recent years, electrostatic precipitators have begun to be installed to remove fly-ash from the emissions. But absolutely nothing is being done to get rid of the sulphur and nitrogen oxides even now.

The sulphur dioxide emissions can be reduced by a desulphurization process using wet or dry scrubbers. Similarly, nitrogen oxides emissions can be reduced by a denitrification process. But none of the power stations is taking any of these processes because of additional cost.

Harm from such activity can be avoided through foresight and imaginative planning. We should learn from the experience of industialised countries and prevent progressive acidification of the environment.

IV. NUCLEAR POWER REACTOR ACCIDENTS

As nuclear power stations accident news come, the horrors of radiation, come to be known. The nuclear power reactor breakdown from chalk river, Canada (1952) to chernobyl, Russia (1986) has forced to rethink countries all over the world, about their nuclear power plans. America has cancelled or deferred the majority of its nuclear power stations. France's massive nuclear plans are running into trouble. In Germany, the threat comes from popular local, opposition unwilling to entrust life to radioactive uncertainty.

According to Greenpeace, there have occurred over 300 nuclear accidents that can be termed "major" and which were life threatening and actually killed people. Accidents involving nuclear dangers are more common than is generally perceived. Most of the nuclear reactor accidents just go unreported until the cover-up of it by the government and the administration is exposed by media and the non-governmental organisations. The elaborate secrecy threat shrouds operations at nuclear installations cannot hide the usually horrific consequences that the environment and sometimes people have to face.

Following are some of the important nuclear power station accidents that have occurred in various countries :

Chalk river nuclear power reactor accident

It was the first nuclear power reactor accident. The reactor NRX was heavy water moderated, located on the Ontario bank of the Ottwa river, about 150 kilometre from Ottawa. The accident occurred on 12 December 1952 due to human error. First four values which kept the air pressure from raising the control rods were opened in error by an operator. The supervisor noted warning lights and rushed to the basement to close the values. Once he had closed the valves, he assumed that the rods had dropped back, but they had not dropped fully - they had dropped far enough to shut off the warning lights.

The supervisor realising that the reactor was still on, called the control room to order the operator to push buttons 4 and 3 to stop the reactor, but mistakenly said 4 and 1. The operator rushed off to do it before he could correct the mistake. Button 1 raised four banks of control rods, causing the reaction to double every 2 seconds. This buildup was noted after 20 seconds and the reactor was screamed. Because of the air pressure problems, the control rods did not go all the way down. After about 44 seconds, the reactor physicist dumped the heavy water to kill the moderation and stop the reaction. This dumped tonnes of radioactive water into the basement. About 3 minutes later, the 4 ton lid blew off the reactor spurting radioactive water and setting off alarms of lethal radiation levels.

The NRx reactor underwent a violent power excursion that destroyed the core of the reactor, causing some fuel melting. Unaccountably, the shut off rods failed to fully descend into the core. A series of hydrogen gas explosions (or steam explosions) hurled the 4 ton gas holder dome four feet through the air where it jammed in the super structure. Thousands of curies of fission products were released into the atmosphere, and a million gallons of radioactively contaminated water had to be pumped out of the basement and disposed off in shallow trenches not far from the Ottawa river. The core of the NR_x reactor could not be decontaminated; the core and calandria of the reactor had to be removed and then burried as radioactive waste. The reactor was largely reconstructed. About 150 United States military personnel, about 170 Canadian military personnel, and about 20 construction company employees joined the 862 staff members to implement the clean-up. It resumed operations 14 months later.

Windscale nuclear reactor accident

The Windscale nuclear reactor was Graphite moderated reactor, air cooled with huge filters on the top of the stack. It was located near the Irish Sea west of the lake district, Liverpool, England. On Monday, October 7, 1957, the blowers were shut down to allow controlled heating of the graphite blocks to achieve a "Winger release, a release of stored up energy in the graphite moderator. The thermocouples were thoroughly tested before the process.

On Tuesday, the operator noted a drop in the temperature of the graphite, when it should have been rising, so they repeated the process. A sudden rise in the temperature of the uranium cartridge was noted. Cadmium control rods were dropped to cool the reactor, but the graphite temperature continued to rise. They got confusing and conflicting instrument readings.

On Wednesday, erratic conditions were noted. Cooling air was started in order to reduce the graphite temperature.

On Thursday, the radiation meters at the top of the stack showed high readings, then dropped back. Then both temperature and radiation readings started to rise. Attempts to cool the reactor failed, and just increased the radiation readings. There were indications of a burst uranium fuel rod with a ten fold increase in ambient radiation. They tried to use an overhead scanning device but it jammed. Two men in protective suits on an elevator device opened an inspection hole to see cherry-red uranium rods and blue flames in the graphite. The problem involved 100 - 200 channels. They tried to punch the rods out, but they were bent and wedged. They disgorged the surrounding channels, applied Carbon dioxide, but still the graphite burned.

On Friday, the decision was made to water down the reactor, ruining it, to avoid a catastrophic release of radiation. The water was kept on until noon Saturday. The milk from 150 surrounding dairy farms was confiscated because of the high Iodine 131 levels. In the affected area of some 200 square miles some cattles were destroyed. Apparently there was no strontium release, and the water supply showed no contamination.

Stationary Lower power - 1 nuclear reactor accident

The SL-1 was a 200 KW nuclear reactor located at Idahofalls in United States. It was designed for electric power production for remote Arctic stations. It was being operated by three men on the night of January 3, 1961. It had a two month history of sticking control rods and the reactor had been shut down for maintenance. The crew was to assemble the control rods devices and prepare for start-up.

Radiation alarms sounded, monitors a mile away gave alarms and health physics people rushed to the reactor. The building was intact and the lights were on, but they measured a level of 25 rads/hr at the entrance, 200 rads/hr as they approached the control room, and 500 rads/hr near the reactor. With protective suits, they rushed to the reactor building and found two of the men, one still alive. They found the third man impaled by a control rod, pinned to the ceiling.

Once the bodies were removed, they measured over 400 rads/hr, too hot for a normal burial.

No meltdown occurred and less than 10 per cent of the radiations was released, but it represented the worst nightmare about nuclear accidents.

Three Mile Island Nuclear accident

The TMI nuclear power generating stations is located on 814 acres on an island in the Susquehanna river some 10 miles southeast of Harisburg, Pennsylvania near some farmland. There are 4 separate Pressurised Water Reactors (PWR) at TMI and it was the second reactor that failed. (Fig. 6.5) At 4 a.m. in the morning of March 28, 1979 the TMI accident was caused by a malfunctioning of the relief valve - RORV, fitted on the top of the equipment called "the pressuriser", which remained open although the instrument showed that it had been closed. Thus, there were causes in both physical hardware (the valve) and the control system software (the erroneous indication). This shut off the water supply to the system that cools down the reactor core. Consequently, too much water escaped through it. When the drain tank was filled, one of its component (rupture disc) broke and thousands of gallons of radioactive water spilled into the basement of the containment building. Water loss from the core left it partially uncovered, and thus the core began to self-destruct. In fact 90 per cent of the reactor core was damaged, 52 per cent of it had melted down. Such a partial melting of the reactor core has been described as the *China Syndrome.* Dumps automatically began removing the contaminated water to waste storage tanks in an auxiliary building, but it overflowed

TMI-2

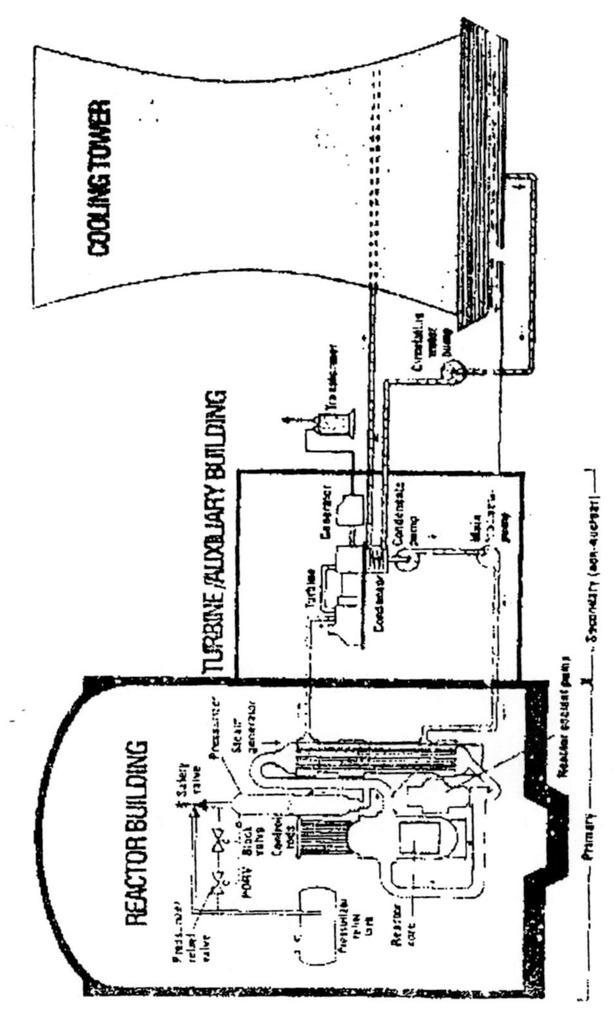
REACTOR BUILDING
TURBINE/AUXILIARY BUILDING
COOLING TOWER
Safety valve
Pressurizer
Steam generator
Block valve
Control rods
PORV
Reactor core
Turbine
Generator
Condenser
Condensate pump
Transformer
Primary

Fig. 6.5

the tanks and spilled on the floor. Since the temperature of the water was high, the gases dissolved in it escaped. These gases carried approximately 2.5 million curries, moved through a stack into the atmosphere and began to increase radiation dose rates outside the power station. It was estimated that 2.5 million curries of radioactive noble gases and 15 curies of radio-iodine was released. These releases resulted in an average dose of 1.4 mrem to the approximately 2 million people around the power station.

It was not clear that how this serious accident occurred ? Probably the operator could not understand the situation quickly enough. They did not check the measurements down-stream from the open valve, which would have told them that the valve had not been closed, and the reactor was losing steam. Not until the next shift came at approximately 6.50 a.m., when the radiation alarm were sounded and the reactor operators declared a state of emergency. In the Senate hearing, the failure to understand the situation and whether the valve had been closed was judged a *Human error.*

Several hundred people at the time of the accident reported nausea, vomiting, hair loss and skin rashes. The next day sickness warned the residents that radiation was being released form the power station. The residents rushed in the opposite direction of the downwind with their belongings. A proclamation was made by the then Governor Mr. Richard Thornburg urging pregnant women and those with small children living within 5 miles of TMI to leave the area, people living within 10 miles radius to stay indoors and calling for the closure of more than 20 local schools. The news of the accident rocked the nation, and its effects were keenly felt by those who lived in the shadow of the great concrete towers of the TMI.

Following the evacuation advisory, approximately 12180 persons living within a 5-10 mile ring evacuated. These figures represent 35 per cent of the total population within 5 mile and 25 per cent of the total population of the 5-10 mile ring. It was further estimated that some 1,44,000 people living within a 15 mile radius evacuated their homes at some point between March 28 and April 3, 1979; this was approximately 39 per cent of the total population (Flynn, 1979). At the same time the Nuclear Regulatory Commission was hedging its bets by saying that there might come a time when everyone had to be evacuated.

However, on April 1, 1979 the crises begun to settle. The officials were able to restore enough coolant to the reactor core to prevent a complete melting and the reactor appeared to be stable (Donnelly and Kramer, 1979). The reactor was shut down. Schools located near the TMI power station were reopened on 4 April 1979. People who left the area began to return home on 6 April, 1979.

Technically of course, TMI nuclear power reactor accident was not a disaster; the major environmental release of radioactive gases and particulates which would constitute a disaster was precluded. The situation may be technically characterised as an emergency, however, and the evacuation which occurred in connection with the nuclear threat may be compared with evacuation in the face of other threats.

Tsuruga nuclear power station accident

The worst accident in the history of Japan's nuclear power industry was revealed on April 18, 1981, occurring at the accident prone Tsuruga plant of the Japan Atomic Power Company (JAPC) in Fukui Prefecture. A whole series of accidents at the Tsuruga clear power plant has been disclosed. In each of the nine accidents revealed so far (Table 6.11), a serious leakage of radioactive materials occurred. Considering the number of workers, a total of 300, exposed to radiation from these accidents at Tsuruga, they were no less grave than the Three Mile Island accident.

At 5 a.m., April 18, the Natural Resources and Energy Agency of the Ministry of International Trade and Industry (MITI), the supervisory organ for commercial nuclear plants held an extraordinary press conference. It announced that radioactive cobalt 50 to 60 times normal in gulf-weed taken from the seabed at the mouth of the Urazoko Bay sewage system which the Tsuruga plant faces. The seabed was highly contaminated with radiation.

However, by this time it was found that the seabed radioactivity by the Tsuruga plant resulted from a March 8 accident in which more than 16 tonnes of radioactive primary cooling water from the reactor overflowed from the sludge storage tank because of a valve left open for hours. The company covered up this incident. 56 workers were exposed in the primitive, week long clean-up. Splashing water was scooped up with dustbins into buckets and mops and tissue paper were used to dry the floor. The water was illegally

Table 6.11 : Accidents Revealed at Tsuruga Nuclear Power Plant of JAPC

Date of accidents	*Date of disclosure*	*Nature of accidents*	*Date of repair & clean-up*	*No. of exposed workers*	*Amount of radiation (Co. statistics)*
1. Dece. 6,'80	April 30,'81	Overflow of radioactive waste water by reporting from filter sludge storage tank	Unknown	Unknown	Unknown
2. June. 1,'81	April 1,'81	Primary cooling water leakage from cracked supply water heater	Welding on Jan. 14,'81	19	Average 7 millirem/day person; 55 millirem/max.
3. Jan. 1,'81	Jan. 1,'81	Trip (shut down) due to rise of gas pressure in containment bldg.	Unknown	Unknown	Unknown
4. Jan. 19,'81	April 25,'81	Three leaks of 2 condensed waste water tanks at new waste storage building	Jan. 24-28,'81	60 total	Average 55 millirem/day person; 92 millirem max.
5. Jan. 19,'81	May 18,'81	Holes in pipes of condenser	Jan. 20-27	30/day 200 total	Unknown
6. Jan. 25,'81 (JAPC staff kills himself)	April 1,'81	2nd crack in supply water heater	Jan. 28-31	76 total	Average 51 millirem/day person; 195 millirem
7. Mar. 8,'81	April 18,'81	2nd overflow of waste water from filter sludge storage tank	Mar. 8-Apr.15	56 total	13 millirem max.
(March 30,'81	Osaka court rejects suit against JAPC by Iwasa, worker irradiated at Tsuruga plant.)				
(April 14,'81	Radiation levels in test sample of gulfweed by Fukui prefecture hygenic institute				
JAPC - April 18, JAPC - MITI 5 a.m., April 19 MITI announcement					
8. Jan.- June.'75	May 14,'81	Condensed waste water tank leakage	June 6-28,'75	37	1118 millirem per person/ max.; 430 millirem per day / person
9. Early '74	May 15,'81	Condensed waste water tank leakage	Mar.-Apr.'74 Aug.-Sept.'74		Unknown

dumped in a manhole connected to the public sewage system. The company has reported excessively low exposure for the workers compared with the levels indicated by the seabed readings. Company officials said the maximum radiation exposure was 13 millirems, but MITI inspectors had to rush from the sludge tank room after only 20 seconds being exposed to 30 millirems. The actual exposure for the clean-up crew must have been much greater. Fish deliveries from the area stopped temporarily because of the disclosures.

Prior to this JAPC had two accidents at this site on January 10 and 24; a feed-water heater cracked causing leakage of primary cooling water between the reactor and turbine. The firm, concealing these accidents, continued operations, repairing the cracks by hammering the steel heater wall to seal it. Disclosure of this came from workers inside the plant. Without their report, the accidents would have remained a secret.

In addition, another Tsuruga accident originally reported as occurring last December, but which the Company said happened in January 19, was soon after disclosed. At that time, highly radioactive waste water leaked from holes in pipes connecting two condensing tanks that are part of a new waste water storage facility. It is feared that 45 workers received large doses of radiation in the clean-up and repairs since it was only safe to be in the area for 5 seconds at a time. The radiation level there at that time was high enough to kill half the people working there in five hours.

A fifth accident has just been disclosed occurring at the same place as the March 8 mishap last December. The details are not yet available.

To date, a known total of 278 persons have been exposed to heavy doses of radiation from the repair and clean-up necessitated by the four accidents. It is notorious for breakdowns accounting for 15 per cent of the total number of Japanese plants. It is also notorious for its careless management of worker's exposure to radiation.

The accidents exposed so far are just the tip of the iceberg. There must be more which have not been revealed and there seems so question that the JAPC is actually trying to conceal an even more serious accident which caused the large amount of radioactive leakage. There are indications that the company tried to erase records of the accidents. JAPC is not qualified to run nuclear plant and has committed crimes against humanity (Jishu-Koza, 1981).

CHERNOBYL NUCLEAR REACTOR DISASTER

On April 26, 1986 at 1.23 O'clock in the morning, one of the four operating reactors units at the Chernobyl power station in Ukraine met an accident. The accident at the Chernobyl was the largest one in the history of this technique. A stream explosion led to complete destruction of the nuclear reactor. The weight of the nuclear fuel of this reactor was as high as hundred ninety two tonnes. The year 1986 was the third year of the exploitation of this fuel, hence the contents of more than three hundred various radionuclides were the highest. The operating reactor went out of control and produced a power peak which was 100 times the normal power within 4 seconds. The temperature of the reactor reached 3000°C and the powerful stream of evaporated, sublimated radionuclides or pulverized fuel material was thrown out from the reactor (Grodzinsky, 1991). The energy released by the stream explosion shifted the 1000 tonne cover plate of the reactor, cutting

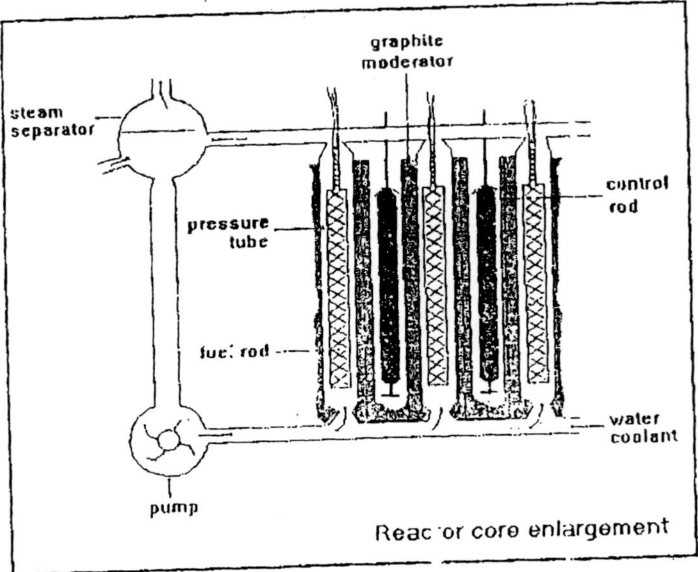

Fig. 6.6 : Showing Chernobyl reactor core which went off during the disaster

all the cooling channels Then another explosion ejected the pieces of the reactor from the damaged building and eventually, due to the influx of air, the graphite which served as moderator in this reactor type, started to burn. This fire extensively damaged the reactor building, equipment in it, the reactor itself and the reactor core. Fire teams came quickly to the site, work was complicated by the fact that neither water nor chemicals could be used.

The Chernobyl nuclear accident led to a severe release of fission products with a high power of hundreds of millions of curies per hour which lasted during ten days. Its peak occurred on the first day, then it was decreasing until it rose again to reach another peak on the tenth day before it dropped sharply. According to the Soviet report to the IAEA in Vienna (IAEA, 1986), the total radioactivity released was in the order of 100 MCI (in the order to 10^{18} Bq). Apparently, radionuclides were released in relative quantities corresponding to their elemental volatility. As Table 6.12 shows, 100 per cent of the noble gases, 10-20 per cent of the iodine, cesium and tellurium but only 6 per cent or less of the more refractory fission product elements left the reactor.

The evaluation of the radiation exposure to population in all affected areas from Chernobyl was initiated by UNSCEAR - United National Scientific Committee on the Effects on Atomic Radiations.

The contribution of $Iodine^{131}$ exposure is now over, as it has an eight day half-life, but that of $Caesium^{137}$ still continues, since it has a 30 year half-life. Thus, a long-term residual exposure to atomic radiations can be expected.

Radioactive contamination in Chernobyl area

Radioactivity releases affected the area beyond the nuclear power station boundaries. Fireman and some nuclear power station personnel were among those affected by radiation. Most residents in adjoining areas were indoors at the time of the accident, and this helped to reduce their exposure.

$Iodine^{131}$ and $Caesium^{137}$ were the most serious among the twenty odd radioactive elements released during Chernobyl disaster. Although no systematic data on radiation levels were made available, some values were given. The maximum level within in 30 km zone had been 10-15 millicem/hour. By 5th May it had dropped to a maximum of 0.15 millicem/hour at the perimeter of the zone. The

Table 6.12 : Core inventory and estimates of total release of radionuclides from Chernobyl nuclear reactor accident (UNSCEAR, 1988)

Radionuclide	*Half-life*	*Inventory* 10^{18}*Bq*[a]	*Percentage released*[b]
^{85}Kr	10.72 yr	0.033	– 100
^{133}Xe	5.25 d	1.7	– 100
^{131}I	8.04 d	1.3	20
^{132}Tc	3.26 d	0.32	15
^{137}Cs	30.00 yr	0.29	13
^{134}Cs	2.06 yr	0.19	10
^{89}Sr	50.50 d	2.0	4
^{90}Sr	29.12 yr	0.2	4
^{95}Zr	64.00 d	4.4	3
^{99}Mo	2.75 d	4.8	2
^{103}Ru	39.30 d	4.1	3
^{106}Ru	368.00 d	2.1	3
^{140}Ba	12.70 d	2.9	6
^{141}Ce	32.50 d	4.4	2
^{144}Ce	284.00 d	3.2	3
^{239}Np	2.36 d	0.14	3
^{238}Pu	87.74 yr	0.001	3
^{239}Pu	24065.00 yr	0.0008	3
^{240}Pu	6537.00 yr	0.001	3
^{241}Pu	14.40 yr	0.17	3
^{242}Cm	163.00 d	0.026	3

Total release : -3.2×10^{18} Bq = 100 Mci.
[a]Decay corrected to 6 May 1986.
[b]Stated accuracy : $\pm 50\%$, except for noble gases.

level of radioactivity in Kiev's water reservoir was within normal limits at all times (Ann, 1986).

The radioactivity releases from the damaged unit were significantly reduced by subsequent shielding and neutron absorbing materials - sand, boron, clay, dolomite and lead dropped from helicopters over the reactor.

Clouds of high radioactivity rose over the Chernobyl area for many days. They caused direct irradiation and internal irradiation due to inhalation as well as considerable contamination of the ground. An area of more than 20,000 km^2 is considered contaminated. This contamination was mainly due to dry precipitation, as little rainfall occurred locally during the period of fission product release.

Evacuation began from the town; of Pripiat on 27th April after radiation levels from the cloud and from the ground had risen as high as 10 millisievert (mSv) h^{-1}. An expected total effective dose 250 mSv-5 times the maximum annual dose for occupational exposure—was considered the lower intervention level for evacuation. Further evacuation had to take place on the following days resulting in a total of about 1,35,000 people from places of Ukraine, from the Gomel area in Bjelorussia and from the Bryansk area in Russian Federation. The area from which a great part of the population was evacuated right in the post-accident period has a radius of about 30 km around the reactor site. However, there were strongly contaminated spots which had to be evacuated later even from greater distances from the reactor.

Four and a half years after the accident, the area around Chernobyl can be divided into three zones according to the level of contamination (Fig. 6.7). Zone I showed a ground radiation of 1-15 Ci/km^2, Zone II had a ground contamination of 15-50 Ci/km^2 and Zone III showed a ground contamination of more than 50 Ci/ km^2.

The elemental composition of the fall out at the site of ground surface fission was found to be in accordance with the volatility of the elements. On May 10, after the bulk of radioactive deposition took place, the radionuclides which contributed atleast 10 per cent of the total radioactive deposition were I^{131}, Ru^{103}, Cs^{137} and Cs^{137}. Radioisotopes of more refractory elements contributed much less e.g., 0.2% Sr^{90} and only $10^{-5\%}$ Pu^{239}. This means that non-volatile elements as far as they were released at all were mainly redeposited close to the source. The estimated fractions deposited in the Chernobyl area have been shown in Table 6.13. Apparently, non-volatile elements did not reach the high altitude after being released from the reactor which allowed the long-range atmospheric transport, and moreover, they formed aerosols of larger diameter favouring faster gravitational setting.

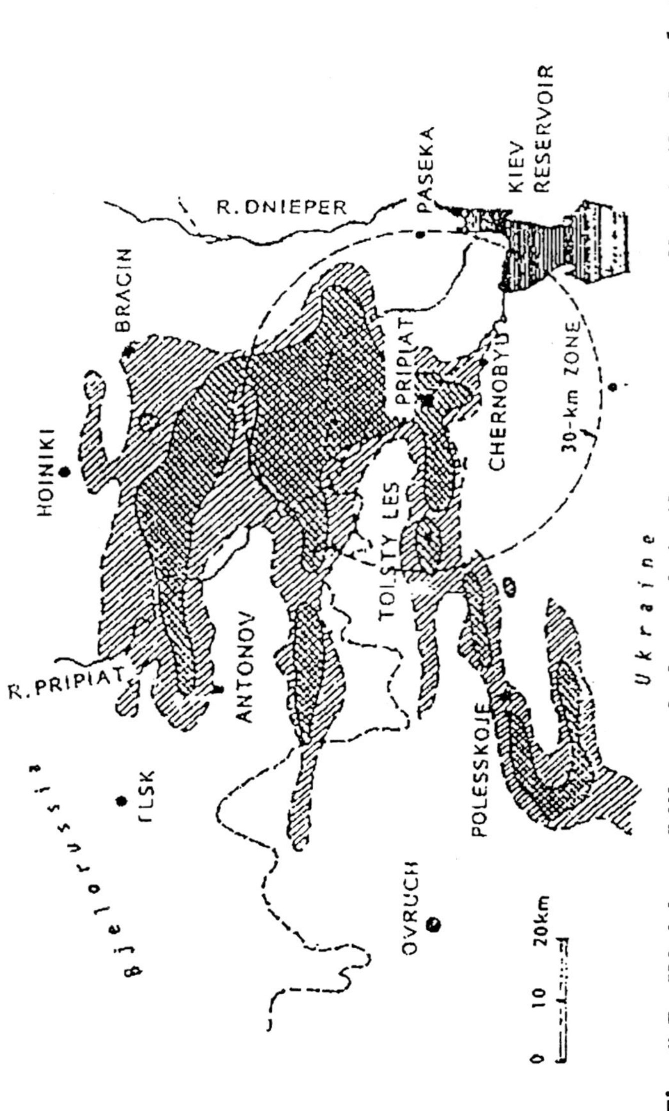

Fig. 6.7 : Division of Chernobyl area into three zones according to the level of radioactive contamination (the hatched areas belong to contamination Zone II, the cross-hatched ones to Zone III) (Kellerer, 1990).

Table 6.13 : Estimated fractions of total releases of long lived radionuclides deposited in the Chernobyl area (WHO, 1987)

Radionuclide	*Estimated fraction of total release deposited near Chernobyl (%)*
^{137}Cs	30
^{134}Cs	30
^{90}Sr	85
^{106}Ru	70
^{144}Ce	85
^{110}Ag	35
^{125}Sb	65
^{240}Pu	
^{239}Pu	
^{238}Pu	
^{241}Pu	
^{241}Am	90
^{242}Cm	
^{243}Cm	
^{243}Cm	

Chernobyl fallout in Europe

Radioactive material from Chernobyl has been distributed over great parts of the world but mainly over Europe. The complex distribution pattern of this fall-out was a consequence of the long duration of the release. Due to violent explosions in the reactor and the heat of the graphite fire, on 26 and 27 April some of the radioactive material was carried as high up as 1200 meter and more. The winds in this altitude came from various easterly directions and led to the dispersion shown in Fig. 6.8. According to the respective meteorological conditions three radioactive plumes were formed in sequence. The first plume reached Scandinavia including Finland on 27 and 28 April and later on affected also Poland and German Democratic Republic, the second plume reached Central Europe (Federal Republic of Germany, Switzerland, Austria, Hungary and Northern Italy) on 29 and 30 April. The third plume arrived in Southern and South eastern Europe (Southern Italy, Greece and parts of Turkey) on 1 to 30 May. As there were no westerly winds during the whole period of fission produce release, Kiew which lies only about 100 km southeast of Chernobyl received radioactive fall-out in the same order of magnitude as Munich, roughly 1500 km North-west of Chernobyl.

The radioactive cloud cases only short-term exposure due to direct irradiation and inhalation. The long-term exposure depends mainly on the deposition of radioactive material on the ground which is a function of weather conditions. Dry deposition leads to fall-out fairly homogeneously distributed over wide ranges whereas rain-storms cause local or regional contamination peaks. Such heterogenous fallout patterns were experienced in some European countries, for example, Bavaria in Southern Germany was heavily and non-uniformly affected due to heavy thunderstorm on 30 April. Other parts were uniformly affected because the fallout there was mainly dry deposition.

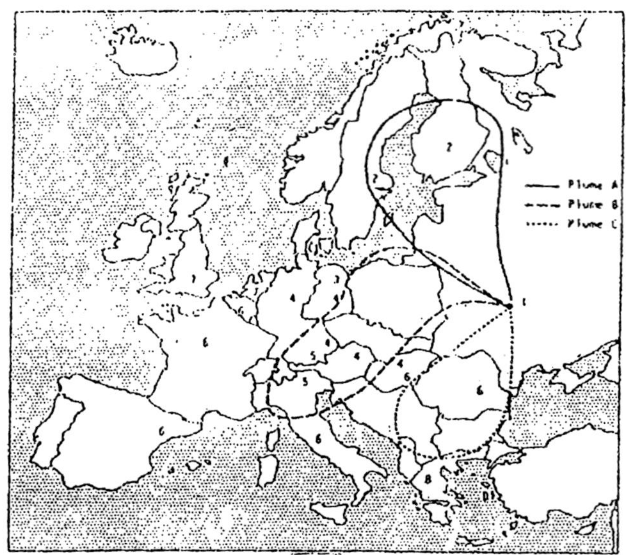

Fig. 6.8 : Descriptive plume behaviour and reported initial arrival times of detectable activity in air. Plumes A. B, and C correspond to air mass movements originating from Chernobyl on 26 April, 27-28 April, and 29-30 April, respectively. The numbers 1 to 8 indicate initial arrival times: 1 (26 April), 2 (27 April), 3 (28 April), 4 (29 April), 5 (30 April), 6 (1 May), 7 (2 May), 8 (3 May). (OCED Nuclear Energy Agency, 1987).

Thus, we see that Chernobyl accident was a true catastrophe. However, after the initial emotions has calmed down, it became clear that it was not a catastrophe of global dimension like pessimists had tried to put it, but rather a catastrophe of regional dimensions.

Ecological consequences of Chernobyl accident

Very limited knowledge about ecological consequences of the accident is available. Priorities in the heavily affected area were not on ecological studies and detectable effects were expected outside the highly contaminated Zone III. Ecological effects occurred only at higher dozes than health effects because ecology is concerned with collective rather than with individual effects (Socolov et al, 1989).

Trees suffered from fall-out because the crown acted as filters. In the neighbourhood of the reactor, they collected 60-70 per cent of the fall-out, and provided some protection for organisms living under their umbrella. The pine forest in an area of about 4 km^2 were found dead by the end of 1988. One year after the accident, damaged forests showed signs of regeneration.

In the vertebrates, deviation from normal blood count, tissue and organ were observed. In the long-run, wild animals will probably increase densely due to absence of humans in a formerly densely populated area. The corresponding harassment factor were no longer effective and crops were left behind providing an unusually rich supply of feed. In fact one year after the accident, for some species over population had been observed rather than a reduction.

Loss of human life

The Chernobyl accident was by far the most severe nuclear accident in a civilian facility which happened since the beginning of the peaceful use of nuclear energy. As a result of explosions two persons lost their lives, one from hot burns and the other as a result of injuries from falling objects. 204 persons including nuclear power station personnel and fire fighters, were affected by radiation, from first degree to fourth degree. All 204 persons were hospitalized in Moscow and treated medically. In some cases bone marrow transplants were performed. The number of additional cancers due to doses received in the region, cannot be predicted.

The chain reactions of uranium fission lasted ten days and radioactive iodine formed in great masses. Therefore, a great impact on thyroid glands was observed. In Ukraine on the population with damaged thyroid glands exceeded four million people. There is a group of about 60,000 child with a high degree of thyroid gland damage. In the city of Prypiat, located not far from the destroyed reactor, the dose load on thyroid glands in children sometimes exceeded 2-3 thousands rads. We can now see an augmentation of the thyroid gland diseases and the first cases of thyroid-gland cancer. The risk of blood cancer has been growing, especially among children. A bad state is certain for the immune system of people, especially in relation to antivirus and anticancerogenic immunity.

Life for many millions of people has become uncomfortable and awful, and a common fear before invisible radioactive rays and radionuclides in accompanied by general people psycho-emotional depression. All people are constantly talking only about the Chernobyl threat and its expected sad consequences.

Since the time of Chernobyl accident, many whole body counts on persons living in the region of Munich were performed. The cumulative frequency distribution of total intake of radiocesium (Cs^{134} + Cs^{134}) during the first year indicate that for an interception factor of 0.1 and 0.2 the radiation uptake was very high. This was due to the fact that assumption was made that all foodstuffs consumed by the population was produced locally. This unrealistic assumption led to an over estimation of the radiation dose for the population of an area of high local contamination.

REVIEW QUESTIONS

1. Discuss the objectives of environmental management. What are the major components of environmental management?
2. Discuss the advantages and limitations of Environmental Impact Assessment (EIA).
3. What are the important aspects in the design of an EIA Process?
4. Discuss briefly the salient features of the Environment (Protection) Act, 1986.
5. What are the basic features of the water (Prevention and Control of Pollution) Act, 1974?
6. Discuss the basic features of the Air (Prevention and Control of Pollution) Act, 1981.
7. What are the basic requirements for ISO 14001?

Chapter 7

POPULATION AND ENVIRONMENT

The word 'population' has been derived from the Latin word *populus* meaning people. Certain formal definitions of population met in the literature are:

1. The whole number of people or inhabitants in a country, section or area (Sociology).
2. The organisms collectively inhabiting an area or region that interbreed frequently (Biology).
3. A group of living individuals set in a frame that is limited and defined in respect of both time and space (Biology).
4. The entire group of organisms from which samples are taken for measurement (Biometry).

Population, that has been defined as a collective group of organisms of the same species occupying a particular place and at a particular time, includes at least four different concepts. These are: number of individuals, likeness of the kind enumerated, aliveness, and limitation of universe in time and space. Number of individuals is an essential theme in all the definitions. Furthermore, properties of population are density, natality, mortality, age distribution, biotic potential, dispersion and growth form.

POPULATION DENSITY

Population density is population size in relation to some unit of space and time. It can be measured in several ways including abundance (the absolute number in population), numerical density (number of individuals per unit area or volume), and biomass density (biomass per unit area or volume). Density of population can refer to the total area (crude density) or the actual area of the habitat (ecological density). It can be measured by applying the formula:

$$D = \frac{nla}{t}$$

where D is population density, n = number of individuals, a = area and t = time. Population density varies with weather conditions, with supply with many other influences. However, there are definite upper and lower limits to species for various population sizes. Thus a large area of forest might show an average of 10 birds per hectare 2000 soil arthropods per square metre, but there would never be as many as 2000 birds per square metre and 10 arthropods per hectare, since upper unit is determined by energy flow, trophic level metabolism of the organism.

NATALITY

Natality can be defined as the force of reproduction in a population. Rate of reproduction is further determined by ***fertility and fecundity***. Fertility refers to actual reproductive performance used on numbers born in a population. Whereas fecundity refers to the capacity reproduction. Ecological or realized natality refers to population increase under an actual or specific environmental condition. Natality can measured and expressed in a number of ways.

ΔN_n = production of new individuals

$$\frac{N\Delta t}{\Delta N_n} = b = \text{Natality rate per unit time per individuals}$$

where,

N = Initial number of individuals
N_n = No. of new individuals added
t = the time
Δ = the changing entity

MORTALITY

Mortality means the death of individuals in the population. Mortality rate is equivalent to death rate in human demography. It is expressed as the number of individuals or a part of population dying in a given period. Ecological or realized mortality is the loss of individuals under a given environmental condition. There is a theoretical minimum mortality for a population which represents the laws under ideal or non-limiting conditions. That is even under the best conditions individual would die of old-age determined by their physiological longevity which is often far greater than ecological longevity.

Specific mortalities at particular ages can be illustrated in the form of life tables. In brief, a life table is the book on death. Deblin and Lotka (1936) and Pearl (1940) have published most useful reviews on life tables as a tool for the study of natural populations.

It is comprised of a series of columns each of which describes something about the mortality relations within population when ages of I_x components are taken into account. The x column first states the age tabulates the survivors; d_x shows the actual number of deaths; d_x is rate at which deaths occur and e_x denotes the life expectation. Thus 1 tables though look tedious are essentially simple. Table 7.1 illustrates mortality statistics for white males in continental United States from 1929 to 1931. It shows that infant mortality is higher during the first year of life. The d_x column points out that out of each 100,000 birth 6,232 will die before the age of 1, and the value of d_x decreases upto 11 years. Thereafter d_x rises at first slowly and later with increasing rapidity. After 50 years the rate accelerates.

Life tables are useful in computing the average longevity of a population for exhibiting age groups in a population, indicating critical stages in species, showing differences between species, showing success of some species in different environments, and for collecting information in game birds and fish exploitation and control of pests.

AGE DISTRIBUTION

A population has individuals from different age groups. Therefore, the natality and mortality is also different for respective ages. Thus age distribution becomes an important phenomenon in population studies.

Three ecological ages have been identified for a population by Bodenheimer (1938). These are pre-reproductive, reproductive and post-reproductive ages. The relative duration of these ages varies in different organisms in population to their life-span. For example, in modern man three ages are relatively equal in length. Many plants and animals have a very long pre-reproductive period. Similarly, insects have a very long pre-reproductive period, a very short reproductive period but no post-reproductive period at all. The may fly and locust are classic examples (Fig. 7.1).

Table 7.1 : Abridged Life Table for White Males in Continental United States 1929 to 1931 (Adapted from Pearl 1940 Originally from Hill 1936)

Age interval	*Of 100,000 males born alive* — *Number alive at beginning of year of age*	*Number dying during year of age*	*Mortality rate* — *Number dying per 1000 alive at beginning year*	*Complete expectation of life* — *Average numbers of years of life remain at beginning of years of age*
x	I_x	d_x	$\frac{d_x}{I_x} \times 1000$	e_x
0–1	100,000	6,232	62.32	59.12
1–2	93,768	931	9.93	62.04
2–3	92,837	483	5.20	61.65
3–4	92,354	331	3.59	60.97
4–5	92,023	285	3.09	60.19
5–6	91,738	243	2.66	59.38
6–7	91,495	208	2.27	58.53
7–8	91,287	179	1.96	57.67
8–9	91,108	156	1.72	56.78
9–10	90,952	142	1.55	55.87
14–15	90,246	172	1.90	51.29
16–19	89,172	268	3.01	46.88
24–25	87,692	321	3.66	42.62
29–30	86,053	346	4.02	38.39
34–35	84,222	410	4.86	34.17
39–40	81,979	522	6.36	30.03
44–45	79,036	691	8.74	26.05
49–50	75,188	900	11.98	22.25
54–55	70,165	1,184	16.87	18.66
59–60	63,496	1,563	24.61	15.34
64–65	54,924	1,960	35.68	12.33
69–70	44.253	2,373	53.62	9.68
74–75	31,986	2,515	78.61	7.43
79–80	19,565	2,344	119.83	5.57
84–85	9,159	1,587	73.33	4.21
89–90	3,068	712	232.11	3.21
94–95	672	211	313.32	2.35
99–100	72	32	438.79	1.62

Furthermore a seasonal change in population can be observed. In a study of a rove of honey-bees, a change from a rapidly growing

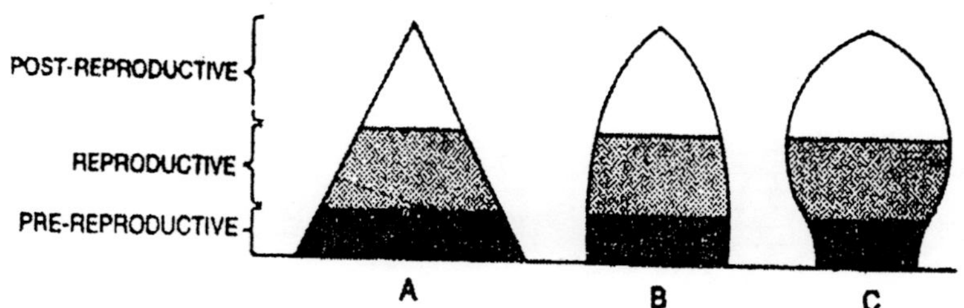

Fig. 7.1 : Age structure in different types of populations: A. expanding population; B. stable population; C. diminishing population.

population as shown by triangular shape in January, to a bell shape showing stable population in March, and then finally an urn shape from July to November shaped progressive dying off (Fig. 7.2).

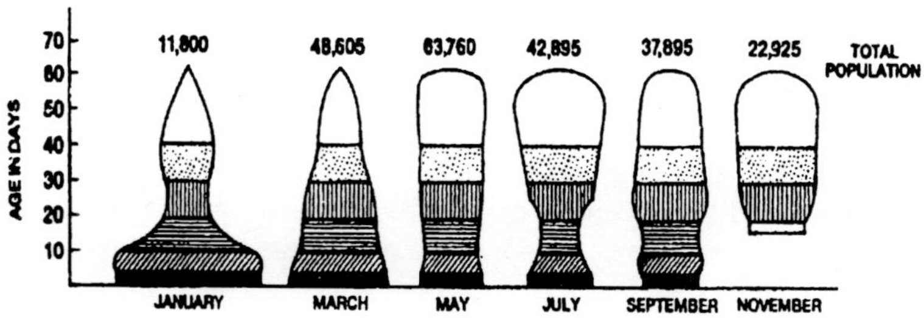

Fig. 7.2 : Changes in age structure in a hive of honey bees over one season.

Age structure of human population in major areas of the world was studied in (1960) as shown in Fig. 7.3. By these bell shapes, it can be seen that the population of Europe, North America and USSR are stable in comparison to rapidly growing population of pre-reproductive and post-reproductive ages noteworthy in this comparison. As medical advances abate the high mortality in the older age-groups and contraception and family planning are more widely practised, the pyramids should show a bell-shape. The age structure may change under a disturbed environment otherwise.

BIOTIC POTENTIAL OF POPULATION(S)

It is denoted as the maximum reproductive power. The constant per cent growth rate of a population under optimum environmental conditions thus represents its biotic or reproductive potential. If it is designated by γ, then

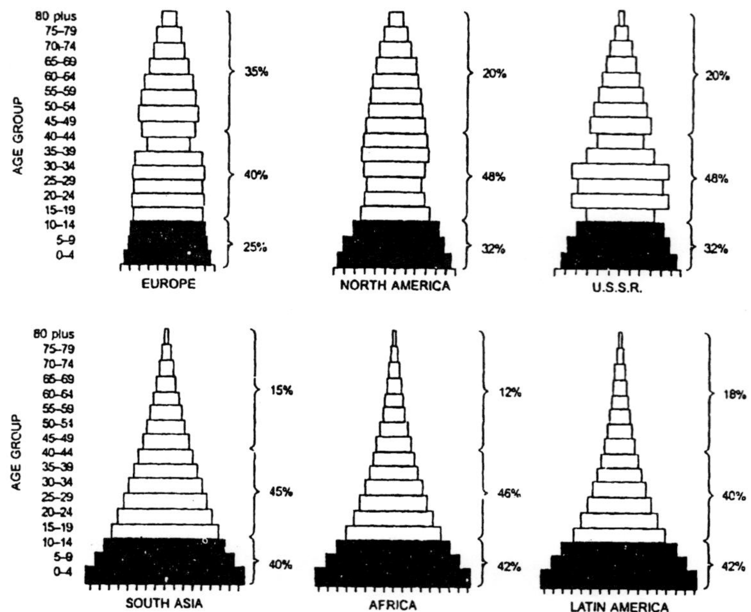

Fig. 7.3 : Age structure in human populations in the major areas of the world. Each segment in the horizontal scale represents one per cent; black bars are the pre-reproductive groups, stippled bars the reproductive group, and clear bars the post-reproductive groups.

$$\gamma = \frac{\Delta A N \Delta t}{N} \text{ [in an unlimited environment]}$$

where N is the number, t is time, A is the area and Δ is a constant. According to biometrician A.J. Lotka (1925) who advanced this mathematical equation, biotic potential altering conditions shall alter which in turn will alter age-structure. When conditions are returned to their previous states, population returns to its previous level and so does the age-structure.

THE GROWTH FORM OF POPULATION

Each population has a characterized pattern of increase which is termed as its growth form.

It increases in size in a characteristic S-shaped or sigmoid fashion. When a population starts growing, first the growing is slow, then it becomes rapid and finally slows down until an equilibrium is reached. If we plot time on x-axis and number of organisms on y-axis on a graph paper, we should get a S-shaped sigmoid curve. Human population and those of yeasts, Drosophila or rabbit in laboratory conditions show a S-shaped growth. However, if the growth stops abruptly, a J-shaped growth curve is obtained.

The level beyond which no major increase can occur is called the saturation level or carrying capacity. The following states have been found to occur in the population growth form:

1. The period of positive growth
2. The equilibrium position
3. Oscillations and fluctuations
4. Decline and extinction
5. Special cases

1. ***The period of positive growth.*** The curve representing this period (Fig. 7.4) is usually sigmoid or S-shaped in form. Literature from population studies shows that Carison (1913), Pearl (1930) and Gause (1934) drew curves for laboratory populations of Drosophila, yeast and Paramecium, respectively. A logistic curve for the growth of yeast cells is shown in Fig. 7.5. Several examples of logistic curve obtained from human populations are available in literature. The curve has been applied to various demographic units like cities, states, countries and the world (Pearl, 1930).
2. ***The equilibrium position.*** Equilibrium state can be defined as that mean numerical stability, i.e. the average size helped by a population very a considerable period of time. As shown by Richards (1932), the growth of the pure strain population of the yeast *Saccharomyces cerevisiae* shows this trend for over 1200 hours of observation. The time is practically horizontal. Davidson (1938) reported equilibrium in the sheep population of Tasmania. Since human populations have not yet attained their maximum growth, it has not been possible to report on their equilibrium state.
3. ***Oscillations and fluctuations.*** Oscillations are the symmetrical departures from equilibrium whereas asymmetrical departures constitute population fluctuations.

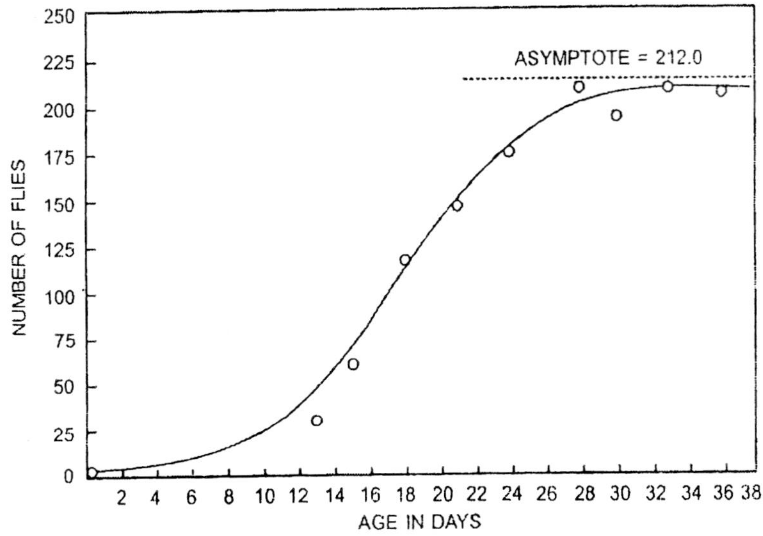

Fig. 7.4 : The logistic growth of a laboratory population of *Drosophila melanogaster*.

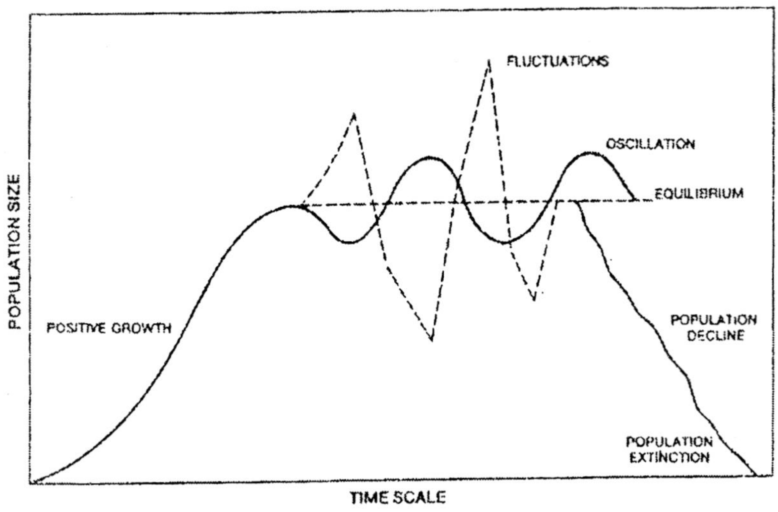

Fig. 7.5 : Stylized representation of the various phases of population growth form in yeast.

Gause (1934, 1935) noted oscillations while rearing a mixed population of yeast *Saccharomyces exigmues* and the ciliate *Paramecium aurelia.* In Smooth oscillations occurred under controlled experimental conditions, the three cycles as shown in the Fig. 7.5 for each species are regular and depart at the same magnitude. Similar studies on insect host (house-fly) and its pupal parasite (Mor-moniella vitripennis) were made by De-Back and Smith (1941).

Fluctuations in the population of a ciliate *Glaucoma piriforinitis* cultured under three nutrient conditions at constant temperature were recorded by Hall and Shottenfeld. The population A was maintained in a solution of casein-peptone, KH_2PO_4 and distilled water. Population B had in addition, a small amount of thiamine. Population C had thiamine at least ten times more than the population B. Each population as shown in the Fig. 7.5 showed fluctuations before decline.

Similarly, fluctuations in the population of Salmon over a long period were observed by Huntsman (1938) who commented that "fluctuations in fish abundance and their causes is a central fisheries problem".

4. ***Decline and extinction.*** A decline in population means consistent and progressive reduction of the populations below the equilibrium or lower than the usual range of fluctuation or oscillation. Whereas extinction can be defined as the final dying out of the group. Although these two forms are separable by definition they actually belong to each other.

There are many examples of known decline and extinction of species like Arizona elk, the great auk, the Labrador-duck, the passenger pigeon and heathen. However, phenomenon was studied by Gause, Nastukova and Alpatov (1934) in the population of two species of Paramecia, viz. *P. caudatum* and *P. aurelia* when cultured in homo-typical and heterotypical conditioned media. After the two form attained equilibrium they were placed as single species into a medium of the two conditions types. The results showed that the population of *P. caudatum* dies out rapidly and disappeared entirely on the 18th day. The rate of decline varies under different conditions.

5. ***Special cases.*** These refer to sudden changes that depart radically from the normal pattern of equilibrium, particularly

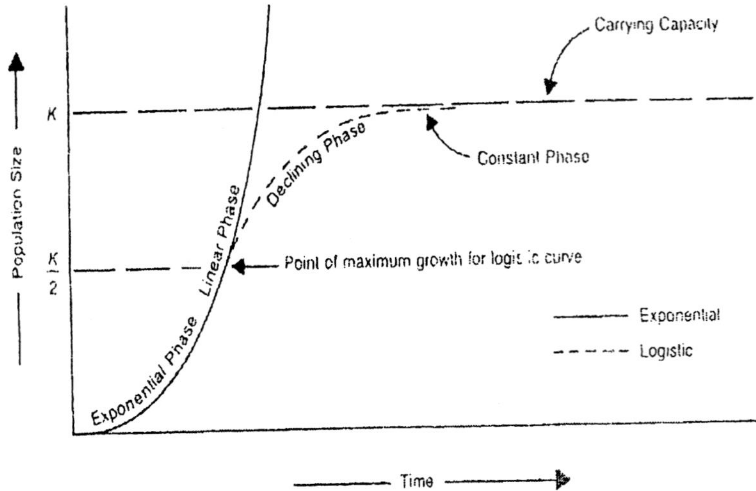

Fig. 7.6 : Exponential and logistic growth. An exponentially growing population increases by a constant percentage every time period. A logistic population initially grows almost at an exponential rate, then slows down and stops growing at its carrying capacity.

to population spurts and crashes. This aspect is important for workers studying the control of economically undesirable insects and mammals as well as epidemics. On the basis of the foregoing discussion, we can conclude that

(i) A population has a certain life history roughly divisible into periods or phases.

(ii) These periods vary in duration and numerical size with the particular species.

(iii) It is meaningful to study population growth form since it provides a numerical measure of the populations past history upto the time of most recent observation.

POPULATION DISPERSAL

Population dispersion is the movement of individuals in or out of the habitat. This movement can be grouped into following three categories.

1. ***Emigration.*** Emigration literally means the one way outward movement. Generally it is the result of overcrowding. This is generally regarded as an adaptive behaviour that regulates the population in particular habitat and avoids its exploitation. Moreover, such movement provides opportunities for inter-breeding with other populations leading genetic diversity and adaptability. Continued emigrations are rare; however, when they occur, the result is depopulation. In such cases equilibrium is maintained by enhancing the reproductive ability as well as decreased mortality among the population. It is exhibited by migratory locust, lemming, grouse, snowy owl, and grey squirrel.
2. ***Immigration.*** Immigration is just opposite to emigration. It can be defined as one way inward movement. As a result, over-population is caused, followed by high mortality among immigrants. It reduces reproductive capacity of the individuals. Rapid or large scale immigration or emigrations are more likely to have pronounced effects than the small or gradual dispersal movements. The introduction of a large number of bluegill fingerlings into a pond where the bluegill population has reached its carrying capacity, resulted into decreased growth rate and smaller size of the fish.
3. ***Migration.*** Migration is the periodic departure and return of a population. It is best developed in insects like the butter flies, dragon flies, fishes like eels, and in several birds and mammals. Migratory movements are controlled by circadian, lunar and tidal rhythms as well as seasonal variations.

The reasons for migration generally are the requirements of food shelter or reproduction. Unexploited habitats and resources are thus utilized by these populations. However, it may cause mortality during dispersion, nevertheless protects the individuals from intraspecific competition. Bird migration still remains a mystery of endurance and navigation.

Dispersal is much influenced by the presence or absence of the barriers and the vagility or inherent power of movement. Vagility is often greater than commonly realized. Although birds and insects are noted for their ability to 'get around', many plants and lesser forms of animals show greater dispersal powers.

POPULATION FACTORS

There seems to be an inherent ability of populations to maintain themselves or to be so regulated as to be maintained. These natural populations are influenced by abiotic and biotic environmental factors collectively called as population factors. These factors are discussed below.

Precipitation (Rainfall)

Available literature shows that rainfall has a direct bearing on locust out-breaks in various parts of the world. The lifecycle of the locust proceeds without essential interruption only so long as the effective rainfall and temperature remain within certain limits. However, are outbreak does not develop unless a number of conditions have beer, satisfied, nevertheless, rainfall remains an important physical factor in a complex system.

Similarly, Errington studied the effect of drought on muskrat population of Iowa. As the drought becomes severe, there is an increase in *intraspecific strife* which usually assumes the form of fighting between old and young of both sexes.

Furthermore, Shelford (1943), reported that rainfall and snowfall are important climatic factors affecting the growth form of populations of collared Lemming in Churchil Manitoba (U.S.A.).

Thus we see that degree of precipitation (rainfall or drought) severely check the growth of population.

Temperature

Temperature constitutes an important abiotic factor affecting growth of the population. Water flea *Moina macrocopa* when abundantly fed produced different sized populations depending upon the temperatures at which they were cultured. Kendeigh (1942) studied nesting losses in fifty; one species of birds and showed that temperature played a vital role in this destruction, Gunter (1941) reported the effect of cold upon marine fish populations in Texas.

Humidity

Humidity is a potent agent in terms of reproduction of different populations. Uvarov (1931) observed the effect of humidity on adult flea populations. As humidity lowers at a constant temperature, the mean longevity of the flea population also decreases.

Storms

Importance of wind as an ecological factor has already been discussed in preceding chapters. An interesting example of the effects of storms on population was reporter by McClure (1943). Practically every storm blew poorly placed nests to the ground. Hail storms killed many adults and cloudbursts dislodged the nests. Doves always tried to protect the youngs but could not be always successful.

pH factor

pH factor is not ecologically so important. However, within certain range of acidity or alkalinity, limiting effects on the population have been observed. Distribution of sessile rotifers in inland lakes was studied by Edmondson (1944). He showed that many rotifers perfected to stay at different corners showing preferences for bicarbonate concentration. Some rotifers tolerated high alkaline concentration while others could not.

Calcium

Sometimes the presence of calcium and magnesium ions in the aquatic environment affect the population density. Jewell (1939) confirmed this aspect in fresh water sponges. He concluded that some sponges are sensitive to calcium bicarbonate while others possess a wide range of tolerance. *Ephydatia* was found absent from waters both high and low in calcium. *Spoilgilla* is extremely sensitive to calcium tension whereas *Heteromyenia* can tolerate an impressive range.

Atmospheric Gases

Atmospheric gases such as oxygen, carbon dioxide, nitrogen as well a biologically inert trace gases do not constitute a limiting factor for the populations of land environments as they remain relatively constant in the atmosphere. However, the situations are quite different in aquatic environments. The amounts of oxygen, carbon dioxide, and other atmospheric gases dissolved in water are variable in different conditions. For example, oxygen is a limiting factor in lakes with a heavy load of organic material. The oxygen supply in water comes from two sources, i.e. by diffusion from the air and through photosynthesis in aquatic plants. Moreover, temperature and dissolved salts greatly effect the ability of water to hold oxygen.

Carbon dioxide is present in water in highly variable amounts. It is extremely soluble in water and gets its supplies from respiration decayed soil or underground sources. It combines with hydrogen molecules of water to form H_2CO_3 which in turn reacts with bicarbonate that keep a neutral pH.

THE REGULATION OF POPULATION

Each population has an optimum size. A population that is self regulating is said to have a density-dependent population regulation and the population which is not self regulating may be subjected to density independent regulation. A real population would grow according to S-shaped curve called logistic curve.

However, the growth must be limited eventually by some resource in its environment. This concept is known as exponential growth and represented by an exponential curve as shown in Fig. 7.7. A balance between logistic growth and exponential growth is called as population's carrying capacity. Carrying capacity of a population may further be defined as the maximum population size that can exist in a habitat or ecosystem over a long period of time without detrimental effects to either population, habitat, or ecosystem (Fig. 7.8).

If the population size exceeds the carrying capacity then deaths exceed births, the growth rate is negative and the population returns to carrying capacity level. On the other hand, if population falls below the carrying capacity, births exceed deaths. The population growth is positive and the population size returns once again to the carrying capacity.

Density Dependent Regulation

Under density dependent population regulation, the rate of population growth is inversely related to population size. Density dependent population regulation implies that there is a feed-back mechanism that the population in someway responds to its own size or density, bringing about changes in its birth and death rates. Density dependent factors are biotic. They depend upon interspecific and intraspecific co-actions. They include competition, predation, reproductive activity, territoriality, emigration, disease and physiological stress.

Density Independent Regulation

In density independent regulation, the mechanism of control has no relationship to population size. These are determined by

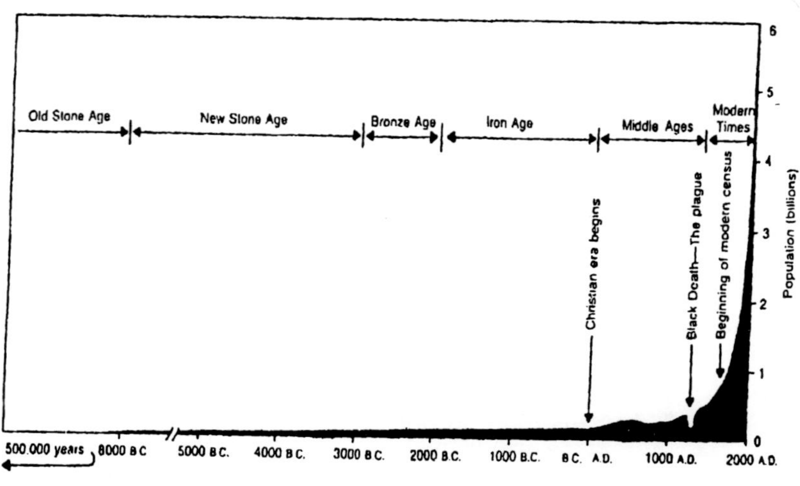

Fig. 7.7: The history of the human population. For most of *Homo sapiens'* time on the Earth, the population was probably less than I million and perhaps only a quarter of a million. It has taken hundreds of thousands of years for the number of people to exceed 1 billion; now, the number will reach 6 billion if current growth rates continue. Note the left-hand side of the graph is not to scale. If the Old Stone Age were to scale, it would stretch 12 m to the left.

physical or abiotic condition of the environment. They include space, prevailing weather and food supply. For example, an insect population in a forest may be subjected to hurricanes. The population size has no effect of the frequency or route of hurricanes. The mortality rate would be 100 per cent regardless of population density.

Kendeigh (1974) suggested that density independent of environmental factors set the ultimate limit of carrying capacity which cannot permanently be exceeded and the density dependent or biotic factors generally stabilize populations at or somewhat below the ultimate level. What stops population growth and what controls abundance have been discussed by Krebs (1972) through following theories.

Biotic regulation. Nicholson (1933) suggested that only factor that can control population is competition. It may be for food, shelter or reproduction.

Climatic regulation. Uvarov (1931) while studying insect populations suggested that climatic factors regulate their population.

Comprehensive theory of natural control. Thompson (1929) suggested that a combination of complex factors regulate the population. A close network of interactions are involved in population control. Schwerdtferg (1941) termed this network as "grandocoen" which he defined as the totality of all factors affecting a population.

Self regulation. Chittay (1960) proposed that all species are capable regulating their population without destroying their renewable resources (Fig. 7.7)

HUMAN POPULATION DYNAMICS

Like every other biological species, *Homo sapiens* can be characterized as a group of populations, each with rates of birth, death and growth and with a population structure, which is the relative proportion or the number of individuals of each age and sex. Like other species, *Homo sapiens* too depends on ecosystems for its existence. We cannot escape these constraints; we are part of the biosphere and must live with its laws.

How many people have lived on the earth? The first estimates of population in Western civilization were attempted in the Roman era. The first human census, however, was made in 1655 in the Canadian colonies, by the French and the British (Desmond, 1962). Sweden began regular census taking in 1750. The first census in the United States was made in 1790 and ever since continued to be recorded every decade. However, by studying modern and primitive people and applying principles of ecology, we can gain a rough idea of the number of people that might have lived on earth during our species history. Edward Deevy (1960) estimated the area of the old world, where early stone age people might have been to live, and calculated that the average early stone age population could have been as low as one quarter of a million less than the population of modern small cities. Early human beings were hunters and gatherers, and people who followed this way of life lived at a density of somewhere between 1 per 130–260 square kilometres in arctic and semi-arid regions, and 1 per 2.6 square kilometres in the most habitable areas.

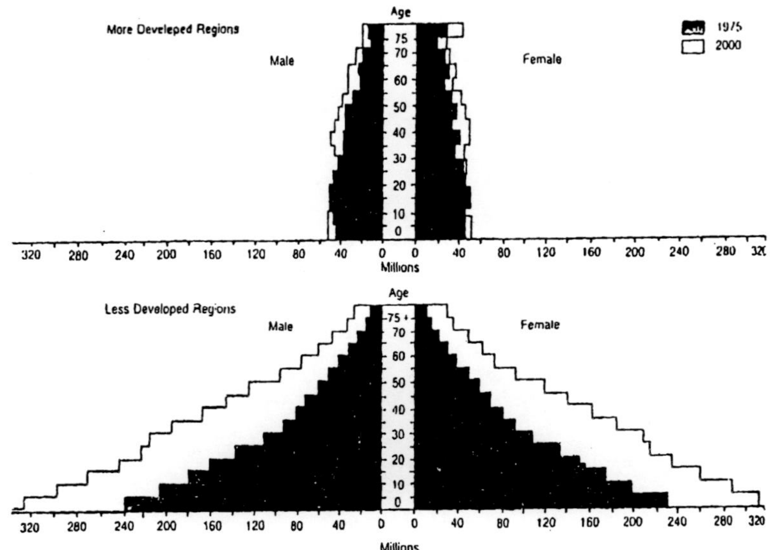

Fig. 7.8 : Age and sex composition of the Earth's human population for 1975 and projected for the year 2000. (From Council on Environmental Quality and the Department of State, 1980).

With the invention of agriculture and the domestication of plants and animals, the population density increased greatly. This phase of more rapid growth saw a general increase in the total number of people. An increase of 0.03 per cent per year would have brought the population from 5 million in 10,000 B.C. to about 100 million in 1 A.D.

POPULATION PROJECTIONS

The Population Reference Bureau projects that the total human population will grow from 4.4 billion in 1981 to 6.1 billion in the year 2000 with an annual increase of 1.7 per cent (Haub, 1981). The age structure in developed countries will move toward a greater percentage of older people. The less developed countries, however, will continue to have a higher percentage of young people and a higher growth rate than the more developed countries. Thus the greater percentage of the world population will live in the less developed countries (Fig. 7.9).

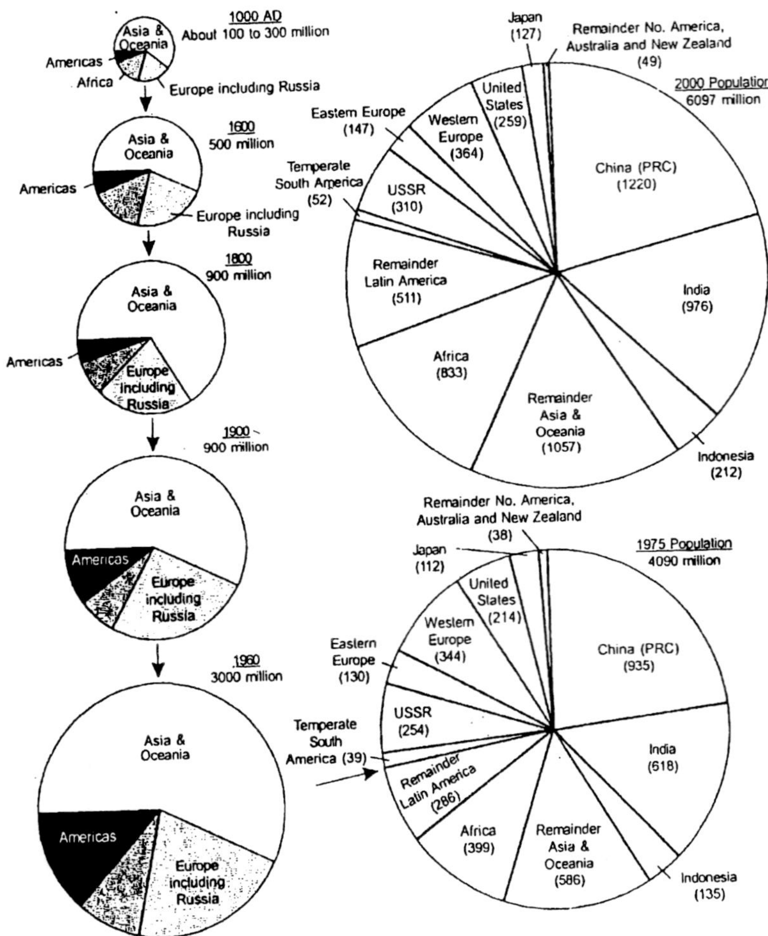

Fig. 7.9 : The figure shows growth of human population from 1000 to 2000 AD. In 1000 AD 60% of the people lived in Asia, 17% in Europe and Russia, 18% in Africa and 4% in the Americas. In 1975, 58% lived in Asia, 18% in Europe, 14% in the Americas, Australia and New Zealand and 10% in Africa. It is estimated that by 2000, 59% will live in Asia, 13% in Europe, 13% in Africa and 15% in Americas and Australia and New Zealand.
(Source: **Council of Environmental Quality, USA).**

Some Generalizations about Human Populations

Human beings cannot escape the laws of population growth. Here we should remind ourselves of the fundamental concept that, 'individual populations are capable of rapid exponential growth, but this is rarely achieved in nature; control of populations is the norm'. The degree of control and causes of control determine many features of ecosystems and the life around us. Modern medicines have decreased death rates, particularly among the young. Although the overall growth rate is approximately 2 per cent per year, the number of individuals added to the world population is huge and beyond our ability to imagine.

Developments in genetic engineering may alter our life or productive life, but that would increase our social problems. If these techniques succeed, the future growth rate of the human population may exceed exponential curve and age structure may incline towards older individuals. The resources available per person will decrease, decreasing further the available opportunities, changing the age structure and population densities. We should remind ourselves of another fundamental concept that 'sustained life on the earth is characteristic of ecosystems, not of individual organisms or populations. For life–including human life–to be sustained, there must be a flow of energy and a compelete cycling of chemical elements necessary for life'.

The scientific and industrial revolutions that began after the European Renaissance provided a third major change; and in our time, the cybernetic, space age revolution is leading to a fourth period in human history. All technology–from the most primitive to most advanced–causes some change in the environment.

Human societies and civilizations have waxed and waned and the reasons for these rises and falls, as given in history, have fascinate people and led them to believe that changes in sedimentation rates in fresh waters appear to correlate with the rise and fall of the civilization.

None of these factors alone is sufficient to cause the rise and fall of human civilizations. People and their societies are too complex to be explained by a single factor. What is necessary to sustain life is scientific question. What effects specific actions have on the life around us and our environment are scientific questions but what we choose is ultimately an ethical question beyond the reach of science.

I. ENVIRONMENT AND HUMAN HEALTH

India has successfully crossed 50 year mark of its independence marching ahead with covetable progress scored through unflagging commitments and tireless efforts of the people and the government in social, economic, scientific and technological areas. A nation which missed three industrial revolutions of the past and could not produce even a needle at the time of independence, and had to start from the scratch, is running neck to neck with most advanced nations of the world in the field of frontier technologies and we are proud to say that in information technology on which the future of human race rests, it is surging ahead, specially in software technology. During the last five decades, India's achievement in science arid technology seem to be very impressive which would reveal expertise built up in space research. Nuclear engineering, production of steel fertilizers, petroleum, chemical, machine tools, thermal power plants, construction of big dams etc. Miraculous achievement has been made in the agricultural production through Green Revolution during last three decades which converted India of the fifties, as an importer of food grains to an exporter. The technological advancement in agriculture is brought about through the increased production of new high yielding varieties of crops by means of the application of chemical fertilizers and pesticides bringing more and more land under crop cultivation for the ever increasing population has no doubt saved mankind from hunger and pestilence. On the other hand various developmental activities such as construction of huge dams unscientific mining of coal for power plants and industrial units have changed the man-nature relationship and they have changed not only the economic and socio-cultural life of the people but also their values, systems, ideas, believes and indeed their entire life style. Destruction of more forests for extension of land for agriculture purpose, for coal mining activities as well as for making multi-storeyed buildings, roads and other constructions has led to the extinction of plant and animal species and is also responsible for ecological imbalance. Apart from these, the indiscriminate disturbance of the freshwater ecosystems leads to the disturbance in corresponding recycling system. The storm of modernization and industrialization has not only uprooted man but in fact has destroyed his habitat and environment too. The increase in the discharge of toxic gases from the Industrial unit and carbon dioxide liberated from animals and human beings and from burning of fossil fuels is as sharp as decrease in release of oxygen by the trees and plants, as a result of which the biosphere equilibrium maintained since time immemorial has been affected.

The relationship between human health and his environment is a two-way process. If one disturbs, other will be automatically disturbed. Therefore, the present work has been undertaken to illustrate the impact of ill environmental health on the health of human beings.

THE BIOSPHERE

It is well known that the earth's surface and the environment surrounding it is important to human health. The nature of the soil, air, water, temperature, barometric pressure, wind, sunshine, cloud, rainfall, humidity and latitude, must all in the last resort determined man's health and welfare. By controlling animal and vegetable life, man supplied himself with the essentials of life, including clothing, housing and food. The physical characteristics of the globe have determined his diet, and hence his health. The monsoons have favoured rice growing, the trade winds have taken merchants and adventures across the oceans, carrying incidentally the seeds of new vegetation from one part to another, and old diseases to new habitats.

The cloudy atmosphere, which keeps out the sunshine, deprives people of sun-made vitamin D, thus accounting for rickets. Both chronic bronchitis and carcinoma of the lung in Europe are, in some measure, due to atmospheric conditions, which, hold down the smoke and turn mist into smog. Where cloudless skies prevail such disabilities would hardly occur, were it not for cigarette smoking, but in strong sunshine fair skins with little pigmentation have an added liability to skin cancer. Rheumatism has for long been related to dampness and cold. Lack of trace elements in the soil causes diseases in animals, cachexia and anaemia in ruminants on soil deficient in cobalt, paralysis (swayback) in sheep on soil with little copper. In man the most pronounced effects of deficiencies are those produced by lack of iodine, iron and copper.

The physical character of the soil or sub-soil, also effects the ease with which water-borne infections spread. The soiling of the

6land and waterways is a more serious matter in some parts than others. Hookworm is inhibited by dryness. Soiling of the land with faeces is particularly liable to result in pollution of water supplies, which is a factor in the spread of hydatid disease.

Climate may affect pathogenicity of bacteria. Amoebae of dysentery devastates the East. Endemic cholera has, for many centuries, existed in India, spreading in pandemic waves across the world. Low lying lands and lakes, rich in organic matter and salts, sheltered from rain and sun, may have favoured the survival of the delicate vibrio outside the human body. Trachoma resulting in much blindness. One of the oldest diseases known to man, is a disease of poverty and promiscuity (Kumar, 1971).

Temperature and humidity, the state of the vegetation, and the abundance or otherwise of animal life, effect the vectors of disease. The mosquitoes, which carry malaria, yellow fever, dengue, encephalitis in many forms, filariasis and other disease, are affected and their disease-carrying qualities altered by a great .variety of climatic conditions. Similar influences may affect other 'carriers', which may have determined the distribution of malaria, yellow fever and filariasis. The flies carrying trypanosomiasis, the ticks carrying rocky mountain spotted fever, and the mites carrying scrub typhus, are all sensitive to climatic changes. And so too is the rat; he may have found living better in India where plague prevailed. The reason why plague occurs mainly in the rainy season may be due to the fact that the rat prefers to remain indoors, where fleas increase in numbers, with the seasonal lowering of temperature and humidity to aid propagation.

The environment effects our output of work, our mental alertness, our desire for change and many other attitudes and responses. A mean temperature of 40°F in winter and 60°F in summer, with a relative humidity of about 60 per cent at noon, and high enough at night to precipitate dew, is the most stimulating for mind and body.

To ensure his well-being man has to produce an ecological balance between himself and his environment. This balance is constantly in danger of being upset from his own activities; Pollution of the environment results from a wide range of human activities;

uncontrolled disposal of human excreta and refuse; smoke from coal or oil burning; fumes from motor vehicles; industrial discharges, such as mercury and other chemicals or the products of nuclear fission, which poison cattle and fish and concentrate undesirable substances in human tissues; misuse or overuse of insecticides and fertilizers. Excessive noise causes deafness and in lesser degree destroys equanimity; the extrovert may be stimulated by the transistor radio, but for the introvert it's an intrusion into privacy which irritates and disturbs. Foodstuffs are contaminated in preparation and processing to cause outbreaks of infection. Population explosion, beyond the capacity of the soil, results in malnutrition; over-consumption of food causes obesity, diabetes and dental decay. Association with the animal world gives rise to zoonotic diseases. The ill-constructed house, overcrowd dwellings and mushroom growths around urban conglomerations cause infections to spread and contribute to mental stress.

CONCEPT OF HEALTH AND HYGIENE

Health signifies a wholeness or soundness of body and mind, but we are confronted with the difficulty of determining its relationship to 'disease'. Health and disease must be intimately related for if disease did not exist it would be irrelevant to talk of health. The two states are contrasted in our minds, as it were the two sides of a coin–so that when one is present the other is absent.

HEALTH

If any one says that only those are healthy who function perfectly in all parts, and that others who function less well are not healthy, he is over-simplifying the definition of health. Also it is clear that the real measure of health is not the Utopian absence of a disease but the ability to function effectively within a given environment. And since the environment keeps changing, good health is a process of continuous adaptation to the myriad microbes, irritants, pressures and problems which daily challenge man. The definition by WHO is 'health' is a state of complete physical, mental and social well-being and not merely the absence of disease or infirmity. Man is a social animal; he cannot easily live for himself alone; there must be harmony with the social environmental, so it can be further defines as below:

Health is a state of feeling well of body, mind and spirit, together with a sense of reserve power; based upon normal functioning of the tissues, a practical understanding of the principles of healthy living, a harmonious adjustment to the environment (physical and psychological), it is a means to a richer life of service

The preservation of health in the group has been defined by WHO as: "The science and art of preventing disease, prolonging life, and promoting health and efficiency through organised community efforts for the sanitation of the environment, the control of communicable infections, the education of the individual in personal hygiene, the organisation of medical and nursing services for the early diagnosis and preventive treatment of disease and the development of social machinery to ensure for every individual a standard of living adequate for the maintenance of health, so organising there benefits as to enable every citizen to realise his birth-right of health and longevity

HYGIENE

In the good olden days man took a fatalistic view of life, death and the disease. Sickness, was something to be endured.

It was not until the middle of the 19th century that man's understanding of nature grew that he could bring the major environmental health hazards to his control. Another half century later achievements of sanitary engineering and preventive medicine in adapting the environment to man were complemented by the success of therapeutic medicine in helping man to conquer some of the diseases.

As a member of society a citizen has an obligation to seek to understand them, and also has the privilege of questioning whether in our effort to add to the comfort and the quantity of life, we may be sacrificing something. Lest the ways in which we are artificially reordering ourselves and our environment may not in some degree diminish our stature in nature and, thereby, rob life of some of its richness and bring sickness. S. S. Bhatti observes in his article that the pace of development is regulated by the animal instincts in man, the process of development itself thrives on a destruction-construction symbiosis. Chandigarh may be cited as an example of disciplined architecture... an environmental conducive to peace... and better quality of life .

HEALTH FOR ALL

The situation today is that nearly one thousand million people are trapped in the vicious circle of poverty, malnutrition, disease and despair that saps their energy, reduces their work capacity and limits their ability to plan for the future. For the most part they live in the rural areas and urban slums of the developing countries. Whereas the average life expectancy at birth is about 70-75 years in the developed countries, it is only 45-55 in most developing countries. Of every 1,000 children born into poverty in the least developed countries, 200 die within a year, another, 100 die before the age of five and only 500 survive to the age of 40.

Health for all is the attainment by all the people of the world of a level of health that will permit them to lead a socially and economically productive life. It will mean that there will be a fair distribution among the population of whatever health resources are available. Essential health care will become accessible to individuals and families in an acceptable and affordable way, and their full involvement.

Alam Ata Conference observed that Primary Health Care is the key to attainment of health for all by the year, 2000. It also identified eight essential elements of primary health care. These are: education concerning prevailing health problems and the methods of preventing and controlling them; promotion of food supply and proper nutrition; an adequate supply of safe water and basic sanitation; maternal and child health care, including family planning; immunization against the major infectious diseases; prevention and control of locally endemic diseases; appropriate treatment of common diseases and injuries; and provision of essential drugs.

It is interesting to note that the various components of primary health care have direct relation to environmental management.

CANCER

The relationship between cancer in different parts of the body and various environmental contaminant and pollutants, have been increasingly observed over a long period. It is now generally agreed that between 60 and 90 per cent of all cancers are directly or indirectly related to the environment, that is, related to factors, in water, air, solar radiation, living and working environments and personal choices of diet and ways of life– such as tobacco smoking and alcohol consumption. The assumption is that all cancers are environmentally caused until it can be proved otherwise.

The environmental component in Communicable disease such as malaria, schistosomiasis, filariasis and trypanosomiasis and the efforts to control these diseases through environmental interventions directed at the habitats of their vector have come in for considerable attention, But the large scale use of pesticides to protest crops and kill disease-carrying insects, and the increasing use, particularly in developing countries, of growth–promoting substances to increase meat production, have led to increasing concern about chronic toxic effects in human beings which might ultimately be related to cancer. Among these growth-promoting substances are antibiotics, hormones and vitamins,

Viewing the world as a whole, the net effects of many known cancer-causing factors remain substantial, though none is as prominent as tobacco. Table 7.2 given enlists causes of cancer.

Table 7.2. : Relative Importance of known Cancer Causing Factors.

Factors	*Percentage of all cancer death*	
	Best estimate	*Range of acceptable estimate*
Tobacco	30	25-40
Alcohol	03	02 -04
Diet	35	10-70
Food additives	<1	5 -2
Reproductive and sexual behaviour	7	01-13
Occupation	4	02-03
Pollution	-2	<1-5
Industrial products	<1	<1-2
Medicines and medical procedures	1	0.5 -3
Geophysical factors	3	02-04
Infection	10?	1-?

OCCUPATIONAL CANCER

It is generally considered to be caused by multiple factors. Certain chemicals on their own are known to be carcinogens. Their effects are potentiated by exposure to other chemical or physical agents. Furthermore, there are a number of chemical agents which become carcinogenic in the presence of other chemicals (co-carcinogents) in the work environment.

Control of all these problems require much more careful observation of the health of workers, coupled with thorough environmental investigation. This, therefore, is one of the areas where WHO has directed attention in the last few years. In December, 1980, WHO experts reviewed the effects of combined exposure in the work environment and identified gaps in our knowledge requiring further research.

NATURE AND ENVIRONMENT

Work, when fully adapted to human capacities and limitations, is a potent factor in health promotion. It is easy to see that good physical performance is always reflected in good physical development, while mental gratification with achievement and the realisation of goals has a positive effect on mental health. It would include: the absence of (or the total control of) chemical, physical and biological hazards, the adaptation of machines to human performance. The provision of adequate lighting, the provision of rest periods, job enrichment to avoid repetitive tasks and boredom, and a balanced physico-social environment coupled with job satisfaction.

The conditions require not only human engineering methods and skilful management but also require evaluation of human capacities, tolerance and susceptibilities.

The utilization of work as a factor in positive health has hitherto been limited and unorganizeu. We observe, however, that people who continue working after their retirement maintain better health than those 'who stop working altogether. We also note that physical disabilities can be overcome through work, hence the well-established practice of occupational therapy and vocational rehabilitation. WHO aims to direct attention to the development of methods by which work would become a valuable contribution to health, rather than a health hazard associated with disease.

RISK AND ACCIDENTS

It is rare to find an occupation which does not entail a potential risk to health and sometimes even more than a risk. Several examples come to mind about farm work in the tropics, where workers may be exposed simultaneously to pesticides and excessive heat. In foundry operation occupational exposure may include various fumes, irritant gases, carbon monoxide, heat and dust. In the chemical industry one may find the combined hazards of various chemicals, a mix lure of solvent gases and vapours. Even in office work where the physical environment may be fully controlled, several psycho-social stress factors may exist–including pressing demands, inter-personal relation and job security.

With an estimated 50 million work-related accidents occurring each year many of them resulting in permanent disability, the war on accidents which began over a century ago is being fought on a global scale. In this war against accidents, some success has been achieved in developed countries. But in the undeveloped lines the reduction of accident frequency and severity rates to the lowest figures attainable by human efforts, lies a long way ahead. It is towards this eventual goal that the worldwide efforts for the improvement of working conditions are underway with the full support of U N agencies.

Apart from the human agony and despair which they cause, these accidents continue to drain away valuable human resources and place a heavy burden on economy of every country. Another disturbing factor, which make the task of prevention all the more urgent in that while the accident frequency may have levelled out in most of the industrialized countries, the rate of fatal accidents in the developing countries like India has doubled or even tripled.

RURAL SANITATION

Most of our villages look like dung heaps. It is very difficult to make people keep their villages and their surroundings neat and clean. The schools, balwadis, other public buildings and houses in the villages are full of dirt and waste of different kinds. The stagnated water acts as the breeding centre for flies and mosquitoes. The open latrine system prevails all-round.

ROLE OF VILLAGE SCHOOLS

Unless villagers are educated on the need for keeping their villages neat and clean, village sanitation would remain a dream. First of all, the surroundings of the schools and public buildings should be kept neat and clean. If our village schools and their surroundings are kept clean and the teachers and children are in a position to remove even the small paper bits in the village streets and throw them in the nearby dustbins, naturally others in the village will follow suit. If the schools have urinals and latrines constructed at a low cost and are used properly, they will serve as models to the community.

In each Panchayat Union, if the schools which are neat and clean are selected and awarded prizes every year, it will create a healthy competition among the school and their pupils to keep their surroundings neat and clean. The Health Department should make it obligatory that there should be latrines and urinals attached to each school.

YOUTH POWER

At present, the energy of the youth in the villages is not properly channellised. Our political parties too are riot effectively directing their workers in the village reconstruction programmes. They don't mind working for cleaning their villages and digging pits for latrines once or twice in a month for a couple of hours. But they are not properly motivated.

WASTE TO WEALTH

People in the village must be told that the waste collected from the dustbins may be converted into good compost manure. This can he sold or can he used for the vegetable garden at the noon meal centres. In the same way the night-soil may also the prepared as good manure. The villagers may not keep their cow-dung and other cattle waste in heaps exposing them to the sum, but he asked to dig compost pits and maintain them properly.

Drainage

In most of the villages no proper system of drainage is in existence. Since this requires a large allocation or government money, we can encourage people to have soak pits in the front yards of their houses for the water to be absorbed. If the waste water finds a place to go into the backyard, it can be utilised for rising a kitchen garden. Whenever the sewage of a particular house dirties the neighbourhood, the village youth leaders have to request the concerned head of that facility to channellise the waste water properly and making the streets and public places untidy.

PERSONAL HYGIENE

Village sanitation can be improved only where people understand the importance of personal hygiene and its interdependence on general cleanliness. If an individual is making his surrounding filthy, then it is his or her health which is affected.

DRINKING WATER

The Health Department has to take effective steps to purify the water in the village wells and tanks to be seen that the drainage water does not mingle with the river water. People should be educated against the use of the river beds as open air latrines.

MASTER PLAN

The Village Planning Committees have to prepare a master plan for improving the health and sanitation of their villages in consultation with experts in this field. Such a plan should be implemented in a phased manner. For implementing this plan villagers should partly depend upon their own resources. There must be a plan to educate the people first. There must be a link between the programmes for the village sanitation and the programmes for the economic development. Improving the village sanitation depends upon both economic uplift and providing health education to the people. Training the children in good health practices and keeping their surroundings neat and tidy is very important. After all the habit formation is ultimately responsible for keeping one-self and one's surroundings neat and clean.

POPULATION

In India an overpopulated country, the growing population is the main cause of environmental destruction. The solution lies not only in containing the population growth, but also by managing the nation's soil and water resources in order to increase production in the fields. It there is poverty and hunger, it is because the managers and scientists of this country have not learnt to use its environment at a high level of suitable productivity. To survive, the poor often has no alternative but to turn up to the forest wealth-to sell the stolen fire-wood. The rich indulge in destruction of the environment to satisfy their greed. The development process would involve not only the restoration and enhancement of nature, but also to achieve small family norm. Women and child are the worst hit in the whole game. So let us study some aspects of environmental health concerning the women and the children.

WOMEN

Poor village women are the worst sufferers of environmental destruction. Every morning they have to go on long march in search of fuel, fodder and water–the women may be old, young or pregnant. The march is becoming longer and more tiresome. They are reaching the limits of their endurance. Water, fuel, fodder, building materials and even food to some extent are gathered freely from the immediate environment but urbanization and advance of cash economy have greatly affected the country's base. Environmental destruction exacerbates women's already acute problems in a way. They do not get any worthwhile health care due to shyness, over-work, and inferior status of women in that society.

CHILD

Until the middle of the 19th century, only half of all the children born in the US reached their fifth year, and as recently as 1900 a new-born infant in the U. S. had less chance of surviving a week than did a man of 90. Today, in the economically developed unions, 97 per cent of new-borns live to adulthood. Few children suffer the effects hunger. Few are infested with worms of other debilitating parasites. The youngsters are spared most of the infectious illnesses small pox, tuberculosis and other fevers, dysentery–that carried off so many in the past. Today, very few mothers die in the process of delivery, although mortality rates were relatively high as recently as the 1930s. Indeed, very few persons of any age or sex die of the acute infections in the developed countries, which used to account for the majority of all deaths. In developing countries like India, despite the advances made in the field of health, the infant and mother death rate is still very high and so is the incidence of preventable communicable disease.

SPORTS AND PHYSICAL FITNESS

There can be no doubt about it : sport offers the best antidote to the tensions and stress that are everyday hazards as our lives become ever more competitive. It offers a respite from our daily cares and contributes a vital element of balance and relation.

Individual sports, such as swimming, athletics or gymnastics offer a school for character formation, where self-discipline and stamina can be developed. Individual exert themselves to improve their own performances however, modest these may be, and to set themselves new goals. Team games like football, rugby, hockey or water-polo, on the other hand, encourage another set of qualities among them—coordination with colleagues, the team spirit and a sense of-belonging, what we call the sporting spirit combines fair play, dexterity, self-control and good manners. It was in this context that Baron Pierre de Coubertin, re-founder of the modern Olympic Games, commented : "The essential thing in life is not so much to win as to fight fair". Sport has a beneficial effect on general hygiene. There is nothing better than a good shower after the sweat of physical effort, and it is just what tired muscles cry out for. Besides all this, athletes noticeably smoke and drink less than those who take no regular physical exercise. At whatever age a person starts, sport has a health-giving effect, although obviously the intensity and the nature of the effort involved varies with age. However, there are some detractors of physical exercise who claim that the benefits which may be derived from sports fall very far short of compensating for the risks entailed. But it is merely to put up a lame excuse for not doing any physical exercise.

ANIMALS

The role of animals as contributors to the environmental preservation and human well-being cannot overemphasized, but while discussing the subject of environmental health, we are inclined to over-react about the risks due to animals and their products. We cannot forget that man's health has always benefited in many ways from the animals.

BIOLOGICAL VALUE

Livestock are an increasing by important source of protein in developing countries. Even "vegetarian" populations are often in fact lacto-vegetarian-adding milk products to their diet. There is no practical substitute for animal proteins in human nutrition. The WHO/FAO/UNICEF has recommended that vigorous steps should be taken to raise the efficiency of producing milk and meat from cattle, meat from sheep and pigs, eggs and meat from poultry throughout the developing world by applying existing knowledge and by introducing new knowledge in these fields. Cattle, buffaloes, horses, camels, yaks and lamas still provide more than 80 per cent of the world's total power. In India, cattle alone account for an irreplaceable 54 per cent of the energy used in agricultural production.

AS COMPANIONS

Man has always kept pet animals among his closed companions and such animal provide people all over the world with various forms of much needed healthful exercise and recreation. Recently, animals have been used as direct aids in the treatment of mental illness. Pets can help human beings through their development problems at various stages of maturity. They can encourage a toddler to become assertive and independent, or help a child to overcome feeling of loneliness or alleviate feelings of guilt. An animal companion frequently provides the lonely person, particularly in cities, with the only remaining possibility for love without rejection. They give timely warning of infectious agents, or the accumulation of toxic materials in the environment. Animals have several advantages as detectors of environmental hazards. Because of their position in a particular food chain, certain species may accumulate toxic substances more rapidly than man; some are more prone than others to become ill or accumulate chemical agents, and can, therefore, be selected as useful detectors. Virtually every human disease from cancer, cardiovascular disease and stroke to degenerative disease associated with aging and psychiatric disorders may be studied in animals.

ZOONOSES

More than 150 zoonoses and food-borne diseases of animal origin are recognized, and their surveillance, prevention, control and eradication are tasks of considerable magnitude in every country. Reservoirs of zoonoses among domestic animals are the greatest source of danger for man, since it is with these that he is in closest contact. Many food-borne diseases due to animal products are the consequence of zoonoses in domestic animals; the infectious agent may be present throughout the chain from animal foodstuffs, animal production and food processing to food preparation. Quite apart from the human suffering caused by zoonoses, these diseases are also responsible for great economic losses.

Among the various human foodstuffs, those of animal origin tend to be epidemiologically the most hazardous. Several factors have contributed recently to the increasing necessity for strengthening food hygiene supervision programme. These include the rapidly increasing world population with its ever greater demands for food: the increase in urban populations with a corresponding decrease in rural populations, which stimulate increased production of processed or semi-processed foods; advances in food technology resulting in new and more "sophisticated" presentation of food, the handling of which may not be properly understood by the consumer; greater use of agricultural and food chemicals; the increase in environmental pollution which acts unfavourably on food quality in general; more active national and international commerce in food, including the transport of basic materials from areas where hygienic standards may be less than satisfactory and greater tourist travel. Food-borne diseases may be caused by various biological agents, such as salmonella, viruses, protozoa, helminths, arthropods, biotoxins (such as paralytic shellfish poisoning), poisonous-chemical (such as residues of heavy metals), and other chemical compounds, drugs or food-drug combinations and radio-nuclides. Both domesticated and wild animals may inflict injuries on humans of varying severity with their teeth, claws, horns, hoofs, spines, and so on. Animal bites may transmit such infections as rabies, pasteurellosis, rat-bite fevers simian herpes virus disease. There is no doubt that they represent substantial costs to the community in the form of treatment of victims, immunization, hospitalization and times lost from work.

High concentration of animals in intensive production units have produced a new series of problems, such as the disposal of animal excreta. Animal faces, which may contain pathogens, are usually converted into a water suspension, called "slurry". Spontaneous generation of heat such as occurs in heaps of solid measure does not occur in slurry, and therefore, pathogenic microbes and parasitic eggs can survive for a considerable time. The best way to dispose of it is to spray it on arable land, but well away from dwellings and attachment are is single infective aerosols may be carried considerable distances by the wind.

A tribal woman even at 80 years of age has her hearing and vision better than those of most teenagers, her teeth sound, her heart strong. Most of her fellow tribesmen are as healthy as she. One explorer noted that life in the tribal villages is so quiet that there is normally less noise than is made by a modern refrigerator. Some tribals enjoy longevity that would be remarkable in the most medically pampered society. Furthermore their declining years are almost free of the usual degenerative diseases of old age. Scientists are still puzzled by the tribal's extraordinary health, but their stable, tranquil environment is almost certainly an important factor. When a tribal moves from home to a big city he is beset by a host of ills he has never known before.

Table 7.3 shows lists of common food additives alongwith the areas of their of their use.

Noise poses quite a different problem. Unwanted sound becomes more pervasive and more intense in urban settings, where transport and industrial sources have particularly high nuisance values. Acute exposure to intense noise may temporarily impair hearing, while repeated occupational exposures to high levels can cause permanent deafness. Increased noise levels arc also associated with cardio-vascular endocrine, respiratory, neurological and psychological changes, some of which are indicators of increased stress. The various levels of acceptable noise that have been proposed in the past are being evaluated and assessed under WHO's programme, taking into account the levels, frequency, direction and duration of exposure.

INDUSTRIALIZATION

Industrialisation has conferred great benefits on mankind. Physical toil has been greatly reduced. People enjoy higher standards

Table 7.3 : List of common Food Additives below along with the Areas of their use

S.No.	*Group and function*	*Examples*	*Uses*
I.	Acidity imparting agents	Acetic acid, citric acid, phosphoric acid	Jams, Jellies, Soft Drinks
2	Aerating or gas imparting agents	Carbon dioxide	Carbonated beverages
3.	Anti-caking and lump preventing agents	Calcium phosphate, starch	Table Salt
4.	Bleaching or colour reducing agents	Chlorine	Wheat flour
5.	Coating or surface glazing agents	Bees wax, paraffin	Chocolates
6.	Colour agents	Turmeric, Saffron	Soft Drinks, confectionery
7.	Emulsifying or oil dispersing agents	Derivatives of fatty acids	Chocolates, Ice-Creams
8.	Flavouring agents	Essential oils	Soft Drinks, chewing-gum, confectionery
9.	Nutrients restoring nutritional value lost in processing	Amino-acids like lysine and various vitamins	Bread, Cereal products
10.	Preservatives preventing spoilage by micro-organisms	Sodium benzoate, Sorbic acid	Jams, Bread, dressed Chicken
11.	Sweetening agents in place of sugar	Saccharin	Diabetic foods.

of living, and developing countries in particular are able to spend more money on development and on health. Unfortunately, the effects on environment have not been beneficial in developing world. These adverse effects include occupational diseases and accidents, pollution, wasteful uses of natural resources and a deterioration of the psycho-social environment. What are these psycho-social consequences of industrialization?

An excessive migration of the rural population to the towns and cities strains these resources of the urban centres. There is often a dense concentration of population. The cacophony of traffic noise, street cries, voices of playing children and music from radio and television sets in the urban environment can be stressful to both ear and mind.

Unlike in a village, where each individual may know everybody else, the social circle of a city dweller can be extremely limited. People in large cities are inclined to be rather curt and impersonal, and may not appear to have any time to spare for others. It is quite easy for a person, especially in a strange city, to feel "lonely in a crowd". All these characteristics of an industrial or urban centre have been said to predispose individuals to such medical problems as insomnia, neurosis, peptic ulcer and hypertension. Social consequences can be equally developing. In many developing countries where industries are mushrooming, alcohol related problems and drug-abuse have increased rapidly. Prostitution, sexually transmitted diseases, illegitimate pregnancy, marital discord and crimes of violence including child-battering, have become rampant.

Rearing of children can also be a problem in an industrialized society, where they become a grievous economic burden. In such instances, many children lack parental love and may drift into juvenile delinquency. Speaking about the aged, their old way of life, in which respect for the aged and dependence on the extended family were prominent features, may have disappeared. In old age, many persons find themselves without work and without a motive for living. This may be one of the main reasons, that the suicide rate among the old has increased.

Mental health in industry appears to depend on the following factors.

(a) Job satisfaction.
(b) Identification and status
(c) Attitudes of fellow-workers and the employer.

AGRICULTURE

The health problems of the workers, entrenched in agricultural environment can be divided into agricultural and special sub-groups. General health problems include those diseases and afflictions that the agricultural worker, in common with everyone else, is exposed to and suffer from. Many of these arise as a result of poor sanitation, inadequate water supply, inadequate accommodation, malnutrition, and a wide variety of communicable disease both parasitic and bacterial, that affect the entire population.

More specific occupational problem occur as a result of workers exposure to parents of disease associated with agriculture, which may be biological, physical or chemical. Biological hazards, obviously include zoonoses–diseases of animals which are transmitted to man in handling animals and animal products. The list of these diseases, include bovine tuberculosis, anthrax and brucellosis. In addition, certain parasite diseases are transmitted as a result of contact with polluted water in farm lands. The common ones are schistosomiasis, ankylostomiasis and teptospirosis. Chemicals are extensively used in agriculture to control insects, fungi, herbs, rodents and so forth which damage crops. These pesticides and insecticides are harmful to man if not used carefully. Contamination of food from the use of empty containers has resulted in human poisoning, and so has the accidental consumption of seeds treated with chemicals (mercurials) as preservatives.

AIR

In the past few decades, man has polluted the atmosphere so heavily that much of the population now breathes a mixture of highly toxic gases with every lungful of air. He has allowed lethal insecticides and weed killers to contaminate his water–a resource from which they cannot be removed by any known method. In fact he has surrounded himself with a new kind of filth which may breed disease as effectively in the long run as the microbial filth of 19th century slums.

Industrial societies are exposing themselves to these man-made poisons in almost total ignorance of their possible delayed or cumulative effects. The dangers of cigarette smoking are now well known. Once radium was added to pep tonics and advertised as a cure for tiredness and arthritis. It was also used to make luminous paint. Some of the first workers employed to paint luminous watch dials habitually pointed their brushes by twirling them between their lips. Some of them died of radium poisoning and bone cancer over a period of a few years. Most had swallowed only a minute speck of radium in all. Every human being depends on the 12,500 quarts of air, he breathes each day-yet man persists in using the sky as a refuse bin. He began to introduce smoke into the air with his first fire. The volume of soot and smoke increased markedly with the advent of soft coal and the birth of industrial cities. In the years that followed, the volume of pollutants in city air has increased faster than ever, while their nature has changed for the worse. Old-fashioned coal and low-grade fuel oil still send soot and sulphur dioxide into the air; now petroleum refineries, chemical industries and motor cars add vast quantity of pollutants.

Leakage of MIC gas, from the tank of Union Carbide Factory at Bhopal on the night of 3-12-1984, converted the entire Bhopal town into a gas chamber, affecting over 2 lac people. This sudden air pollution brought blindness, respiratory distress, sudden death and so many other miseries instantaneously. The death toll may have been as high as 10,000–mostly poor children and women. This has been described as one of the biggest failures of our time to prevent air pollution.

WATER

When health workers in Rajasthan asked village women to name their single greater need, they answered, "Water". That response would have been echoed throughout India and the rest of underdeveloped world, where water supplies are not only insufficient but also impure, and often carry the microbes of such diseases as cholera, typhoid and amoebic dysentery. In rural areas, one well or water-hole may serve an entire community as well as cattle. Of the 250 million people living in the cities of Africa and Asia, only one in five has piped water in his home and fewer than half have water within half a mile. Where water lines have been laid, the effects on health have been immediate and dramatic in one rural section of India, the death rate from cholera fell by 74 per cent in five years.

FOOD

Adulteration of food with chemicals of various kinds, to make a fast buck is widely known. But inadvertent contamination of the food-stuffs with microbes, parasites, residual pesticides and weedicides is not fully appreciated. Above all the fad of mixing food additive is a bane of modernization. The bread you eat daily or the soft drink you relish contain such substances as sodium propionate, sorbic acid, potassium bromate, phosphoric acid, citric acid and caramel colours. In the broadest sense the word "food additives" includes food colours which impart different colours and food preservatives which extend the storage life of food.

NATURAL VS. SYNTHETIC

It is estimated that there are about substances which are added to food. Depending on their source, these substances may be classified as natural or artificial. Some of the natural additives are various gums like arabica, ghatti and karaya gums, colouring materials like saffron, turmeric, flavouring materials like lemon grass, rose and cinnamon oils. Artificial or synthetic food include phosphoric acid, saccharin and potassium metabisulphite. Bread, biscuits, cakes, confectionery jams jellies, marmalades, ketchup, various soft drinks, alcoholic drinks, tinned products, cheese, sweets etc. are only a few of the list of products containing additives.

IMPROVING NUTRITIVE VALUE

Addition of iodine to common salt in the Himalayan region for reducing the incidence of goitre (swollen thyroid gland) in humans is a good example. Similarly bread (specially that manufactured by the organized sector) is enriched with several vitamins and minerals. Additives such as various vitamins are used to compensate for losses processing. Often, vitamin C is added to fruit and vitamin A to skimmed powder.

FLAVOUR

Flavour is a combination of taste, feeling and odour on receptors in the mouth and the nose. Traditionally, several ingredients were being added to food to improve the flavour. Flavouring materials are added to give the flavour or modify the particular flavour or to mask the original flavour. A good example for flavour is sugar in coffee. Flavouring is widely used in bakery products, confectionery and soft drinks. Concentrated fragrance is the essence.

INFECTIOUS DISEASES

In the underdeveloped countries, infectious illnesses are a major threat. Their cost is high not only in suffering, but in the permanent damage they can do. In some rural sections of India, trachoma has weekended the vision of 80 to 90 per cent of the population; in some African villages, blindness is so common that ropes are strung to guide women on their way to the village well and bamboo poles laid to guide men planting in the fields. The first step in the control of many diseases is to find their carriers, e.g., flies and mosquitoes, etc., and then trace the paths by which they travel. These two activities are a major concern of medical research in the new nations.

TIME POLLUTION

The ease with which men jet from one time zone to another may also have undesirable effects on health due to environmental change and stress involved. An airline pilot who flies from New York to Tokyo is still on New York time biologically speaking when he arrives at his destination. It is several days before his pulse rate, body temperature, digestive processes and secretion of adrenal hormones adjust. After several such trips he may develop various unpleasant symptoms; he may loss or gain weight, and suffer from insomnia and a general sense of tension and irritability.

STRESS ENVIRONMENT

During a car through heavy traffic, working at a frustrating job, watching a child struggle with illness, quarrelling with one's mate the stresses of life take infinitely varied forms. And they can pose just as much of a challenge to health as bacteria, viruses, malnutrition, or chemical and physical forces. Each man reacts in his own way. The family quarrel that triggers a heart attack in one may only make another resentful, while for a third it may even serve as a goad to useful and productive work. Whatever the response, it involves the whole person, both body and mind play a part in dealing with the stresses of life. The stress reactions have both physical and psychological aspects. The physical effects that result are dictated in part by the stimulus itself and in part by past experience. Everyone sneezes if sufficient plant pollen is introduced into his nasal passages–this universal physical response has the effect of eliminating the pollen from the body. But in some people this response is very much exaggerated. Because of their genetic endowment and the effects of previous exposures, these people are allergic. They find pollen so threatening that they over-react to it, throwing up a defence that produces all the symptoms of a disease-stuffed noses, itching eyes, sneezing, weeping.

EDUCATION

In the undeveloped countries, superstition, ignorance and fear are as much a threat to health as are the microbes of disease. In villages the centuries-old custom of rubbing the navels of the new-born with vegetable oil often results in death from tetanus. It could not be possible, if many primitive peoples were more afraid of the hypodermic needle than they were of disease. Similarly vaccination against poliomyelitis, diphtheria, tetanus, whooping cough, measles, mumps, rubella and rabies can prevent these dreadful disease. This is possible only, if immunization education is widely disseminated and emulated.

HOSPITALS

Once upon a time, hospitals rarely had running water, and what water they had was usually contaminated. Garbage, human wastes and assorted offal were dumped into a pit in the courtyard. Surgeons wiped their instruments on their trousers; bad-clothes were rarely. Hospital infection was rampant upto one-third of all women giving birth died of puerperal fever, a form of blood spesis. As late as mid-century only a few visionaries recognized any connection between filth and disease.

Unless people take the right precautions, hospitals still risk becoming breeding-grounds for disease. This was stated by Miss Nightingale, who was horrified by what she saw in the cholera and dysentery- infested hospital. In her "Notes on Nursing", she wrote. "There are five essential points in securing the health: Pure air, Pure water, Efficient drainage, Cleanliness and Light". By pure air and light she meant what today we call a good environment. She was saying just what WHO says today—that health is not just hospitals, doctors and drugs, but everything that we understand by "the quality of life" and the environment.

At Budapest's St. Rochus Hospital, Ignaz Semmelwels was given freedom to enforce his antiseptic practices in a ward assigned to him. Although the rest of the hospital remained typically dirty (and the incidence of hospital infection remained typically high), but in this ward fever and infection virtually disappeared among patients. This was a turning point, towards new era, in hospital's environment, bringing cleanliness and freedom from infections.

WHO PROGRAMMES

During the last two decades, several WHO programmes were engaged in evaluating health hazards from various agents. High priority went to food additives, pesticide, residues that might reach us through the food chain, the quality of drinking water, occupational exposure, air quality in the cities and the cancer risks from certain chemicals. Conference on the Human Environment, held in Stockholm in 1972, highlights the need for an expended programme. While addressing this conference, Smt. Indira Gandhi, the then Prime Minister of India observed: "Life is one and the world is one and all these questions are interlinked. The population explosion, poverty, ignorance and disease, the pollution of our surroundings, the stockpiling of - nuclear weapons and biological and chemical agents of destruction are all parts of a vicious circle. Each is important and urgent. The extreme forms in which questions of population or environmental pollution are posed obscure the total view of political, economic and social situations. We believe that planned families will make for a healthier and more conscious population. But we know also that no programme of population control can be effective without education and without a visible rise in the standard of living. Our own programmes have succeeded in the urban or semi-urban areas. To the very poor, every child is an earner and a helper. We are experimenting with new approaches and the family planning programme is being combined with those of maternity and child welfare, nutrition and development in general".

WHO's new programme was initiated in 1973 and has four fundamental objective.

(a) to assess existing information on the relationship between exposure to environmental pollutants and man's health, and to provide guidelines setting the exposure limits consistent with health protection;

(b) to identify new or potential pollutants; to identify gaps in our knowledge about the health effects of agents likely to be increasingly used in industry, agriculture, in the home or elsewhere; to encourage countries to harmonize their methods of assessing harmful factors so that the research results can be compared on an international scale.

II. HUMAN RIGHTS

The concept of Human Rights is as old as political philosophy. The New Lexicon Webster's Dictionary of the English language describes the "Human Rights" expression as under :

Human rights program enunciated at Helsinki.

"The right to be free from Governmental violations of the integrity of the person."

"The right to the fulfilment of such vital needs as food, shelter, health care and education and the right to enjoy civil and political liberties" released January 1978 by U.S. State Department and urged on all nations" Helsinki Agreement.

The Lexicon Universal Encyclopaedia describes the "Human Rights" expression as follows:

"Human Rights are basic political and social conditions–variously defined–to which every individual is entitled as a human being. Originally they were called natural rights or the rights of man, and included the rights to life, liberty; and the pursuit of happiness cited in the U.S. Declaration of Independence over the years the concept of human rights has been broadened to include rights to social benefits such as social security, rest and leisure, and education.

In the Encyclopaedia of the United Nations and International Agreements, the meaning of "Human Rights" is given thus :

"An International term which is not defined In any Act of International Law introduced by U.S. Declaration of Independence, 1776, and the declaration of the Rights of Man and Citizen, 1789, of the French Revolution. It was adopted by the U.S. Constitution : and extended by the 16th Constitutional Amended 1913 and is the subject of international declarations, the first being the declaration of Human Rights and duties worked out in 1929 by the New York Institute of International Law and submitted to Inter American Legal Committee. In Art. 1, it proclaimed that it is a duty of every state to recognize equal rights of an individual to life, freedom and property and to fully grant and protect these rights on its entire territory regardless of nationality, sex, race; language or religion; and in Art. 2, that it is a duty of every state to recognize equal right an individual to the free execution, whether in public or private, of any faith, religion or worship, the practice of which is not in violation of public order and good manners. This declaration together with the resolution of Inter-American Conference in Chapultepec, March 8, 1945 on the need to establish international protection of human rights, laid the foundations for a draft of the universal declaration of Human Rights submitted to the U.N. Human Rights Commission."

Earlier, the U.N. Charter introduced an unspecified term "Fundamental Human Rights" of international character and set up a special U.N, Commission for its promotion of human rights. The obligation of protecting human rights was introduced into peace treaties concluded February 10, 1947 with Bulgaria (Arts. 2, 3, 4), Finland (Arts. 6, 7, 8) Romania (Arts. 3, 4, 5), Hungary (Arts. 2 and 4) and Italy (Art. 15) as well as on May 15, 1955 with Austria (Arts. 6, 7, 8). Work on the universal declaration of Human rights went on in the years 1946-48 concurrently in the U.N. Human Rights Commission and the International Legal Commission which in March, 1948, Won approval of the 9th Inter-American Conference in Bogota for the declaration of American Human Rights and Duties (specifying 28 rights and 10 duties) and served as a basis for an inter-American convention (the convention consisting of 88 Articles was debated internationally in the years 1950, 1953, 1954 and 1959 as well as elaborated on for many years by the Human Rights Commission set up by the OAS). The U.N. Human Rights Commission prepared December 10, 1948 the Human Rights, Universal declaration of which enumerates human rights alone and only mentioned (Art. 29) human duties "to the community in which alone the free and full development of his personality is possible." Though not an international treaty but a U.N. General Assembly Resolution, the declaration bears significantly on the shaping of modern norms of international law which found two-fold reflection in international agreements concluded outside the U.N. and referring to the declaration (e.g., the 1951 Peace Treaty with Japan); in the international conventions concluded with a view to protecting particular human rights, as well as in facts and declaration, presented here in chronological order :

The convention of prevention and prosecution of the crime of **Genocide, 1948.**

The European Convention of protection of human rights and the fundamental freedoms adopted by members of European Council on November 5, 1950.

1. The convention on the status of refugees 1951.
2. The convention on political rights of women, 1952.
3. The protocol on the modification of the 1926 convention on slavery 1953.
4. The convention on the status of stateless persons, 1954.
5. The supplementary convention on elimination of slavery, 1956.
6. The convention on citizenship of the married women, 1957.
7. The convention on reduction of the number of apatridos (stateless person), 1961.
8. The convention on marriage (terms, minimum age, registry), 1962.
9. The convention on elimination of all forms of racial discrimination, 1965.
10. The international pacts of Human Rights: Civil and Political.
11. Rights and Economic, Social and Cultural Rights, 1966.
12. The protocol on the status of refugees, 1966.

Added to these Acts in the U.N. General Assembly resolution of Nov. 20, 1959 called "Declaration of the Rights of the Child" stipulating for legal protection the rights of the child, as well as the resolution of Nov. 20, 1963 called "Declaration on elimination of all forms of racial discrimination" the fruit of which was an international convention to this effect of December 22, 1965.

In keeping with U.N. General Assembly Res. 2081/XX of Dec. 20, 1965, the year 1968 was declared the International Year of Human Rights. An international conference of Human Rights was held in Tehran April 22/May 13, 1968 and concluded with the adoption of 28 resolutions and (unanimously) the so-called Human Rights Tehran proclamation. The International covenant of Economic, Social and Political Rights introduced (Arts. 16 and 22) a monitoring system for human rights based on report submitted by states parties to the covenant. An eighteen Nation Human Rights Committee considered the reports, including communications alleging human rights violations. The committee held its first and second sessions in 1977.

The position of the Warsaw pact countries is stated in the declaration of consultative political committee signed on November 23, 1978 at Moscow whose provisions concerning human rights were as follows :

"The U.N. Charter obliges all countries to contribute to the respect and effectuation of freedoms for all regardless of race, sex, language or religion. Accordingly the socialist countries, showing initiative and consistent efforts, have added considerably to the preparation and adoption of the most important international agreements and treaties in this respect: human rights pacts, conventions on prevention of genocide, liquidation of all forms of racial discrimination and many others; they implement in practice all the provisions of these treaties and agreements."

The U.S. Department of State in Jan. 1978 has released an official U.S. Government definition of Human Rights :

"Freedom from arbitrary arrest and imprisonment, torture, unfair trial, cruel and unusual punishment, and invasion of privacy, rights to food, shelters, health care, and education; and freedom of thought, speech, assembly, religion; press, movement and participation In government."

The concept of Human Rights is as, old as the ancient doctrine of "natural rights" founded on natural law. The expression "human rights" is of recent origin, emerging from (post-second world war) international charters and conventions. The expression "Human Rights" has not been specifically defined in any declaration or covenant of the United Nations. Human Rights are generally defined as "those rights which are inherent in our nature and without which we can not live as human beings". The recognition of these natural rights of human beings as ancient as the human civilization.

The expression "Human Rights" had its origin in international law, appertaining to the development of the status of an individual in the of international legal system, which was originally confined to the relation between states, who were regarded as the only person in international law. For all practical purposes, the genesis of this international aspect of human rights is not older than the second world war, though the concept of an individual, having certain inalienable rights as against a sovereign state had its origin in the dim past, in the somewhat nebulous doctrines of natural law and natural rights.

The most important question is –What are human rights? The United Nations has described human rights thus :

Human Rights can be generally defined as those rights which are inherent in our nature and without which we can not live as human beings.

Human Rights and Fundamental Freedoms allow us to fully develop and use our human quality, our intelligence, our talents and our conscience and to satisfy our spiritual and other needs. They are based on mankind's increasing demand for a life in which the inherent dignity and worth of each human being will receive respect and protection.

Human Rights are some times called "Fundamental Rights" or "Basic Rights" or "Natural Rights." As Fundamental or Basic Rights these are those rights which must not be taken away by any Act of legislature or government and which are often set out in the Fundamental Law of the land, i.e., Constitution. As natural rights, these are seen as belonging to men and women by their very nature. Another way to describe them would be to call them as common rights, for these are the rights which all men or women in the world should share, just as the common law in England, for example the body of rules and customs which unlike local customs governed the whole country.

To speak of human rights requires a conception of what rights one possesses by virtue of being human. Of course, we are not speaking here of human rights in the self evident sense that those who have them are human but in the sense that in order to have them one need only be human. Human Rights have been identified as those rights which are "important, moral and universal". They may also be called as "Rights" in some moral order or perhaps under some natural law.

Human Rights are those minimal rights which every individual must have against the state or other Public authority by virtue of his being a 'member' of the human family' irrespective of any other consideration.

The first documentary use of the expression 'human rights' is found in the charter of the United Nations, which was adopted (after the Second World War) at Sen Fransisco on June 25, 1945 and ratified by a majority of its signatories in October that year. The preamble of this charter, which was drawn up to prevent a recurrence of the destruction and suffering caused by the second world war, by setting up the international organisation called the United Nations, declared that the United Nations shall have for its object inter-alia, "to reaffirm faith in fundamental human rights .....", and Art. 1, thereafter stated that the 'purposes' of the United Nations shall, among others.

"To achieve international co-operation... in promoting and encouraging respect for *human rights and for fundamental freedoms* for all without distinct as to race, sex, language or religion.

Thus, the preamble to the charter in most sonorous terms recognises the human rights and the dignity worth of the human being, which were most flagrantly violated during the two world wars. It may be pointed out here that the charter of the United Nations though makes reference to "human rights" and "fundamental freedoms" in the preamble and various other provisions, yet it did not define what is meant by "human rights" and "fundamental freedoms".

"Human Rights", as the term is most commonly used, are the rights which every human being is entitled to enjoy and to have protected. The underlying idea of such rights fundamental principles that should be respected in the treatment of all men, women and children exists some form in all culture and societies. The contemporary international statement of those rights is the Universal Declaration of Human Rights. Many Northern Governments, however, as well as the News media, have come to use the term 'human rights' in a very narrow sense, referring only to the civil and political rights set out in the declaration. As a result the term 'human rights' is often broadly misused to mean only civil and political rights and to exclude from consideration as a matter of rights, fundamental issues such as the right to food, health, education and social security.

Human Rights are something more than a set of do's and don'ts. They imply ethical values and attitudes to human relationship.

They come to life and grow–only if they relate to the local culture, traditions and philosophy.

Most societies have always had some form of legal recognition and protection of human rights, but definitions as to what constitutes human rights (and who should have them) have varied considerably. In the West, Human Rights were made explicit and systematically elaborated and codified in the 17th and 18th century, although individual rights were hinted at as early as the Magna Carta (1215). The English Petition of Right (1628) and the English Bill of Rights (1689) were early documents establishing civil liberties. Philosophers such as John Locke in Civil Government (1690) and Baron De Motesquieu in the spirit of the laws (1748) gave these and future documents an intellectual basis. The idea of individual rights and freedom was further embodied in the U.S. Declaration of Independence (1776) in the Bill of Rights, the first ten amendments to the U.S. Constitution (1791); and in the French Declaration of the Rights of Man and of the Citizen (1789). These statements of human rights included such principles as every person's right to be free from coercion by other, including governments; freedom of speech; freedom of the press; freedom of religion; freedom to join other in pursuing political goals; due process; representative government; and the right to private property.

The widespread misinterpretation of the term, 'human rights' has resulted in charges that 'human rights' and the universal declaration itself consist of nothing more than bourgeois rights' or 'western rights.' Some governments had begun to argue that the very idea of universal human rights conflicts with the very specific characteristics of local or regional cultures and customs, some argued that human rights must be recognised as different religious contexts, some held that under the banner of protecting universal rights, northern governments would continue to focus on civil and political rights in their dealings with southern nations while refusing to change financial and other practices that deny the economic, social and cultural rights of people in such nations, the conventional wisdom had been that human rights were 'indivisible', meaning that respect for civil and political rights could not be divorced from the enjoyment of Economic, Social and Cultural Rights.

Expressed the other way round, authentic economic and social development could not exist without the political freedom to participate in that process, including the freedom to dissent. Here too, views diverged, some governments had argued that strict measures curbing political freedoms were necessary to get their economies going, some argued that priorities must be established. What was the point of taking about the establishment of courts and reforming the prison system when the pressing issue was ending starvation and seeking relief from crippling foreign debt? Stemming in part from the one-sided interpretation of the term 'human rights'. The concept of 'development' had also come to be regarded as a human aspiration separate from the achievement of human rights, despite the fact that at least half of the 30 articles of the Universal Declaration of Human Rights specify the economic, social and cultural rights which constitute the core of much of the world's development efforts. The various international practices that had grown up as a result also gave rise to increasingly heated debates. These focused particularly on two aspects of international relations most commonly referred to in diplomatic jargon as 'conditionally' and 'the right to development'.

According to western political and philosophical thinking human rights are innate in individuals and are an intrinsic factor in the 'quality of the human persons.' They are, to quote the words of President Jefferson, "inherent and inalienable rights of man," and hence a state that violates them in its laws and its actions breaches one of the very prerequisites of civil co-existence between state and may be legitimately brought to account.

Human Rights to the people are believed to be an extremely modern concepts of recent origin, some ascribe their origin to the French Declaration of the Rights of Man and Citizen of 1789, the second article in their declaration says–

"The aim of all political association is the conservation of the natural and inalienable rights of man. These rights are: liberty, property, security and resistance to oppression."

The French Declaration announced a number of other rights for people and they were :

- All men are born and remain free and equal in their rights.
- All men are equal before law.

- All men should have the freedom from arrest.
- All men are innocent until they are proved guilty etc.

The declaration made it clear the definition of human liberty, i.e., freedom to do anything which is not harmful to others. Englishmen, however, consider their 'MAGNA CARTA' of 1215 as the foundation of their liberties. They say that the Magna Carta guaranteed to them freedoms from imprisonment or from dispossession of property and from prosecution, etc., unless by the lawful judgment by the law of the land."

In India the Protection of Human Rights Act 1993 came into existence on 28th September, 1993. The provide for the constitution of a National Human Rights Commission, State Human Rights Commission and Human Right Courts for better protection of Human Right and for matter connected therewith or incidental thereto. The Act defined "Human Rights" to mean the rights relating to life, liberty, equality and dignity of the individual guaranteed by the constitution or embodies in the International covenants and enforceable by courts in India.

Thus, it is abundantly clear that Section 2(d) of the Protection of Human Rights Act, 1993 defines Human Rights as in this Act unless the context otherwise requires human rights means the rights relating to life, liberty, equality and dignity of the individual guaranteed by the constitution or embodies in the International covenants and enforceable by courts in India.

Article 19 of the Constitution of India deals with protection of certain rights regarding freedom of speech, etc. Article 20 of the Constitution of India deals with protection in respect of conviction for offences Article 21 of the Constitution of India dealing with protection of life and personal liberty says that no person shall be deprived of his life or personal liberty except according to procedure established by law.

The relevant international Covenants are : (a) the Universal Declaration of Human Rights, 1948, (b) the International Covenant on Civil and Political Rights, 1966, (c) the International Covenant on Economic, Social and Cultural rights, 1966 and (d) the Optional Protocol, 1966. Clearly an International instrument can not simply list of the human rights which it protects. The scope of the right must be defined. For example, what is the relationship of the basic principle of the right to life and such associated issues as the right to an abortion or suicide or to voluntary euthanasia or capital punishment? But beyond that, human rights can not be seen in isolation. One man may wish his right to freedom of speech to allow him to libel his neighbour; his neighbour is not likely to agree. Human rights have to be interpreted with some degree of relativity to ensure fairness to all and to the interests of society as a whole. In this respect, the general principles which temper the impact of substantiative human rights in their individual application are vitally important in determining the ultimate scope of human rights protection. All the major international instruments have qualifying provisions and anyone studying their provisions or seeking to claim rights under them must have regard to this fact.

Thus, we find that the term "Human Rights" has not been specifically defined in any declaration or covenant of the United Nations. It has been defined in general terms in various declaration or covenants. The definition of term still eluded all. Evolution and crystallisation of the concept took a long time. Initially there was confusion between the natural rights propounded by political philosophers in the by gone ages and the concept of Human Rights.

The first documentary use of the expression 'human rights' could be seen in the charter of the United Nations which declared promotion and fostering of human rights as one of the basic goals of the United Nations, the charter of the United Nations does not define the contents of human rights. The framers of the charter left the task to the organisation itself. The internationai Bill of Human Rights which was subsequently drawn up comprises.

1. The Universal Declaration of the Human Rights, 1948.
2. International Covenant of Economic, Social and Cultural Rights, 1966.
3. International Covenant on Civil and Political Rights, 1966.
4. The Optional Protocol, 1966.

The sum and substance of the aforesaid deliberation is that all Human Rights derive from the dignity and worth inherent in the human being and human person is the central subject of Human Rights and Fundamental Freedoms.

In simple terms, whatever adds to the dignified and free existence of a human being should be regarded as Human Rights. In general terms, we can also describe Human Rights are those minimal rights that every individual must have by virtue of his being 'a member of human family' irrespective of any other consideration. The sweep of the Human Rights concept is expanding day by day. Broadly, they fall into categories of

(a) Civil and Political Rights.
(b) Economic, social and Cultural rights, and
(c) Group Rights

The horizon of Human Rights is expanding day by day and the people all over the world are able to enjoy their basic rights and assured of dignified existence. Broadly speaking an emphasis on the concept of Human Rights is a step in the right direction leading to the evolution of the World Public Order. Twentieth Century witnessed two global wars displaying unabashed lust for power and dominance destroying eternal values and respect for human beings. The crux of the matter is that in our era of human rights as a political and legal topic, the subject of Human rights and fundamental freedoms has no parallel today. At present, there is no shadow of doubts in our society concept of Human rights has established its strong foothold. The importance of the study of Human Rights is constantly growing.

III. VALUE EDUCATION

WHAT ARE VALUES?

Value literally means something that has a price, something precious, dear and worthwhile; hence something one is ready to suffer and sacrifice for. In other words values are a set of principles or standards of behaviour. In the words of John Dewey, "the value means primarily to prize, to esteem, to appraise and to estimate. It means the act of achieving something, holding it and also the act of passing judgement upon the nature and amounts of values as compared with something else."

Values are regarded desirable, important and held in high esteem by a particular society in which a person lives. Thus values give meaning and strength to a person's character by occupying a central place in his life. Values reflect one's personal attitudes and judgements, decisions and choices, behaviour and relationships, dreams and vision. They influence our thoughts, feelings and actions. They guide us to do the right things.

Values are the guiding principles of life which are conducive to all round development. They give direction and firmness to life and bring joy, satisfaction and peace to life. Values are like the rails that keep a train on the track and help it move smoothly, quickly and with direction. They bring quality to life.

People, especially the young, are confused about their values and I value system. They are facing value conflicts and dilemmas. This is mainly due to the dramatic and far reaching socio-cultural and political changes that are taking place in our country and in other parts of the world. Besides these, there is breakdown of traditional values without proper replacement, lack of adequate role models, conflicting ideologies and double standards practised by people in position of power and influence. At the same time there is new awareness among people about human dignity and rights, a greater concern tor the poor and the oppressed, the sick and the old.

In the light of all these, it is but natural that people arc confused about proper and sound values. Many young people openly reject some of the traditional values and question dogmatic beliefs held sacred for centuries. Often it happens that such people do not find replacements for the traditional values and hence a kind of vacuum is created in their life. This is not desirable because in the absence of values, they have no principle or foundation on the basis of which they can face life situations, make choices and decisions. A life without proper values will become chaotic and disastrous. It will be a boat without rudder To guide our life in the right path and to embellish our behaviour with good qualities we need values.

From the above discussion, a broad and simple working concept of value could be evolved. Any human activity, thought or idea, feeling, sentiment or emotion which could promote self-development of the individual in all its dimensions could be said to constitute a value. The other complementary function of a value is it should also contribute to the welfare of the larger social unit such as the family, the community and the nation of which the individual is a member.

WHAT IS VALUE EDUCATION?

Value education means inculcating in the children a sense of humanism, a deep concern for the well-being of others and the nation. This can be accomplished only when we instill in the children a deep feeling of commitment to values that would build this country and bring back to the people pride in work that brings order, security and assured progress.

Through value education we like to develop the social, moral, aesthetic and spiritual sides of a person which are often undermined in formal education. Value education teaches us to preserve whatever is good and worthwhile in what we have inherited from our culture. It helps us to accept respect the attitude and behaviour of those who differ from us. Value education does not mean value imposition or indoctrination.

Value education has the capacity to transform a diseased mind into a very young, fresh, innocent, healthy natural and attentive mind. The transformed mind is capable of higher sensitivity and a heightened level of perception. This leads to fulfilment of the evolutionary role in man and in life.

HOW ARE VALUES CLASSIFIED?

Values, in general, could be classified broadly under five headings: personal, social, moral, spiritual and behavioural.

Personal Values

Personal values refer to those values which are desired and cherished by the individual irrespective of his or her social relationship. The individual determines his own standards of achievement and attains these targets without explicit interaction with any other person. Examples : ambition, cleanliness, contentment, courage, creativity determination, dignity of labour, diligence, excellence, honesty, hope, maturity, regularity, punctuality, self-confidence, self-motivation, simplicity etc.

Social Values

Social values refer to those values which are other oriented. They are concerning to society. These values are cherished and practised because of our association with others. Unlike personal values, the practice of social values necessities the interaction of two or more persons. Social values are always practised in relation to our neighbour community society nation and the world is countability, brotherhood, concern for environment, courtesy, dialogue, dutifulness, forgiveness, freedom, friendship, gratitude, hospitality, justice, love, magnanimity, patience, repentance, responsibility, service sharing, sportsmanship, sympathy, team spirit, tolerance etc.

Social values are local, parochial and temporary in application. They may change with the change of social circumstances. These are external relationships of the individual with society. These values may also be considered as historically conditioned socio-political expressions of religion.

Moral Values

Moral values refer to those values which are related to an individual's character and personality conforming to what is right and virtuous. They reveal a person's self-control. Examples : honesty, integrity, sense of responsibility, compassion etc. The realm of moral values is rather a debatable one. For example : According to the sage Yagnavalkya, as a moral value 'dharma' signifies the cultivation of the virtues of non-injury, sincerity, honesty, cleanliness. Control of the senses, charity, self-restraint, love and forbearance. It may be observed that this list includes both social values (values that refer to the good of others) and individual values (values that serve to develop one's own charact r and will).

SPIRITUAL VALUES

We define ethical value as the perception of the 'within' in man, it arises from the inner depth dimension of man; it bestows the capacity to see the false as the false and the true as the true; it is the key to the integration of man. The ultimate ethical value is called spiritual value. Spiritual value is the awareness itself.

All knowledge is structured in consciousness. This recognises that consciousness alone can be totally self-reliant so that pure mind or non-dual state of mind is presented as an object to pure consciousness, then it reveals itself ; or in other words, non-dual mind perceives the spiritual value. Virtues that are associated with spiritual values are : purity, contentment, austerity, scriptural study, devotion to God, spiritual wisdom, dispassion, self-discipline, control of the senses, endurance, piety, sublimity, japa, meditation, tranquillity etc.

Behavioural Values

Behavioural values refer to all good manners that are needed to make our life successful and joyous. They are those values which we exhibit by our conduct and behaviour in our daily life. Behavioural values will adorn our life and spread cordiality, friendliness, and love all around.

Keeping in view the nature of professional requirements values can be classified into several categories such as economic, social, political, spiritual, modern, aesthetic, religious, material, academic, socio-political, global, environmental, cultural, moral, professional values etc.

What is Moral Education?

Moral education in its broad sense includes not only inculcation of moral and ethical values but also all spiritual, humanistic, scientific, .aesthetic and sporting values. Thus comprehensively moral education means value oriented education.

Ethical values arc essential to make an individual a good and useful member of society and a good person himself and these values are essential for sustaining societies. These values in the past were referred to as 'dharma' in Indian philosophy. Broadly speaking 'dharma' is right action or right conduct.

Is moral education different from religious education?

The term moral education and religious education are used interchangeably because religious preachings and practices have been the sources of social and moral values. But instead of drawing fundamental similarities in the social and moral values of different religions, religious education of the denominational type tends to be restricted to an indoctrination or beliefs and preachings of particular religion. Being moral and being religious are not synonymous expressions. Being moral refers to being able to understand the principles of right and wrong and act in the right way. Being religious refers to having belief in a supernatural power or God. It is not logically necessary for a moral person to have faith religion One could be moral without committing oneself to any religious faith. This is more so in a pluralistic society like ours.

Should moral education be taken to be essentially a matter of developing character ?

Character which refers to appropriate conduct and behaviour is undoubtedly a very important aim or moral education. But this does not mean that moral education has no knowledge component in it. Actually, ability to make moral judgement based on sound reasoning is also an equally important aim of moral education. This ability has to be deliberately cultivated. Further, character is not just a matter of conforming to certain virtues or norms. Blind adherence to custom or mechanical conformity with tradition is not a real indicator of moral character.

Is moral education essentially a matter of educating the feelings and emotions?

Morality is not just developing the right feelings and emotions. For, moral development involves both thinking morally and behaving morally. Even effective outcomes are developed through cognitive components. Moral thinking is a distinct type of thinking characterised by the exercise of rational choice. The separation of the cognitive and effective aspects of moral a development untenable. A moral person not only does the right thing but also knows the reason as to why he does it. Moral education is thus a complex process which involves developing the ability to think morally, the ability to do right things, and also the ability to feel the right emotions.

The development of these abilities should be pivoted around the most fundamental values preached by Bhagavan Sri Sathya Sai Baba. Those values are truth (Sathya), righteousness (dharma), love (prema), peace (shanthi) and non-violence (ahimsa). To quote Bhagavan,

"Prema filled with thought is Sathya
Prema in action is Dharma
Prema with feelings is Shanthi
Prema with understanding is Ahimsa."

How does Moral Education differ from Value Education?

Value education is more wider, practicable and adoptable than religious education or moral education, as no specific faith or religion is reflected through ethical, moral, social, cultural or spiritual values (Gupta, 1988).

Value education is an approach to inculcate broad-mindedness, tolerance and proper social and emotional qualities. Understood in its broader sense, value education aims at training the young in the entire realm of values–physical, emotional, intellectual, imaginative, aesthetic, democratic, scientific, social, moral and spiritual–which can be pursued by any individual irrespective of his practice of any religion or no religion at all.

What is the difference between Moral, Spiritual and Religious Values?

According to professor Kireet Joshy (1986), the distinction between moral (naitik), spiritual (adhyatmik) and religious (dharmik) values in Indian thought is as follows : The word morality connotes a pursuit of the control and mastery over impulses and desires under the guidance and inspiration of a standard of conduct in consideration of man's station and duties in the society or in consideration of any discovered or prescribed intrinsic law of an ideal morality is often conceived as a preparation for spirituality, Spirituality, on the other hand begins when one seeks whatever one conceives to be the ultimate and absolute for its own sake unconditionally and without any reserve whatsoever, Morality is often limited to duties. Spiritually is fundamentally a search of the knowledge of the highest and the absolute by direct experience and manifestation of this search in every mode of living, thinking and acting.

The religious values are different from these two. A specific religious belief which is so exclusive that one cannot accept it if he does not hold that belief. Every specific religion has, as its essential ingredient, certain prescribed acts, rituals and ceremonies. A religious authority to which religious matters are referred and the decision of which is final.

Both moral and spiritual values can be practised irrespective of whether one believes in one religion or in no religion. Morality and spirituality can be independent of rituals and ceremonies or any acts specified by any particular religion.

Which were the Supreme Values in Ancient Indian Life and Education?

Truth, Beauty and Goodness were the supreme values of ancient India and they served as the guiding lights for men in their lives. With Dharma (right conduct) as the base, Artha and Kama were held as instrumental values to attain Moksha, that is, spiritual freedom. The concept of 'Purusharthas' formed the spiritual rind moral basis for life manifested in the four-fold stages of life– brahmacharya, grihastha, vanaprastha and sanyasa.

The doctrine of pancha koshas mentioned in Taittiriya upanishad represents a hierarchy of values. As elaborated by Oad (1981), happiness was achieved stage by stage. Starting from the lowest happiness experienced by the gross body, progressive realisation of the ultimate value 'anand' forms the very essence of good life.

The purpose of education in ancient India had been to teach the youngsters these values. The school and the society were the two agencies involved in developing character shaping personality, inculcating social values, promoting social efficient and the national culture. Righteousness and piety were the most cherished values. There are multiple origins of values but Indian values owe a great deal to culture specially philosophy and religion. Culture is one of the significant sources of values. Dharma and truth may be considered as basic tenets of Indian culture.

Value—Oriented Thoughts from Ramayana, Mahabharata and Bhagvadgita

The character of Srirama was the embodiment of truth and dharma (duty). Dasaratha stood by his word solemnly given to his wife Kaikeyi. Sita who had to leave Ayodhya, wept more for Kausalya, her mother-in-law than for her husband, Srirama. That was the mother-in-law and daughter-in-law relationship.

Yudhistira said "Truth is my mother, Jnana is my father, Daya is my friend, peace is my wife and Mercy is my son". These qualities are the sine qua non of Indian culture. Indra calls upon Yudhistira to abandon a dog that had followed him faithfully and to mount the celestial chariot that would take him to heaven. Yudhistira declined the offer saying "Never let me be joined unto that glory for whose sake a loyal dependent must be abandoned."

Bhagavad Gita enjoins the performance of one's duty without expecting a reward. All the teachings of Lord Krishna are nothing but values to be practised in life.

Harischandra sacrificed everything for the sake of truth (Satya). This great man courted misery, suffered uncounted difficulties, humiliations and the pangs of separation for the sake of truth. A single lie would have saved him from all the sufferings but he preferred to say truth only.

Unselfishness, sacrifice and renunciation are important components of Indian culture. Respect for women and looking upon other women as their mothers arc the noble characteristics of our culture.

As well all know our country has its own identity based on great values. Every Indian can be proud of the identity, cultural heritage and glory of our motherland. Indian philosophy occupies a proud place in values. The modem educationists have been thinking of introducing value based education in the curriculum. But value based education for having a better society was an age old practice in the ancient Indian educational system, which has deplorably been forgotten in recent times. Gross materialism has become the order of the day.

What did Commissions, Committees, Policies say about Value Education?

All-round development of the child–its head, heart and hand, is emphasised in Basic Education proposed by Gandhiji. He identified certain values as the basis for the establishment of a new social order in India. They are Truth, Non-violence, Democracy, Sarvadharma Sambhava, Equality, Self-realisation, Purity of ends and means, Self-discipline and Cleanliness.

Prior to independence, the ruling British Government stuck to a policy of strict religious neutrality. The Central Advisory Board of Education (1943-46) recommended that provision of spiritual and moral instruction for building up of the character of the young should be the responsibility of the home and community. The university education commission (1948) recommended moral and spiritual instruction at the university stage too. Moral and religious instruction from the elementary to the university stage was recommended by Sriprakasa Committee on Religious and Moral Education (1959). The Committee on Emotional Integration (1961) recommended that every student who takes up science should have some background in the humanities and should study a compulsory paper on Indian culture heritage, just as students in humanities should have some knowledge in general science.'

'The idea of national unity and the unity of mankind should be introduced in the curriculum with due regard to the children's age and understanding.'

The Indian Education Commission (1964 -66) recommended instruction on moral, social, and spiritual values at all levels of study.

Character–building as an object of education was stressed in the curriculum framed by the National Council of Educational Research and Training in 1975. Further, curriculum was sought to be related to national integration, social justice, productivity, modernisation of the society and cultivation of moral and social values.

The UNESCO in its report of the International Commission in 1972 suggested that educational systems should encourage the promotion of the values of world peace, international understanding and unity of mankind.

In 1981, the first National Moral Educational Conference was organised in our country. The important resolutions of the conference were : (1) A course on moral education in all classes upto the high school stage in all schools should be run inculcating moral and human values in students to make them better and more useful members of society; (2) Moral education should be a separate subject; (3) The content of moral education should include common ethical teachings of all great religions highlighting their unity; (4) The syllabus for moral education should include stories, illustrations and events mainly from our own country and its literature from various religions, so that it leads to national integration; (5) In higher classes, a course in comparative religion should be introduced as it will promote social harmony a liberal attitude and a less fanatical approach to religion.

The National Policy on Education (1986) stressed the need for readjustments in the curriculum in order to make education a forceful tool for the cultivation of social and moral values. It also emphasised the combative role of value education in helping to eliminate obscurantism, religious fanaticism, violence, superstition and fatalism. It was also commanded that primary emphasis was to be laid on the inculcation of positive content based on our heritage, national goals and universal perceptions.

Various committees and commissions set up by Government of India before and after independence have been highlighting the urgent need for incorporating appropriate programmes in our educational system that would directly or indirectly develop among the students an integrated growth of body, mind and spirit. But the majority of educational institutions have failed in evolving an integrated approach in the curricular and co-curricular programmes for the all round development of human personality which is impossible without imbibing effectively the ideas of value. Lured by commercialism we follow money-making education ignoring man-making education.

From the above survey, it is clear that schools have all along considered the training ground for the development of values and desirable habits in children. In older times, religious institutions were the centres of education and religious instruction dealt with moral and spiritual values too, along with the social life of the people. Gradually, the shifts took place from religious education to secular education, because of many factors such as loss of faith in education, scientific development, plural societies etc. Besides moral values, the social, democratic, cultural and scientific values also gained importance in education. Emphasis is now on 'Value Education' for the proper development of the human personality. Transmission of values is inherent in the theory of all round development of human personality which is a prominent aim of education.

What is the Condition of the Present Society?

We are creating generation of youth who are neither Indian nor Western, with the result that they find themselves caught in a dilemma. We are building a purely economic society which seeks security in money and not in concern for social harmony and social well-being. Wherever we go we hear people talking of corruption, This has become so widespread that it is at the root of many other evils like injustice, exploitation and violence.

It is a fact that religious fundamentalism, language and regional chauvinism and caste and communal feelings are raising their ugly heads and threatening the very existence of our nation.

Religion when practised in its true spirit makes for high character. But we have allowed it to make us selfish and negligent. We can see evidence of this even when people go to temple to offer worship. They would readily break a queue to have 'darshan' of the deity and create chaos.

We try our best to build private property and protect but we have little regard for public property which in fact is the property of all of us This is the hall-mark of a people whose conduct is not based on an appreciation of values. The primary values we have to develop to promote the strength and greatness of our society, our country, is to show an abiding concern for others.

Increasing concern is expressed every day about the general deterioration in values in contemporary social life. It is always the acts of immorality that have shown the seeds of unhappiness, hatred, jealousy, enemity in the society. The same immorality is exhibited in the form of mass copying, indiscipline and turmoil in the educational institutions. There is need for appropriate educational action to meet these challenges. The National Educational Policy (1986) has stated "The growing concern over the erosion of essential values and an increasing cynicism in society has brought to focus the need for readjustments in the curriculum in order to make education a forceful tool for the cultivation of social and moral values".

What is the place of Values in the Present Educational System?

The present educational system, however, through its mechanistic approach has added to human psychological problems which through its formal and non-formal agencies has developed an apparatus of structures, content and processes to transmit knowledge without much concern or commitment to respond to the task of communicating and inculcating values necessary for creating ideological climate congenial to appreciation of cultural heritage and taking pride in one's history and having confidence in human capabilities to overcome material, social, religious and spiritual problems of living.

Our educational institutions at all levels are labouring hard for their survival and continuance under serious problems and it is felt that such problems as exist in these institutions, are a reflection of common problems in the community. The students of today have been deprived of their roots with the result that they have failed, by and large, to identify themselves with the community.

The dismal picture of our educational institutions and of social living within them and outside, makes it obligatory to redefine the role of education and employ it for building up an edifice of attitudes, values and moral standards.

Education does not consist merely in the imparting of text-book knowledge or preparing the child for a career, however important these goals are. The moral and spiritual foundation on which we structure our educational system alone can mould the future destiny of India. As Swami Vivekananda said 'we need a man-making and nation building education'.

Why is Value Education Indispensable?

An educational system, if it really aims at making human life peaceful and happy, ought to pay undeviated attention, special are and constant focus on thoughts, motives, attitudes, actions and finally values in the life of human beings.

The present educational system, with all its complexities and intricacies, has proved to be deficient in so far as it neglects, or does not give the deserving importance to values in human life. Thus, human sufferings and sorrows are for ever on the increase in spite of the phenomenal explosion of knowledge. Values have become the neglected lot in the current educational system and consequently the maxim "Education Changeth man" ceases to be meaningful or has to almost lost its 'value'.

"Education without vision is waste; education without value is crime; education without mission is life burden". Education in our life enables us to become comfortable and to look after our family well. But so far as the social progress is concerned, value-based education is an unavoidable necessity. "A nation with atomic power is not a strong nation; but a nation with people with strong character is indeed a strong nation. If a nation is to be strong, then the character of the people of that nation needs to be elevated. For this purpose, value-based education is an indispensable device".

How are Values Inculcated ?

The inculcation of values is by no means a simple matter. There is no magic formula, technique or strategy for this. Value education in all its comprehensiveness involves developing a sensitivity to values, an ability to choose the right values, internalising them, realising them in one's life and living in accordance with them. Therefore, it is not a time-bound affair. It is a life long quest.

In inculcating values, all human faculties such as head, heart and hand should play a role. Thus value education covers the entire domains of learning, the cognitive, affective and psychomotor.

Inculcation of values is influenced by a complex net work of environmental factors such as home, school, peer group, community, the media and society at large. Home takes the highest position in the hierarchy followed by school. As the home, so the society and within the home, as the parents so the children, and within the school, as the teacher, so the taught, are common sayings.

In the pursuit and promotion of values, the teacher has the most vital role to play. It is the teacher who is the guide, friend and philosopher and the first interaction of children, after the parents, is with the teacher. Teachers with vision, dealing with curricular subjects such as languages, science, social science, music, art, work experience and curricular activities such as NCC, Scouts and

Guides, Community Service, Red Cross, field Trips, Sports and Games can develop suitable strategies and methods which would enable proper transmission of values.

Value education can be achieved directly, indirectly or incidentally. Direct value inculcation refers to deliberate, systematic instruction given during the time of formation. Indirectly, value inculcation can be imparted through the regular subjects or curriculum and co-curricular activities. Incidental value inculcation can be given through events and incidents related to good values occurring around us thus relating value inculcation to concrete situations.

There is an urgent need for adopting such methods which promote value education, through the use of various curricular and co-curricular activities in the entire educational programme. There is also a growing awareness among the educationists that ear marking one period exclusively in the school time-table for value education and allocating this work only to one teacher will not be very helpful because values cover the whole gamut of curricular co-curricular activities of schools.

Since every person belongs to the family of humanity, there are certain basic values which are accepted universally. Without these basic values, the character would be lacking in certain primary traits. The basic values are essential to a profound character just like the foundation to the building. Without the foundation, the building would not stand, so also without essential basic values, we cannot build a sound character.

What should be the Content and Approaches of Value Education?

(1) Integrating Values with subject Areas and Educational Programmes

Values could be integrated properly with different subject areas and educational programmes. Through physical education emphasis on health, strength, agility, grace and beauty can be laid. Through sports, the qualities of courage, hardihood, energetic action, initiative, steadiness of will, rapid decision and action, the perception of what is to be done in an emergency, sportsmanship, leadership etc., can be developed. Besides, one would also develop right attitudes, friendliness, self-control, acceptance of victory or defeat, supremacy of a judge or refers, discipline, obedience, order, team spirit and working for a common goal. Likewise, work experience which now constitutes one of the areas of core-curriculum, will help in perfecting skills, utilising materials, tools and processes of works and will promote spirit of 'love of work' and dignity of labour, more so, in a social setting. Responsibility towards and identification with the community can be aroused and awakened by organising work experience activities in social settings. Students can be encouraged to live in harmony with nature, appreciate art, music and can dream of the country and the world that future would shape, of course, with some degree of realism. Examples of national leaders, greatmen of art and culture and of science can be motivating forces in the life of students. In fact, with the help of all educational programmes and subjects of study, values can be included with ease and in a natural setting.

(2) Value Education as Separate Curriculum

People say that 'values cannot be taught but caught'. Against this belief educationists strongly advocate that values could be taught with sufficient care and caution. The NCERT in its publication "Documents on Social, Moral and Spiritual Values in Education" (1979, p. 56) has drawn up 84 values to be inculcated through education (Vide Appendix). All these 84 values are subsumed in the three categories of values. They are: social, ethical and spiritual values. A lot of exercise is required for dividing the values to be imparted at pre-primary, primary, secondary, collegiate and university levels. A lot of literature is available on which values have to be taught at different levels upto secondary schools.

The college student has a great responsibility as he is one among two out of a hundred who are literate. It is this stage of education that is beset with grim problems. These problems have vitiated higher education and made it meaningless and farcical. The objective for inculcating social and ethical values is to develop positive personalities who manifest good citizenship, responsibility, integrity, duty consciousness, service mindedness, patriotism, courage, knowledge of religion and ethics, acceptance of other religions, consideration for others, leadership qualities, willingness to adjust, cooperation, sound health and interest in education and love of knowledge for the sake of knowledge. Education for ethical and social values in college level can be both indirect and direct.

The indirect is the integrated approach with no subject or syllabus. Values are taught along with all the subjects, like Mathematics, Physics and Humanities through (a) daily assembly with readings from scriptures and holy books, (b) talks by eminent religious/spiritual leaders (at least monthly once), (c) regular participation in sports, games, yogasanas and meditation, (d) seminars on ethical topics with students' participation and recognition of students with outstanding moral values.

The direct approaches ate through: (a) institution of formal course—a foundation course spread over three years i.e., first year to third year of the degree course, (b) including topics in different subjects—integration of topics, (c) co-curricular activities. Several factors of personality development can be introduced with the following as foundation course in the curriculum: 1. Spiritual and cultural heritage of India and modern challenges, 2. Relevance of the teachings of great saints of India to modern times, 3. Health education, 4. Yogasanas, 5. Transcendental meditation and 6. Moral values in the society in restructured pattern.

The content of value oriented post-graduate education should include (a) a yearning for knowledge and capacity to utilise it for the good of the society, (b) democratic education, (c) a course in ethics, (d) spiritual education and (e) provision for activities involving values. Once the syllabus for the three year degree course is finalised, the syllabus for the postgraduate courses could be developed by way of including higher level of values that are not taught at the lower levels of courses.

How should Teacher Training Programmes be Designed?

Once the content, areas are delineated the next and most important task remains with the teacher who has to act in diverse learning situations with purpose and precision. The curriculum needs to be designed in such a way that an integration is achieved among the theory courses and values are woven with skill dominated and attitude building areas.

In-service education has to reorient teachers for up dating their know-how in adopting most appropriate teaching strategies with special focus on value inculcation.

The programmes of training need to be so planned as to create conditions to motivate teachers to innovate, devise appropriate methods of communication and activities relevant to needs and capabilities of the community and its concerns with tradition of high intellectual and spiritual attainment.

The curriculum of teachers' training needs to be revised in the light of the new policy thrusts with an emphasis on the integration of education and culture, work experience, physical education, the study of Indian culture and the problems of unity and integration of India.

The pre-requisite to value oriented teacher education is the identification of content areas where values can be integrated and interwoven so as to give prospective teachers value-orientations. It would also necessitate establishing linkages between the pedagogical courses, working with community and the affective domains—interests, values and attitudes.

The teachers will have to be given focused training regarding the evaluative practices that would follow to ensure whether the values have been internalised. A rational system of evaluation as against the conventional tests has to be developed.

The preparation of teachers does not end up with the pre-service education. In-service programmes need greater care in their organisation through universities, NCERT, NCTE, SCERTs and DIETs for enabling them to learn the methods and techniques of inculcating different desirable values among the students.

What are the strategies for the Teaching of Values?

For value orientation, there is necessity to fully utilize the life setting and life experience of students through a wide mix of pedagogic strategies wherein teacher occupies a position of key resources.

Instead of teaching moral maxims, the students should be led to make moral choices. The intellectualization of moral education puts one in a quandary as to how to relate moral education with other intellectual studies.

For a question whether values can be taught as a separate subject or these need to be taught through the instrumentality of a number of subjects, the answer is that both ways may be adopted as per the demands of the situation.

It is said that values are caught but not taught. Modern educationists are of the opinion that values are caught as well as taught. Though values are intimately related to volition and affection, yet cognition plays a role in the training of volition and affection.

The secret of teaching values lies in inspiring and kindling the quest by one's own example and mastery of knowledge of values. Only education in human values can help retain humanness, promote purity and thereby help human beings to unfold the hidden divinity. A congenial social environment of cumulative examples winsome altitudes and salutary interactions between teachers and teachers, teachers and students, teachers and parents can promote and foster the growth of values. It is well known that the right discipline in an institution consists not in external compulsions but in healthy moral influences and habits of the mind which lead spontaneously to desirable behaviour.

The saying, "as the teacher, so the taught" and "as the school, so the students" is something which cannot be disputed. Teachers' responsibility in this national task is tremendous. They can make a modest beginning by helping and guiding children to keep their school premises clean, instilling in them a sense of punctuality and a sense of duty. Many of these elementary qualities can be inculcated by the teachers themselves through personal example. There is a good deal that can be done by the united efforts of teachers and parents and we must therefore have everywhere parent-teacher associations.

We must employ simple methods, that methods that the Buddha employed to teach Gautam. Women are the guardians of culture. Women have to play an important role in the value oriented education of the children. Women have the capacity not only to convert a house into a sweet home but also to transform a country into a strong nation. Mahatma Gandhi once said that if men are good teachers, women are better teachers.

IV. AIDS

AIDS, the acquired immuno-deficiency syndrome (sometimes called "slim disease") is a fatal illness caused by a retrovirus known as the human immuno–deficiency virus (HIV) which breaks down the body's immune system, leaving the victim vulnerable to a host of life-threatening opportunistic infections, neurological disorders, or unusual malignancies (Among the special features of HIV infection are that once infected, it is probable that a person will be infected for life. Strictly speaking, the term AIDS refers only to the last stage of the HIV infection. AIDS can be called our modern pandemic, affecting both industrialised and developing countries.

PROBLEM STATEMENT

World

Recognized as an emerging disease only in the early 1980s, AIDS has rapidly established itself throughout the world, and is likely to endure and persist well into the 21st century. AIDS has evolved from a mysterious illness to a global pandemic which has infected tens of millions in less than 20 years.

According to the recent estimates, by the end of 1999 about 34.4 million people were living with HIV/AIDS. There were 5.4 million new infections and 2.8 million deaths during 1999, bringing the cumulative death toll to 18.8 million (Table 7.4).

Trends in AIDS incidence show important differences between regions of the world. More than 95 percent of new cases remain in developing countries. In Africa, HIV infection in women now outnumber those in men. The number of AIDS deaths in industrialized countries has recently been falling due to combined antiretroviral therapies introduced during the past few years. In developing countries where the vast majority of those infected with HIV live, the number of new cases continue to increase. Beyond the death toll and human suffering, AIDS continues to roll back hard won development gains in many regions of the world. The trend in HIV infection will have a profound impact on future infant, child and maternal mortality, life expectancy and economic growth.

In South-East Asia Region, the number of reported cases continue to increase and is likely to do so well into the early part

Table 7.4 : Global summery of HIV / AIDS epidemic by end of 1999

People newly infected with HIV in 1999	Total	5.4 million
	Adults	4.7 million
	Women	2.3 million
	Children < 15 years	620000
Number of people living with HIV/AIDS	Total	34.3 million
	Adults	33.0 million
	Women	15.7 million
	Children <15 years	1.3 million
AIDS deaths in 1999	Total	2.8 million
	Adults	2.3 million.
	women	1.2 million
	children < 15 years	500,000
Total number of AIDS deaths since the beginning of the epidemic	Total	18.8 million
	Adults	15.0 million
	women	7.7 million
	children < 15 years	3.8 million
Total number of AIDS orphans since the beginning of the epidemic		13.2 million

of 21st century. The potential for continued spread of HIV/AIDS in Asia and Western Pacific is real and requires determined and sustained prevention efforts. Several countries have already experienced intense epidemic in certain population or in the population at large. In these countries, including Combodia, India, Myanmar, and Thailand, AIDS has imposed new demands on health care systems.

In 1984, Thailand was the first country in the region to report a case of AIDS. In most other countries, HIV infection was not diagnosed till 1986 or later. Since then HIV infection has spread rapidly and WHO estimates that currently there are about 5 million HIV infected people in the region (Table 7.5), an alarming 15 percent of the world's total. Besides persons with high risk behaviour, HIV infection rates have now begun to increase in the general population as well. As of January 2000, over 135,000 cases of AIDS have been reported. India, Thailand and Myanmar report the majority of cases with HIV/ AIDS in the region. The analysis of the recent regional data shows that 91 percent of the AIDS cases are in the age group of 15-49 years and 4.6 percent are children. The male to female ratio is 4 to 1. Heterosexual contact is the predominant mode of spread (85%), followed by injecting drug use (7%) and mother to child transmission (50;0). Table 7.5 shows the magnitude of the problem with HIV/AIDS in SEAR countries.

Table 7.5 : AIDS and HIV infection in South -East Asia Region

(As on 1.1.2000)

Country	*Reported AIDS cases*	*Estimated HIV infections*	*HIV rate per 100,000 population*
Bangladesh	12	21,000	16
Bhutan	1	<100	<16
DPR Korea	0	<100	<1
India	8491	3,500,000	418
Indonesia	255	25,000	12
Maldives	5	<100	<25
Myanmar	2854	440000	760
Nepal	261	25000	66
Sri Lanka	77	7500	32
Thailand	123,55	950000	1345
Total	135,311	5,000,000	>358

While the AIDS epidemic in the Region continues to grow, there are now a number of excellent examples which show that progress is being made in responding to the pandemic in SEAR. The 100 per cent condom use programme in Thailand has received worldwide attention. The effectiveness of this programme can be assessed by the declining HIV incidence among military recruits from 3.6 % in 1993 to 2.1% in 1995. At the same time STDs are at a lower rate than ever before–In Calcutta, India, the Sonagachi health care and education project among sex workers has become a model of successful–peer education. The latest data shows that HIV prevalence continues to be low–around.1.4% and STDs are declining. Needle exchange programme and community-based treatment approach for injecting drug users in Myanmar and Nepal have been effective in bringing about behavioural changes and reducing HIV infection rates.

INDIA

Now into its second decade, India's epidemic is marked by heterogeneity–not a single epidemic but made up of a number of distinct epidemics, in some places within the same state. It continues to be strongly driven by heterosexual transmission, with HIV infection moving steadily beyond its initial foci among sex workers and their clients into the wider population.

India's epidemic seems to be following the so-called type 4 pattern, first described in Thailand. The epidemic shifts from the highest risk group (commercial sex workers, drug users) to bridge population (clients of sex workers, STD patients and partners of drug users) and then to general population. The shift usually occurs where the prevalence in the first group reaches 5%. There is a time-lag of 2-3 years between the shift from one group to the next. The trends indicate that HIV infection is spreading in two ways; from urban to rural areas and from individuals practising high risk behaviour to the general population. Data from antenatal clinics indicate rising HIV prevalence among women, which in turn contribute to increasing HIV infection in children.

Estimates at the national level are–about 3.7 million people are suffering from HIV infection at the end of 1999. Sero-surveillance findings from the states show that by June 2000, out of 3662969 persons screened for HIV, 9845 were HIV positive with seropositivity rate of 26.88 per thousand. 1143 persons tested positive for HIV in the month of June 2000. The cumulative number of AIDS in the country has risen to 12389. This includes 9757 males and 2632 females, 150 new cases of AIDS were detected in the month of June 2000. The distribution of likely source of infection is shown in Table 7.6.

Data from various sentinel sites in Maharashtra shows that over the years, HIV infection has increased sharply among commercial sex workers and has reached a level of 60% in Mumbai. It has rapidly progressed in STD clinic attenders (14- 60%) and is steadily progressing in low risk population (Over 2% among women attending antenatal clinic). Among injecting drug users the infection has spread very rapidly in Manipur with HIV prevalence of more than 70% and is spreading in Nagaland. States such as Tamil Nadu, Andhra Pradesh, Karnataka and Manipur are also reporting high

Table 7.6 : Probable route of HIV infection in India

Category	*HIV*		*AIDS*	
	Number of cases	*%*	*Number of cases*	*%*
Sexual	48466	49.23	10006	80.77
Through blood and blood products	6517	6.62	673	5.43
Through infected syringe and needle	3918	3.98	645	5.21
Perinatal	353	0.36	102	0.82
Other	39197	39.81	963	7.77
Total	98451	100	12389	100

levels of infection (between .1-2 per cent in antenatal women). It reflects the broadening of the epidemic across the southern and western states of India as well as the continued concentration of HIV, through injectable drug users in the North -East. In other parts of the country, where the overall level of HIV are still low, high levels of STDs make for a continuing potential for the epidemic to become generalized among all sexually active adults. Differences across states may just be matter of moment.

In India the epidemic continues to shift towards women and young people with 21.4% of all HIV estimated to be women with an accompanying increase in vertical transmission and paediatric HIV. Migration (both within and between the states) is a potential source of spread of disease from urban to rural population. Indications are that a sharp increase in injecting drug use is underway with drug users switching from smoking to over the counter injecting drugs. Important new elements are entering the picture. The burden of AIDS cases is beginning to be felt in states affected early in the epidemic. Mumbai in Maharashtra and Imphal in Manipur report high hospital bed occupancy rates from HIV–associated illness. Home-based care is still in its infancy in most parts of the country. Today greater awareness of HIV has triggered a mixed response from society. While individual examples of support and care are increasingly seen, HIV–positive individuals in general have been poorly received in the society. Assess to health care is emerging as a critical first point of contact with services for HIV positive individuals and their families and a site of frequent discrimination.

Fig. 7.10 shows the diversity of the epidemic in different states of India.

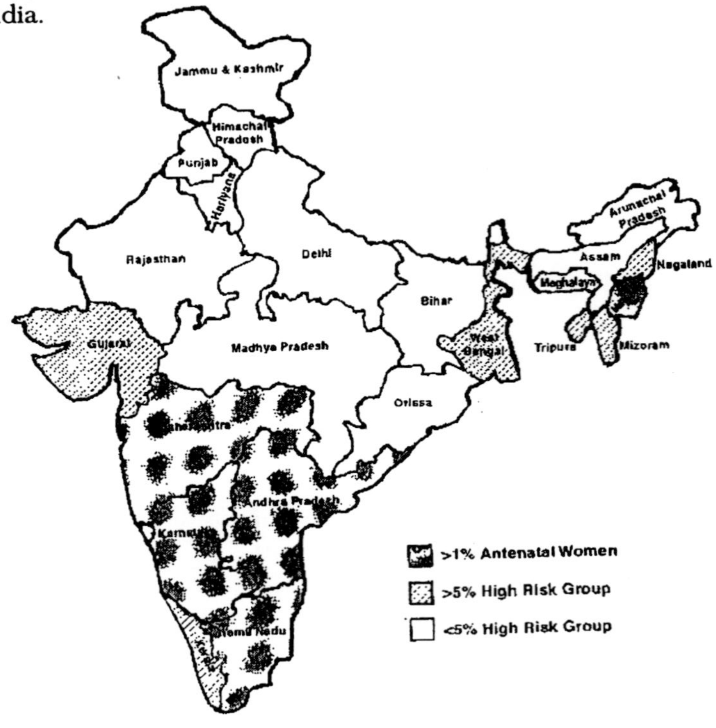

Fig. 7.10 : Adult prevalence - HIV

Based on the analysis of existing sentinel surveillance data, the HIV prevalence in adult population can be broadly classified into three groups of states/UTs in the country.

1. Group I : A "generalised epidemic", where the HIV infection has crossed 5% mark in high risk group, is 1% or more in antenatal women and spreading from urban to rural areas. These states include Maharashtra, Tamil Nadu, Karnataka, Andhra Pradesh and Manipur.
2. Group II : A "concentrated epidemic", where HIV is still concentrated among high risk group and has crossed 5% mark but the infection is below 1% in antenatal women. It includes states like Gujarat, Goa, Kerala, West Bengal and Nagaland.
3. Group III : A "low–level epidemic", when none of the sub groups of population including sex workers, injecting drug users

has reached 5% level of HIV and is less than 1% in antenatal women. This group includes the remaining states.

It can be seen that majority of states in the country including some of the most populous states such as Uttar Pradesh, Madhya Pradesh, Bihar, Rajasthan etc. are still in the low level of endemicity.

The sentinel surveillance data from the antenatal clinics from 7 metropolitan cities in the country shows that HIV infection has crossed 2.5% in Mumbai, it is more than 1% in Hyderabad, Bangalore, and Chennai and is below 1% in Calcutta, Ahmedabad and Delhi (Fig. 7.11).

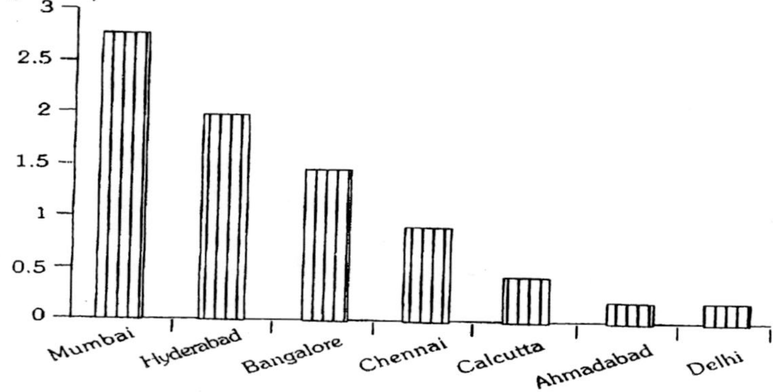

Fig 7.11 : Sentinel Surveillance for HIV Infection. HIV prevalence among antenatal women in metropolitan cities—1998

The major opportunistic infection in the AIDS is tuberculosis. Thus it may emerge as dual epidemic of TB and AIDS in future.

EPIDEMIOLOGICAL FEATURES

1. Agent factors

(a) Agent : When the virus was first identified it was called "lymphadenopathy—associated virus (LAV)" by the French scientists. Researchers in USA called it "human T-cell lymphotropic virus III (HTLV-III)". In May 1986, the International Committee on the Taxonomy gave it a new name: human immuno—deficiency virus (HIV).

The virus is 1/10,000th of a millimetre in diameter. It is a protein capsule containing two short strands of genetic material (RNA) and enzymes. The virus replicates in actively dividing T4 lymphocytes and like other retroviruses can remain in lymphoid cells in a latent state that can be activated. The virus has the unique ability to destroy human T4 helper cells, a subset of the human T-lymphocytes. The virus is able to spread throughout the body. It can pass through the blood-brain barrier and can then destroy some brain cells. This may account for certain of the neurological and psychomotor abnormalities observed in AIDS patients. HIV mutates rapidly, new strains are continually, developing. There are two types of HIV—the most common HIV 1 and a more recently recognized virus (in West Africa) called HIV 2.

The virus is easily killed by heat. It is readily inactivated by ether, acetone, ethanol (20 per cent) and beta-propiolactone (1:400 dilution), but is relatively resistant to ionizing radiation and ultraviolet light.

(b) Reservoir of Infection : These are cases and carriers. Once a person is infected, the virus remains in the body life-long. The risk of developing AIDS increases with time. Since HIV infection can take years to manifest itself, the symptomless carrier can infect other people for years.

(c) Source of Infection : The virus has been found in greatest concentration in blood, semen and CSF. Lower concentrations have been detected in tears, saliva, breast milk, urine, and cervical and vaginal secretions. HIV has also been isolated in brain tissue, lymph nodes, bone marrow cells and skin. To date, only blood and semen have been conclusively shown to transmit the virus.

2. Host factors

(a) Age : Most cases have occurred among sexually active persons aged 20-49 years. This group represents the most productive members of society and those responsible for child-bearing and child-rearing. Children under 15 make up less than 3 per cent of cases.

(b) Sex: In North America, Europe and Australia, about 70 per cent of cases are homosexual or bisexual men. In Africa, the picture is very different; the sex ratio is equal. Certain sexual practices increase the risk of infection more than others, e.g., multiple sexual partners, anal intercourse, and male homosexuality. Higher rate of HIV infection is found in prostitutes.

(c) High Risk Groups : Male homosexuals and bisexuals, heterosexual partners (including prostitutes), intravenous drug abusers, transfusion recipients of blood and blood products, haemophiliacs and clients of STD.

(d) Immunology : The immune system disorders associated with HIV infection/AIDS are considered to occur primarily from the gradual depletion in a specialised group of white blood cells (lymphocytes) called T-helper or T-4 cells. The full name of T - helper cell is CD4 + T-lymphocyte and is also commonly known as CD4 + cell. These cells play a key role in regulating the immune response.

HIV selectively infects T-helper cells apart from several other cells in the immune system such as B-cells, microphages and nerve cells. When the virus reproduces, the infected T-helper cells are destroyed. Consequently people with AIDS tend to have low overall white blood cell count. Whereas healthy individuals have twice as many "helper" cells as "suppressor" cells, in the AIDS patients the ratio is reversed. A decreased ratio of T-helper to T-suppressor cells may be an indirect indicator of reduced cellular immunity. One of the most striking features of the immune system of patients with AIDS is profound lymphopenia, with a total lymphocyte count often below 500 c. mm. It is the alteration in T-cell function that is responsible for the development of neoplasms, the development of opportunistic infections, or the inability to mount a delayed-type hypersensitivity response. The lack of an obvious immunological response by the host to the virus is one of the problems confronting scientists. That is, those with antibodies to HIV, usually have too few HIV antibodies, and these antibodies are also ineffective against the virus.

MODE OF TRANSMISSION

The causative virus is transmitted from person-to-person, most frequently through sexual activity. The basic modes of transmission are:

(a) Sexual transmission

AIDS is first and foremost a sexually transmitted disease. Any vaginal, anal or oral sex can spread AIDS. In the USA, over 70 %

of the cases were in homosexual or bisexual men. In contrast, in equatorial Africa, AIDS is acquired mainly through heterosexual contact (infected man to woman; infected woman to man). Every single act of unprotected intercourse with an HIV-infected person exposes the uninfected partner to the risk, of infection. The size of the risk is affected by a number of factors, including the presence of STD, the sex and age of the uninfected partner, the type of sexual act, the stage of illness of the infected partner, and the virulence of the HIV strain involved. A European study of 563 heterosexual couples in which only one partner was infected at the start, suggests that chances of transmission of HIV infection from male to female is twice as likely as from female to male. Generally, women are more vulnerable to HIV infection because a larger surface is exposed, and semen contains higher concentration of HIV than vaginal or cervical fluids.

Anal intercourse carries a higher risk of transmission than vaginal intercourse because it is more likely to injure tissues of the receptive partner. For all forms of sex, the risk of transmission is greater where there are abrasions of the skin or mucous membrane. For vaginal sex the risk is greater when woman is menstruating.

Exposed adolescent girls and women above 45 years of age are more prone to get HIV infection. In teenagers the cervix is thought to be less efficient barrier to HIV than in mature genital tract of adult women. The thinning of mucosa at menopause is believed to lessen the protective effect. The production of mucus in the genital tract of adolescent girls and in postmenopausal women is not as prolific as in women between these life stages and this may also enhance their susceptibility to HIV infection.

An STD in either the HIV-negative or the HIV-positive partner facilitates the transmission of HIV. If an STD, such as syphilis, chancroid or herpes, cause ulceration in the genital or perineal region of the uninfected partner, it becomes far easier for HIV to pass in to his or her tissues. An STD causes inflammation. T–cells and monocytes/macrophages, get concentrated in the genital area. In a person already infected with HIV, some of these key cells of the immune system will be carrying the virus–which magnifies the risk of transmission to the uninfected partner.

As for HIV–infected people, they are more infectious to others in the very early stages, before antibody production i.e. during the "window period", and when the infection is well advanced, because levels of virus in the blood at that time is higher than at other times.

(b) Blood contact

AIDS is also transmitted by contaminated blood—transfusion of whole blood cells, platelets and factors VIII and IX derived from human plasma. There is no evidence that transmission ever occurred through blood products such as albumin, immunoglobulins or hepatitis vaccines that meet WHO requirements. Contaminated blood is highly infective when introduced in large quantities directly in to the blood stream. The risk of contracting HIV infection from transfusion of a unit of infected blood is estimated to be over 95 per cent. Since the likelihood of HIV transmission through blood depends on the "dose" of virus injected, the risk of getting infected through a contaminated needle, syringe or any other skin-piercing instrument is very much lower than with transfusion. Nevertheless, among drug users who inject heroin, cocaine or other drugs, this route of transmission is significant because exposure is repeated so often, in some cases, several times day. As a result, needle-sharing by drug users is a major cause of AIDS in many countries both developed and developing and in some it is the predominant cause. Any skin piering (including infections, earpiercing tattooing acupuncture or scarificance) can transmit the virus if the instruments used have not been sterilized and have previously been used on an infected person. It may be mentioned that transfusion of blood and blood products has played a minor role in the spread of AIDS in developed countries.

(c) Maternal—foetal trasnmission: mother-to-child transmission

HIV may pass from an infected mother to her foetus through the placenta or to her infant during delivery or by breast-feeding. About one-third of the children of HIV-positive mothers get infected though this route. The risk of infection transmission is higher if the mother is newly infected or if she has already developed AIDS. Infants and children progress rapidly to they already account for about 20 per cent cases to date.

There is no evidence that HIV is transmitted through mosquitoes or any other insect, casual social contact with infected persons including within households, or by food or water. There is no evidence of spread to health care workers in their professional contact with people with AIDS.

Incubation period

While the natural history of HIV infection is not fully known current data suggest that the incubation period is uncertain, (from a few months to 6 years or even more) from HIV infection to the development of AIDS. The virus can lie silent in the body for many years. The percentage of people infected with HIV who will develop clinical disease AIDS and another 25-30 percent AIDS related complex. However it is estimated that 75 per cent of those infected with HIV will develop AIDS by the end to ten years.

Clinical manifestations

The clinical features of HIV infection have been classified into four board categories :

I. Initial infection with the virus and development of antibodies.
II. Asymptomatic carrier state.
III. AIDS-related complex (ARC).
IV. AIDS.

(I) Initial infection

Except for a generally mild illness fever sore throat and rash that about 70 per cent of people experience a few weeks after initial infection with the virus most HIV–infect people have no symptoms for the first five years or so. They look healthy and feel well although right from the start they can transmit the virus to others. Once infected, people are infected for life. Scientists have not found as yet a way of curing them, or making them uninfectious to others.

HIV antibodies usually take between 2 to 12 weeks to appears in the blood-stream, though they have been known to take longer. The period before antibodies are produced is the "window period" during which although the person is particularly infectious because of the high concentration of virus in the blood he will test negative on the standard antibody blood test. Though the body's immune system reacts to the invasion of HIV by producing antibodies, these do not inactivate the virus in the usual way.

(II) Asymptomatic carrier state

Infected people have antibodies but not overt signs of disease expect persistent generalised lymphadenopathy. It is not clear how long the asymptomatic carrier state lasts.

(III) AIDS-related complex

A person with ARC has illnesses caused by damage to the immune system, but without the opportunistic infections (Fig. 7.12) and cancers associated with AIDS, but they exhibit one or more of the following clinical signs; unexplained diarrhoea lasting longer than a month, fatigue, malaise, loss of more than 10 per cent body weight, fever, night sweats, or other milder opportunistic infections such as oral thrush, generalised lymphadenopathy or enlarged spleen. Patients from high-risk groups who have two or more of these manifestations (typically including generalised lymphadenopathy), and who have a decreased number of T–helper lymphocytes are considered to have AIDS-related complex. Some patients with AIDS- related complex subsequently develop AIDS.

(IV) AIDS

AIDS is the end-stage of HIV infection. A number of opportunist infections commonly occur at this stage (Fig. 7.12) and/or cancers that occur in people with otherwise unexplained defects in immunity. Death is due to uncontrolled or untreatable infection. *Tuberculosis* and *Kaposi* sarcoma are usually seen relatively early. Serious fungal infections such as Candida oesophagitis, *Cryptococcus* meningitis and *penicillosis,* and parasitic infections such as *Pneumocystis carinni* pneumonia or *Toxoplasma gondii* encephalitis tend to occur, when T–helper cell count has dropped to around 100. People whose counts are below 50 have the late opportunistic infections such as *cytomegaloviral retinitis.*

Many people with AIDS are affected by a wasting syndrome that is known, especially in Africa, as "slim disease". It involves chronic diarrhoea and severe weight loss. Another condition, seen

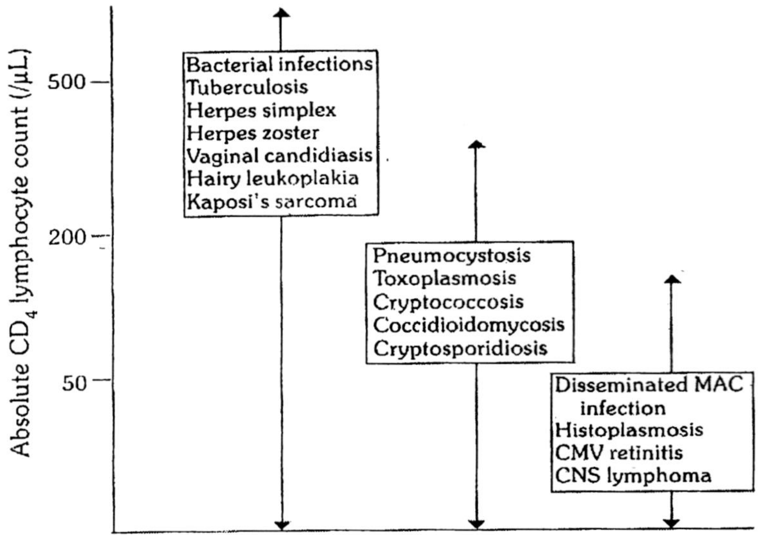

Fig. 7.12 : Relationship of CD_4 count to development of opportunistic infection

world-wide, is AIDS encephalopathy or AIDS dementia, which is caused by HIV crossing "blood-brain barrier". In its late stages, AIDS encephalopathy resembles senile dementia or Alzheimer's disease. AIDS dementia appears to result not from opportunistic infection, but from the action of the virus itself.

TUBERCULOSIS. An alarming factor in the AIDS epidemic is the increasing link between HIV infection and tuberculosis. In countries where tuberculosis is endemic, many people are infected in childhood. When the immune system breaks down, as in HIV infection, tuberculosis becomes active and the person becomes contagious to others. Studies in Rwanda, the USA, Zaire and Zambia found that HIV-positive individuals were 30-50 times more likely to develop active tuberculosis than HIV-negative people. As a consequence, AIDS is reviving an old problem in developed countries e.g. in the USA, where there was sudden increase in tuberculosis cases. The situation in developing countries is still worst.

The emergence of drug resistance makes it essential that antibiotic sensitivity be performed on all positive cultures. Drug therapy should be individualized. Patients with multi drug resistance should receive atleast three drugs to which their organism is sensitive.

PERSISTENT GENERALISED LYMPHADENOPATHY. Lymph nodes are larger than one centimetre in diameter, in two or more sites other than the groin area for a period of at least three months.

KAPOSI SARCOMA. A tumour featuring reddish brown or purplish plaques or nodules on the skin and mucous membranes. Endemic in Africa prior to HIV, it used to affect mainly older men. With HIV infection it affects a wider age range and both sexes, and is characterised by lesions in the mouth or gut; or lesions are generalised (in two or more places) or rapidly progressive or invasive.

OROPHARYNGEAL CANDIDIASIS. Caused by a common yeast fungus, oral thrush presents with soreness and redness, with white plaques on the tongue, and in the mouth and throat; and sometimes a white fibrous layer covering the tonsils and back of the mouth. Infection of the oesophagus presents with pain behind the breastbone.

CYTOMEGALOVIRUS RETINITIS. Inflammation of the eye retina which may lead to blindness.

PNEUMOCYSTIS CARINII PNEUMONIA. Symptoms can include a dry, non-productive cough; inability to take a full breath and occasional pain on breathing; and weight loss and fever.

TOXOPLASMA ENCEPHALITIS. Protozoal infection in the central nervous system, presenting with focal neurological signs such as mild hemiplegia or stroke, resulting from damage to part of the brain, seizures or altered mental status.

HAIRY LEUKOPLAKIA. White patches on the sides of the tongue, in vertical folds resembling corrugations.

CRYPTOCOCCAL MENINGITIS. A fungal infection in the central nervous system which usually presents with fever, headache, vomiting and neck stiffness.

HERPES ZOSTER OR SHINGLES. Viral inflammation of the central nervous system, presenting with localised pain and burning sensations, followed by vesicle eruption (skin blistering) and ulceration.

SEVERE PRURIGO OR PRURITIC DERMATITIS. Chronic skin inflammation in the form of a very itchy rash of small flat spots developing into blisters.

SEVERE OR RECURRENT SKIN INFECTIONS. Warts; dermatophytosis or ring-worm; and folliculitis (inflammation of hair follicles).

DIAGNOSIS OF AIDS

CLINICAL

I. WHO case definition for AIDS surveillance

For the purposes of AIDS surveillance an adult or adolescent (> 12 years of age) is considered to have AIDS if at least 2 of the following major signs are present in combination with at least 1 of the minor signs listed below, and if these signs are not known to be due to a condition unrelated to HIV infection.

Major signs

- weight loss $\geq$ 10% of body weight
- chronic diarrhoea for more than 1 month
- prolonged fever for more than 1 month (intermittent or constant)

Minor signs

- persistent cough for more than 1 month[a,b]
- generalized pruritic dermatitis
- history of herpes zoster[b]
  oropharyngeal candidiasis
- chronic progressive or disseminated herpes simplex infection
- generalized lymphadenopathy

The presence of either generalized Kaposi sarcoma or cryptococcal meningitis is sufficient for the diagnosis of AIDS for surveillance purposes.

a. For patients with tuberculosis, persistent cough for more than 1 month should not be considered as a minor sign
b. Indicates changes from the 1985 provisional WHO clinical case definition for AIDS ("Bangui definition")

The clinical definition is relatively *specific* (if used correctly), meaning that the vast majority of people diagnosed as having AIDS will have been correctly assessed. However, studies show that the definition is relatively *insensitive,* meaning that only half the patients who have severe illness related to HIV infection are included. This is because not all HIV-related opportunistic diseases are in the AIDS definition. Tuberculosis is widely recognised as the commonest opportunistic disease associated with HIV in Africa. But because TB causes wasting, cough and fever in most patients, the AIDS clinical case definition cannot reliably distinguish between HIV-positive and HIV-negative TB patients.

The clinical case definition was developed to enable reporting of the number of people with AIDS for the purposes of public health surveillance, rather than for patient care. However, for the purposes of individual case management, it is useful to be able to diagnose whether illnesses may be related to HIV infection (symptomatic HIV infection) because:

- clinical manifestations can be a reliable indicator of underlying HIV infection;
- over-use of HIV testing is avoided, testing is used to confirm suspected HIV infection, rather than as a diagnostic tool in the first instance;
- a patient with suspected HIV infection can be counselled about having an HIV test, the implications for them and .. their sexual partners, self-care and nutrition;
- many HIV-related illnesses can be treated, improving the patient's quality of life;
- certain drugs (such as thiacetazone) cause severe side effects in people with HIV infection, and should not be prescribed for them.

II. Expanded WHO case definition fo AIDS surveillance

For the purposes of AIDS surveillance an adult or adolescent (>12 years of age) is considered to have AIDS if a test for HIV antibody gives a positive result, and one or more of the following conditions are present:

- 10% body weight loss or cachexia, with diarrhoea or fever, or both, intermittent or constant, for at least month, not known to be due to a condition unrelated HIV infection

- cryptococcal meningitis
- pulmonary or extra-pulmonary tuberculosis
- Kaposi sarcoma
- neurological impairment that is sufficient to prevent independent daily activities, not known to be due to a condition unrelated to HIV infection (for example, trauma or cerebrovascular accident)
- candidiasis of the oesophagus (which may be presumptively diagnosed based on the presence of oral candidiasis accompanied by dysphagia)
- clinically diagnosed life-threatening or recurrent episodes of pneumonia, with or with out aetiological confirmation
- invasive cervical cancer

Major features of this expanded surveillance case definition are that it requires an *HIV serological test,* and includes a broader spectrum of clinical manifestations of HIV such as tuberculosis, neurological impairment, pneumonia, and invasive cervical cancer. The expanded definition is simple to use and has a higher specificity.

LABORATORY DIAGNOSIS

SCREENING TESTS: As antibodies to HIV are far easier to detect than the virus itself, their presence or absence in blood-stream is the basis for the most widely used test of HIV infection. A person whose blood contains HIV antibodies is said to be HIV-positive, or seropositive, meaning that he or she is infected with HIV. There is now a wide range of screening tests based on detection of HIV-antibodies. To be reliable at screening test must be ***sensitive*** enough to identify all "true positives", while being ***specific*** enough to record few "false positives". The ideal test needs both the attributes.

At present, to ensure accuracy, two different tests are commonly applied. At first a sensitive test is used to detect the HIV-antibodies, while a second ***confirmatory test*** is used to weed out any false positive results. The first kind of test is normally the *ELISA*. The confirmatory test, usually a ***Western Blot*** is a highly specific test; it is based on detecting specific antibody to viral core protein and envelop glycoprotein. This is a more difficult test to perform and requires trained and experienced laboratory workers to interpret the test.

VIRUS ISOLATION: A test for the virus itself would eliminate the painful uncertainty of AIDS infection. HIV can be recovered from cultured lymphocytes. This type of testing is ; very expensive and requires extensive laboratory support.

The current trend in HIV-antibody tests is towards simple, cheap, reliable kits whose results can be read on the spot without much waiting and without the need for laboratory back-up.

Nonspecific laboratory findings with HIV infection may include anaemia, leukopenia (particularly lymphocytopenia) and thrombocytopenia in any combination, polyclonal hypergammaglobinemia. Cutaneous energy is frequent early in the course and becomes universal as the disease progresses.

Several laboratory markers are available to provide prognostic information and guide therapy decisions. The most widely used marker is the absolute D_4 lymphocyte count. As the count decreases, the risk of opportunistic infection increases. People with healthy immune system usually have more than 950 CD_4 cells/μL ill of blood. The number falls over the course of HIV infection. People with AIDS usually have CD_4 cell count below 200 (USA makes CD cell count below 200 in an HIV-infected person a definition of AIDS). The trend of the count is much more important than any single reading. The frequency of performance of counts depends on the patient's health system. Patients whose count is substantially above the threshold for antiviral therapy (500 cells/μl) should have count performed every 3 months, This is necessary for evaluating the efficacy of antiviral therapy and for initiating Pcarinii prophylactic therapy when the count falls below 200 cells/μl Some studies suggest that the percentage of CD_4 lymphocytes is more reliable indicator of prognosis than the absolute count because the percentage does not depend on calculating a manual differential, Risk of progression to AIDS is high with percentage of CD_4 lymphocyte less than 20.

The laboratory tests and their significance are summarized in Table 7.7.

The WHO has pointed out the danger of compulsory testing programmes in their tendency to social rejection of HIV-carrier and the resulting social and psychological consequences. Diagnostic

Table 7.7 : Laboratory findings with HIV infection

Test	*Significance*
HIV enzyme linked immunosorbent assay (ELISA)	Screening test for infection sensitivity >99.9% to avoid false positive results, repeatedly reactive result must be confirmed with western blot.
Western blot	Confirmatory test for HIV specificity when combined with ELISA > 99.99% indeterminate result with early HIV infection, HIV-2 infection automune disease pregnancy and recent tetanus toxoid administration.
CBC	Anaemia, neutropenia and thrombo-cytopenia common with advanced HIV infection
Absolute CD_4 lymphocyte count	Most widely used predictor of HIV progression. Risk of progression to an AIDS opportunistic infection to malignancy is high with CD_4 < 200 cells/ μL.
CD4 lymphocyte percentage	Percentage may be more reliable than the CD_4 count. Risk of profession to an AIDS opportunistic infection or malignancy is high with percentage <20%.
HIV viral load tests	These tests measure the amount of actively replicating HIV virus. Correlation with disease progression and response to antiretroviral drugs.

testing may be useful in gauging the magnitude and course of the epidemic.

Control of AIDS

There are four basic approaches to the control of AIDS:

Prevention

(a) Education

Until a vaccine or cure for AIDS is found, the only means at present available is health education to enable people to make life-saving choices (e.g., avoiding indiscriminate sex, using condoms), There is, however, no guarantee that the use of condoms will give full protection. One should also avoid the use of shared razors and toothbrushes. Intravenous drug users should be informed that the sharing of needles and syringes involves special risk. Women suffering from AIDS or who are at high risk of infection should avoid becoming pregnant since infection can be transmitted to the unborn or newborn. Educational material and guidelines for prevention should be made widely available. All mass media channels should be involved in educating the people on AIDS, its nature, transmission and prevention; this includes international travellers.

(b) Prevention of bloodborne HIV transmission

People in high-risk groups should be urged to refrain from donating blood, body organs, sperm or other tissues. All blood should be screened for HIV 1 & HIV 2 before transfusion. Transmission of infection to haemophiliacs can be reduced by introducing heat treatment of factors VIII and IX. Strict sterilization practices should be ensured in hospitals and clinics. Presterilized disposable syringes and needles should be used as far as possible. One should avoid injections unless they are absolutely necessary.

2. Antiretroviral treatment

At present there is no vaccine or cure for treatment of HIV infection/AIDS. However, the development of drugs that suppress the HIV infection itself rather than its complications has been important development. These antiviral chemotherapy, while not a cure, have proved to be useful in prolonging the life of severely ill patients.

The availability of agents that alone and in combination suppress HIV replication has had a profound impact on the natural history of HIV infection. Patients who achieve excellent suppression of HIV generally have stabilization or improvement of their clinical course which results from partial immunologic reconstitution and a subsequent decrease in complications of immunosuppression. Concept about the timing of such therapy have changed considerably. The recognition of continuous viral replication during early HIV infection and the considerable risk of disease progression in individuals with even low levels of circulating HIV RNA has

Table 7.8 : Antiretroviral therapy

Drug	*Dose*	*Common side effects*	*Monitoring*
Nucleoside analogs			
Zidovudine (AZT) (Retrovir)	500-600 mg orally daily in two or three divided doses	Anaemia, neutropenia, nausea, malaise, headache, insomnia, myopathy	Complete blood count and differential (every 3 months once stable)
Didanosine (ddI) (Videx)	12-30 mg orally twice daily (for pill formulation)	Peripheral neuropathy, pancreatitis, dry mouth, hepatitis	CBC and differential, aminotransferases, K^+, amylase, triglycerides, biomonthly neurologic questionnaire for neuropathy
Zalcitabine (ddC) (Hivid)	0.375-0.75 mg orally three times a day	Peripheral neuropathy, aphthous ulcers, hepatitis	Monthly neurologic questionnaire for neuropathy, aminotransferases
Stavudine (d4T) (Zerit)	40 mg orally twice daily	peripherl neuropathy, hepatitis pancreatitis	Monthly neurologic questionnaire for neuropathy, aminotransferases, amylase
Lamivudine (3TC) (Epivir)	150 mg orally twice	Rash, peripheral neuropathy	No additional monitoring

Table 7.8 : *Contd.*

Drug	*Dose*	*Common side effects*	*Monitoring*
Protease inhibitors			
Saquinavir (invirase)	600 mg orally three times daily	Gastrointestinal distress, headache	No additional monitoring
Ritonavir (Novir)	600 mg orally twice daily or 400 mg orally twice daily in combination with other protease inhibitors	Gastrointestinal distress, peripheral paresthesias	Bimonthly aminotyransferases, CK, uric acid, triglycerides
Indinavir (Crixivan)	800 mg orally three times daily	Kidney stones	Bimonthly aminotyransferases, bilirubin level
Nelfinavir (Veracept)	750 mg orally three times daily	Diarrhoea	No additional monitoring
Nonnucleoside reverse transcriptase inhibitors (NNRTIs)			
Nevirapine (Viramune)	200 mg orally daily for 2 weeks, then 200 mg orally twice daily	Rash	No additional monitoring
Delavirdine (Rescriptor)	400 mg orally three times daily	Rash	No additional monitoring

given impetus to the concept of treating the majority of infected individuals with antiretroviral therapy even if the CD_4 lymphocyte count is preserved. Although the long term benefit of this strategy has not been proved, it is reasonable to offer immediate antiretroviral therapy to all patients. An approach to antiretroviral therapy is shown in Table 7.8.

Combination Therapy

There is now little debate about the necessity for combining drugs from all available classes in order to achieve long term suppression of HIV and its associated clinical benefit, Only combination of the drugs have been able to decrease HIV viral load by 2-3 logs and allow suppression of HIV RNA to below , the threshold of detection for longer than 2 years in some individuals. Data from clinical trials have demonstrated the superiority of regimens that contain either non non-nucleoside or protease inhibitor agents in addition to nucleoside agents alone.

There is a consensus about the desirability of treating aggressively once a decision to start antiretroviral therapy has been made rather than adding drugs subsequently over months or year. Fig. 7.13 outlines the current approaches to first line and salvage antiretroviral therapy.

Monitoring HIV-infected patients with serial T-helper cell count and other clinical and laboratory markers is standard practice in developed nations.

Post exposure prophylactic treatment: Post exposure prophylaxis (PEP) for HIV refers to antiretroviral drug treatment started within hours following accidental exposure to the virus. Four weeks of treatment with AZT monotherapy after accidental needle stick exposure to HIV among health care workers decreases the chance of their becoming infected by 79%, according to the results of a recent study done by US government. Given this high rate of success in preventing HIV infection among health care workers, the question naturally arises as to whether PEP also can abort HIV infection among injecting drug users. Although the answer is not yet known, some clinicians are already prescribing 2 to 3 drug therapy to patients exposed to HIV through sex or injection drug use. The objective of the PEP is to prevent HIV infection of cells. The following treatment is recommended by the US Centre for Disease Control and Prevention for health care workers accidentally

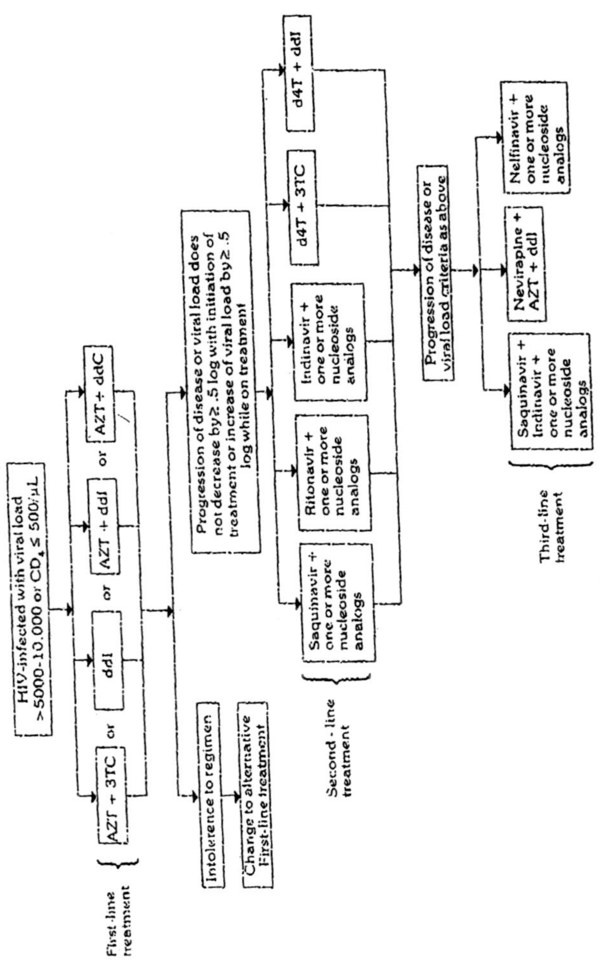
HIV-infected with viral load >5000-10,000 or CD4 ≤ 500/μL
AZT + 3TC
or
ddI
or
AZT + ddI
or
AZT + ddC
First-line treatment
Intolerence to regimen
Change to alternative First-line treatment
Progression of disease or viral load does not decrease by ≥ .5 log with initiation of treatment or increase of viral load by ≥ .5 log while on treatment
Saquinavir + one or more nucleoside analogs
Ritonavir + one or more nucleoside analogs
Indinavir + one or more nucleoside analogs
d4T + 3TC
d4T + ddI
Second - line treatment
Progression of disease or viral load criteria as above
Saquinavir + Indinavir + one or more nucleoside analogs
Nevirapine + AZT + ddI
Nelfinavir + one or more nucleoside analogs
Third-line treatment

exposed to HIV.

a. Double combination of treatment with AZT (200 mg 3 times daily) plus Lamivudine (3 TC) (150 mg twice daily) for 4 weeks.
b. If the "source" individual has advanced AIDS, the protease inhibitor nelfinavir (750 mg 3 times daily) should be added to the AZT /3TC regimen.
c. If the "source" individual has failed on AZT/3TC therapy (e.g., no benefit or intolerance), stavudine (d4T) plus ddl should be used instead of AZT/3TC.

3. Specific prophylaxis

Until more effective antiviral therapy becomes available, the main aim of existing therapies will be to treat the manifestations of AIDS. Primary prophylaxis against *P. carinii* pneumonia should be offered to patients with CD_4 count below 200 cells/ μl. The regimens available are trimethoprim-sulfamethoxazole, aerosolized pentamidine and dapsone. Patients who develop *P. carinii* infection on a particular prophylactic regimen should be switched to the other drug or should receive a combination regimen.

M. *avium* complex occurs in at least one-third of AIDS patients. Rifabutin has been shown in a randomized trial to decrease the incidence of disseminated *Mavium-intracellulare* in persons with less than 200 CD 4 cell/Ill. Clinicians should make certain that patients do not have active *M. tuberculosis* infection before starting Rifabutin. Prophylaxis against *M. tuberculosis* is 300 mg isoniazid daily for 9 months to one year. It should be given to all HIV-infected patients with positive PPD reactions (defined for HIV-infected patients as more than 5 mm in induration). Kaposi's sarcoma might be treated in some stage with interferon, chemotherapy or radiation. Cytomegalovirus retinitis can be controlled by ganciclovir, cryptococcal meningitis with fluconazole. Esophageal candidiasis or recurrent vaginal candidiasis can be treated by fluconazole or ketoconazole. Herpes simplex infection and herpes zoster can be treated with acyclovir or foscamet.

4. Primary health care

Because of its wide-ranging health implications, AIDS touches all aspects of primary health care, including mother and child health, family planning and education. It is important therefore that AIDS control programmes are not developed in isolation. Integration into country's primary health care system is essential.

NATIONAL AIDS CONTROL PROGRAMME

With the spread of AIDS from one country to another it became necessary to initiate a national control programme. The Government of India in 1985 constituted a task force to look into this matter. It began by pilot screening programme of high-risk population. National AIDS Control Programme was launched in 1987. In the year 1992, the Ministry of Health and Family Welfare set up a National AIDS Control Organization as a separate wing to implement and closely monitor the various components of the programme. The government of India launched a 5 year HIV/AIDS Control Project from September 1992 to September 1997 as 100 per cent centrally sponsored project for all states/UTs. The project was later on extended upto March 1999.

The national strategy has the following important components: establishment of surveillance centres to cover the whole country; identification of high risk groups and their screening; issuing specific guidelines for management of detected cases and their follow-up; formulating guidelines for blood banks, blood product manufacturers, blood donors, and dialysis units; information, education and communication activities by involving mass media; research, reduction of personal and social impact of the disease; control of sexually transmitted diseases and condom programme.

The aim of the first phase of the programme was to prevent further transmission of HIV, to decrease morbidity and mortality associated with HIV infection and to minimize the socio-economic impact resulting from HIV infection.

Blood safety programme: Blood transfusion services have been considered as an integral part of the health care system. Blood Transfusion Councils have been set-up at national and state level. Professional blood donation has been prohibited in the country since 1st January 1998. Only licensed blood banks are permitted to operate in the country and voluntary blood donation is encouraged. The strategy is to ensure safe, collection, processing, storage and distribution of blood and blood products. Zonal blood testing centres have been established to provide linkage with other blood banks affiliated to public, private and voluntary sector. The function of these Zonal centres is to test the blood sample from the blood banks attached to them and to report the results of HIV tests same day. As per national blood safety policy, testing of every unit of blood is mandatory for detecting infections like HIV, hepatitis B, malaria and syphilis.

In the country 1233 blood blanks have been licensed to supply blood, 815 blood banks have been modernized with provision of adequate facilities of equipment and development of appropriate man power, 40 blood component separation units have been established, besides the district level and small blood banks with independent HIV testing facilities. 154 Zonal Blood Testing Centres and 9 HIV Reference Centres are functioning in the country. HIV test kits are supplied upto district level blood banks.

Policy on *HIV testing:* HIV testing is carried out on a voluntary basis with appropriate pre-test and post-test counselling. The basis and objectives of testing are to monitor the trend of HIV infection in a population or sub-group; to test blood, organ or tissue for ensuring safety of the recipients; and to identify an Individual with HIV Infection on voluntary testing basis.

Often these objectives can not be met by single testing strategy. If surveillance is taken as an example, the objective is to obtain an estimate of prevalence of HIV in a certain group of population (e.g., STD patients attending STD clinics). In case there is referral from persons being tested the prevalence is under estimated or over inflated. On the other hand HIV testing could be carried out without identification of name of samples collected for other purposes for example VDRL in STD clinics. The objective of surveillance may be fulfilled in this example whereas the positive persons are not identified. Therefore, different objectives require separate testing procedures (e.g. unlinked anonymous or voluntary testing), and choice of test (e.g. two ERs for surveillance or 3 ERs for diagnosis of symptomatic HIV).

STD Control Programme: STD control is linked to HIV/AIDS control as behaviour resulting the transmission of STD and HIV are same. HIV is transmitted more easily in the presence of another STD. Hence, early diagnosis and treatment of STD is now recognized as one of the major strategies to control spread of HIV infection. Patients with STD form special group for health education and individual counselling.

STD control programme has been in operation in India since 1946. The programme is based on diagnosis and treatment of STD and relies on the health seeking behaviour of individuals with STD. The facilities currently providing STD control are: 5 Regional STD Reference Centres, the Skin-Leprosy-STD clinics in medical colleges and 504 STD clinics usually located at district hospitals.

Colidom *Promotion:* Among the probable source of HIV infection in India, heterosexual promiscuity constitute the major route, as almost 75 per cent HIV infections occur due to unprotected and multipartner sexual contacts. This type of transmission can be prevented by consistent use of good quality condoms. While the use of condom is easy, making a programme to cover the whole country needs careful planning on certain issues. These issues are mainly related to following questions: (a) how to sensitize people for using condom not only as a family planning method but also as the best preventive step against HIV and STD; (b) how to convince the commercial sex workers and their clients about the importance of use of condom as a means for preventing HIV and STD transmission; and (c) how to make available low cost and good quality condoms for people at the time and place when they need it most.

The three major areas in which NACO has made significant progress in relation to condom programming are a. quality control of condoms, b. social marketing of condoms, and c. involvement of NGOs and private voluntary organizations in the programme.

NACO has brought in the quality control specification parameters as prescribed by WHO. The unlubricated condom NIRODH has already been phased out and manufacturers have started adhering to the new specifications. Social marketing has been accepted as the most effective strategy for condom promotion. This strategy not only helps to increase the acceptability but also provides easy access to the users while improving the sustainability of condom provision. The free distribution of condom has its drawbacks, as it is very difficult to continue the supply line and also it creates doubts about the actual use of freely distributed condoms. It has therefore been emphasized now to increase the social marketing of condoms as a priced item but at subsidized rate. To make condom easily available to regular visitors to brothels, vending machines have been installed on a pilot basis in Mumbai, Pune, Chandigarh , etc. As it is realized that the target group for condom promotion for disease protection is not the same as the group for family planning, market research is necessary to identify the different potential user group.

HIV Surveillance: In the year 1985 ICMR started screening of blood from high risk groups at National Institute of Virology at Pune and Christian Medical College at Vellore to determine if HIV was present in India. As the first case of HIV was diagnosed in 1986 at Chennai, the surveillance activity was extended by establishing 62 sero -surveillance centres and 9 HIV Reference Centres in the country for identification of geographic spread of HIV and determination of major modes of transmission. During 1993 a Sentinel Surveillance System was introduced with a pre-defined scientific protocol. The objective was to identify trends of seropositivity in specific high risk groups as well as low risk groups. During 1997, 115 additional sentinel sites were established to carry out sentinel survey in the country during February-March 1998.

Different types of surveillance activities are being carried out in the country to detect the spread of the disease and to make appropriate strategy for prevention and control viz., area specific targeted intervention and best practice approach. The types of surveillance are: a. HIV Sentinel Surveillance, b. HIV Sero-Surveillance, c. AIDS Case Surveillance, d. STD Surveillance, e. Behavioural Surveillance, and f. Integration with surveillance of other diseases like tuberculosis etc.

HIV *Sentinel Surveillance:* After the establishment of the fact that HIV infection is present in wide geographic area, the objective of surveillance was redefined to monitor the trends of HIV infection. The objective of the surveillance is best achieved by annual cross -sectional survey of the risk group in the same place over few years by unlinked anonymous serological testing procedures by two ERS (i.e., when HIV testing is carried out without identification of name of samples collected for other purposes e.g., VDRL in STD clinics. The objective of surveillance may be fulfilled in this example whereas the positive person is not identified). The number of samples to be screened must represent the risk group under study and the sample size is determined accordingly. Clinical based approach for such collection has many advantages including the procedure for collection of samples which should be carried out on the above lines to avoid "selection bias" and "participation bias".

To start with, the HIV sentinel surveillance for HIV was taken up from 1994 in 55 sentinel sites attached to the existing surveillance centres and were increased to 180 in 1998. The high risk groups of population is represented by the patients attending STD clinics and intravenous drug users while low risk population includes mothers attending antenatal clinics. Each sentinel site conducts regular round of surveillance every year with a sample size of 250 cases of high risk groups and 400 samples of low risk groups. If the sample size is not adequate then the number of samples collected upto 3 months is taken as adequate sample size. After each round of surveillance, data collected is compiled and analysed by NACO. The drawback in this surveillance is that there is no rural representation. Thus there is a need to expand the present surveillance system so that each state will have adequate number of sentinel sites representing urban as well as rural population.

Information, Education, Communication and Social Mobilization

The objective of the IEC strategy in the national AIDS control programme are: to raise awareness, improve knowledge and understanding among the general public about AIDS infection and STD, routes of transmission and method of prevention; to promote desirable practice such as avoiding multiple sex partner, use of condom, sterilization of needles/ syringes and voluntary blood donation, to mobilize all sectors of society to integrate messages and programme on AIDS into their existing activities; to create a supportive environment for the care and rehabilitation of persons with HIV/AIDS.

In the present day awareness campaign through multimedia has made easy the efforts to reach at larger segment of people. The print media, electronic media, press campaign, interpersonal publicity and field publicity holds the key to success. A massive media campaign was launched by NACO in 1996 through well designed generic materials. Posters, pamphlets, booklets, news papers, advertisements, film clippings, TV spots, radio spots, wall paintings and cinema slides were prepared in Hindi and all regional languages. IEC programming cannot exist in isolation.

The ***National AIDS Control Programme Phase II*** was officially launched on the 15th December, 1999. It is being implemented by the National AIDS Control Organization with support from World Bank, USAID and the Department For International Development. The new programme has the features which will amount to a paradigm shift in the nation's response to prevention and control of HIV/AIDS at all levels. Based on the epidemiological data obtained from annual sentinel surveillance, the country has been divided into three categories. Maharashtra with a prevalence of 2-2.4 per cent in the age group of 15-49 years, Andhra Pradesh, Karnataka, Manipur and Tamil Nadu with a prevalence of 1-2 per cent in 15-49 years age group, and the rest of the country where the prevalence rate is less than one per cent in the same age group.

The objective under the Second Phase is to restrict the future spread of the infection to less than 5 per cent in Maharashtra, less than 3 per cent in Andhra Pradesh, Karnataka, Manipur and Tamil Nadu and less than one per cent in the rest of the country.

The objective is to be realised by : (a) Raising the level of awareness on STD/HIV in rural areas and other vulnerable groups of population; (b) Encouraging health seeking behaviour in general population for reproductive tract infections and STD; (c) Making people-aware about services available in the public health system for the management of RTI/STD; (d) Provision of facilities for early detection and prompt treatment; and (e) Implementing focused IEC strategy.

V. WOMEN AND CHILD CARE

REPRODUCTIVE AND CHILD HEALTH PROGRAMME

Reproductive and child health approach has been defined as "people have the ability to reproduce and regulate their fertility, women are able to go through pregnancy and child birth safely, the outcome of pregnancies is successful in terms of maternal and infant survival and well being and couples are able to have sexual relations free of fear of pregnancy and of contracting disease".

The concept is in keeping with the evolution of an integrated approach to the programme aimed at improving the health status of young women and young children which has been going on in the country namely family welfare programme, universal immunization programme, oral rehydration therapy, child survival and safe motherhood programme and acute respiratory infection control etc. It is obviously sensible that integrated RCH programme would help in reducing the cost inputs to some extent because overlapping of expenditure would not be necessary and integrated implementation would optimise outcomes at field level.

The RCH programme incorporates the components relating child survival and safe motherhood and includes two additional components, one relating to sexually transmitted disease (STD) and other relating to reproductive tract infection (RTI). Fig.7.14 represents the various components of RCH programme.

Family Planning	Child Survival and Safe Motherhood Component
Client approach to health care	Prevention/Management of RTI/STD AIDS

Fig. 7.14 : RCH Package

The main highlights of the RCH programme are :

1. The programme integrates all interventions of fertility regulation, maternal and child health with reproductive health for both men and women.
2. The services to be provided will be client oriented, demand driven, high quality and based on needs of community through decentralised participatory planning and target free approach.
3. The programme envisages upgradation of the level of ~ facilities for providing various interventions and quality of care. The First Referral Units (FRUs) being set-up at sub-district level will provide comprehensive emergency obstetric and new born care. Similarly RCH facilities at PHCs will be substantially upgraded.
4. It is proposed to improve facilities of obstetric care, MTP and IUD insertion in the PHCs. Also for IUD insertion at sub-centres.
5. Specialist facilities for STD and RTI will be available in all district hospitals and in a fair number of sub-district level hospitals.
6. The programme aims at improving the out-reach of services primarily for the vulnerable group of population who have been, till now, effectively left out of planning process, e.g. special

programme will be taken up for urban slums, tribal population and adolescents; NGOs and voluntary organizations will be involved in a much larger way to improve out-reach and make it people's programme; practitioners of Indian System of Medicine will be trained and research and development in Indian System of Medicine will be supported to improve range of RCH services; the Panchayat Raj functionaries will have a central role in determining the need of the local population for RCH services generally and for contraceptives particularly under the target free approach. The Panchayat will also be the agency for implementing the programme and for extending financial support to women for taking them to specialist at sub-divisional hospitals for delivery.

The RCH programme is based on a differential approach. Inputs in all the districts have not been kept uniform. While the care component are the same for all districts, the weaker districts will get more support and sophisticated facilities are proposed for relatively advanced districts. On the basis of crude birth rate and female literacy rate, all the districts have been (divided into three categories. Category A having 58 districts, category B having 184 districts and category C having 265 districts. All the districts will be covered in a phased manner over a period of three years. The programme was formally launched on 15th October 1997.

RCH interventions at district level will be as follows:

Interventions in All Districts

- Child Survival interventions i.e. immunization, Vitamin A (to prevent blindness), oral rehydration therapy and 'prevention of deaths due to pneumonia.
- Safe Motherhood interventions e.g. antenatal check up, immunization for tetanus, safe delivery, anaemia control programme.
- Implementation of Target Free Approach.
- High quality training at all levels.
- IEC activities.
- Specially designed RCH package for urban slums and tribal areas.
- District sub-projects under Local Capacity Enhancement.
- RTI/STD Clinics at District Hospitals (where not available)

- Facility for safe abortions at PHCs by providing equipment, contractual doctors etc.
- Enhanced community participation through Panchayats, Women's Groups and NGOs.
- Adolescent health and reproductive hygiene.

Interventions in selected States/Distts.

- Screening and treatment of RTI/STD at sub-divisional level.
- Emergency obstetric care at selected FRUs by providing drugs.
- Essential obstetric care by providing drugs and PHN/Staff Nurse at PHCs.
- Additional ANM at sub-centres in the weak districts for ensuring MCH care.
- Improved delivery services and emergency care by providing equipment kits, IUD insertions and ANM kits at sub-centres.
- Facility of referral transport for pregnant women during emergency to the nearest referral centre through Panchayat in weak districts.

Maternal health care was a part of Family Welfare Programme from its inception. Interventions were introduced as vertical schemes, namely the National Nutritional Anaemia Control Programme, TT immunization of pregnant women (part of immunization programme) and Dais training programme etc. Family Planning remained a separate intervention. In 1992, the *Child Survival and Safe Motherhood Programme* integrated all .the schemes for better compliance. This programme had the following components:

a. Early registration of pregnancy;
b. To provide minimum three antenatal check-ups;
c. Universal coverage of all pregnant women with TT immunization;
d. Advice on food, nutrition and rest;
e. Detection of high risk pregnancies and prompt referral;
f. Clean deliveries by trained personnel;
g. Birth spacing, and;
h. Promotion of institutional deliveries.

The current RCH Programme has integrated these services and the major intervention are essential obstetric care, 24 hour delivery services at PHCs/CHCs, emergency obstetric care, Medical Termination of Pregnancy, prevention of reproductive tract infection (RTI) and sexually transmitted diseases (STD), and District Surveys.

Essential obstetric care

Essential obstetric care intends to provide the basic maternity services to all pregnant women through (1) early registration of pregnancy (within 12-16 weeks), (2) provision of minimum three antenatal check ups by ANM or medical officer to monitor progress of the pregnancy and to detect any risk/complication so that appropriate care including referral could be taken in time, (3) provision of safe delivery at home or in an institution, (4) provision of three postnatal check ups to monitor the postnatal recovery and to detect complications.

This component in RCH programme is more relevant for Assam, Bihar, Rajasthan, Orissa, Uttar Pradesh and Madhya Pradesh as most of the deliveries In these states are conducted at home in unclean environment causing high maternal morbidity and mortality.

Emergency obstetric care

Complications associated with pregnancy are not always predictable, hence, emergency obstetric care is an important intervention to prevent maternal morbidity and mortality. Under the CSSM programme 1748 Referral Units were identified and supported with equipment kit E to kit P. However, these FRUs are not fully operational because of lack of manpower and adequate infrastructure,' Under the RCH programme the FRUs will be strengthened through supply of emergency obstetric kit, equipment kit and provision of skilled manpower on contract basis etc. Traditional Birth Attendant still plays an important role during deliveries in our society, Under the CSSM programme Dai training was a uniform country-wide activity, However, it was observed-that the delivery practice adopted vary from state to state, e.g. in Kerala and Goa more than 90 per cent deliveries take place in health institutions, whereas in most northern states majority of deliveries take place at home. Therefore, the current RCH programme decided to decentralise this activity by involving NGOs to make it more local specific.

24-Hour delivery services at PHCs/CHCs

To promote institutional deliveries, provision has been made to give additional honorarium to the staff to encourage round the clock delivery facilities at health centres.

Medical Termination of Pregnancy

MTP is a reproductive health measure that enables a woman to opt out of an unwanted or unintended pregnancy in certain specified circumstances without endangering her life, through MTP Act 1971. The aim is to reduce maternal morbidity and mortality from unsafe abortions. The assistance from the Central Government is in the form of training of manpower, supply of MTP equipment and provision for engaging doctors trained in MTP to visit PHCs on fixed dates to perform MTP.

Control of reproductive tract infections (RTI) and sexually transmitted diseases (STD)

Under the RCH programme, the component of RTI/STD control is linked to HIV and AIDS control. It has been planned and implemented in close collaboration with National AIDS Control Organization (NACO). NACO will provide assistance for setting up RTI/STD clinics upto the district level. The assistance from the Central Government is in the form of training of the manpower and drug kits including disposable equipment. Each district will be assisted by two laboratory technicians on contract ~ basis for testing blood, urine and RTI/STD tests.

Immunization

The Universal Immunization Programme (UIP) became a part of CSSM programme in 1992 and RCH programme in 1997. It will continue to provide vaccines for polio, tetanus, DPT, DT, measles and tuberculosis. The cold chain established so far will be maintained and additional items will be provided to new health facilities.

Drug and equipment kits

The drug and equipment kits supplied at various levels are as follows :

At sub-centre level

- Drug kit A
- Drug kit B
- Mid-wifery kit

Sub-centre equipment kit

At PHC level

- PHC equipment kit

At CHC/FRU level

- Equipment kits from kit E to kit P

In addition, a drug kit for essential obstetric care will be supplied to PHCs in category C districts.

Essential newborn care

The primary goal of essential newborn care is to reduce perinatal and neonatal mortality.. The main components are resuscitation of newborn with asphyxia, prevention of hypothermia, prevention of infection, exclusive breast feeding and referral of sick newborn. The strategies are to train medical ; and other health personnel in essential newborn care, provide basic facilities for care of low birth weight and sick new borns in FRU and district hospital etc.

Oral Rehydration Therapy

Diarrhoea is one of the leading cause of child mortality. Oral Rehydration Therapy Programme started in 1986-87 is being implemented in a phased manner. Supplies of ORS packets to the states are being organized by Central Government. Twice a year 150 packets of ORS are provided as part of drug kit : supplied to all sub-centres in the country. The programme emphasises the rational use of drugs for the management of diarrhoea. Adequate nutritional care of the child with diarrhoea and proper advice to mothers on feeding are two important areas of this programme.

Acute respiratory disease control

The standard case management of ARI and prevention of deaths due to pneumonia is now an integral part of RCH programme. Peripheral health workers are being trained to recognise and treat pneumonia. Cotrimoxazole is being supplied to the health workers through the CSSM drug kit.

Prevention and control of vitamin A deficiency in children

It is estimated that large number of children suffer from sub-if clinical deficiency of vitamin A. Under the programme, 5 doses of vitamin are given to all children under three years of age. The first dose (1 lakh units) is given at nine months of age along with measles vaccination. The second dose (2 lakh units) is given along with DPT/OPV booster doses. Subsequent three doses (2 lakh units each) are given at six months intervals.

District Surveys

There is no regular source of data to indicate the reproductive health status of women. The RCH programme conducted district based rapid household survey to assess the reproductive health status of women. The survey was conducted in two phases covering 252 districts. The outcome of the survey is shown in Table 7.8.

Table 7.8 : Key indicators of reproductive health in India (Survey report September to December 1999)

Indicator	*Percentage*
CMW age 15-44 years	
* With any ANC	67.2
* With full ANC	14.8
Institutional delivery Safe delivery	
Women having	
* Pregnancy complications	34.8
* Delivery complications	36.4
* Post delivery complications	42.0
CMW age 15-44 years	
* With symptoms of RTI/STD	28.8
* Aware of AIDS	41.1
Males age 20-54 years	
* With symptoms RTI/STD	12.7
* Aware of AIDS	57.4

CMW refers to currently married women who had the last live/still birth since January 1995.

Safe delivery refers to institutional delivery and home delivery attended by doctor/nurse/ANM.

Full ANC refers to 3 ANC check up + at least one TT injection + 100 tablets of IFA.

Pregnancy complications : Percent women who reported any of the following complication : swelling of hands and feet, visual disturbance, bleeding, convulsions, weak or no movement of foetus and abnormal presentation.

Delivery complications : percent of women who reported any of the following complications : premature labour, obstructed labour, prolonged labour.

Post delivery complications : percent women who reported any of 'to the following complications : high fever, lower abdominal pain, foul smelling vaginal discharge, excessive bleeding and dizziness or severe headache.

REVIEW QUESTIONS

1. Define population. Give a concise account of different properties of a population.
2. Give a detailed description of factors that control a population.
3. Describe growth form of a population. How is it regulated?
4. Discuss the problems related with human population dynamics
5. Explain the following:

   (i) Age structure of human population
   (ii) Population dispersal
   (iii) Natality and mortality
   (iv) Density dependent and density independent regulation.